MATHEMATICS OF
ENGINEERING SYSTEMS

D0870240

MATHEMATICS OF ENGINEERING SYSTEMS

FRANCIS H. RAVEN

Associate Professor of Mechanical Engineering
University of Notre Dame

International Student Edition

McGraw-Hill Book Company

New York   St. Louis   San Francisco   Toronto   London   Sydney

Kōgakusha Company, Ltd.

Tokyo

MATHEMATICS OF ENGINEERING SYSTEMS

INTERNATIONAL STUDENT EDITION

Exclusive rights by Kōgakusha Co., Ltd. for manufacture and export from Japan. This book cannot be re-exported from the country to which it is consigned by Kōgakusha Co., Ltd. or by McGraw-Hill Book Company or any of its subsidiaries.

I

COPYRIGHT © 1966 BY McGRAW-HILL, INC. ALL RIGHTS RESERVED. THIS BOOK, OR PARTS THEREOF, MAY NOT BE REPRODUCED IN ANY FORM WITHOUT PERMISSION OF THE PUBLISHERS. LIBRARY OF CONGRESS CATALOG CARD NUMBER 65-27680

TOSHO PRINTING CO., LTD., TOKYO, JAPAN

To my beloved wife Therese
and God's Blessings
Betty, Ann, Paul, and John

PREFACE

Because of the complex nature of today's engineering problems, the methods of the engineering profession have become increasingly analytical. This has resulted in a marked increase in the mathematical content of engineering courses. The purpose of this text, which is intended for the junior or senior year, is to explain the application of advanced mathematical methods to the solution of engineering problems. Our experience at Notre Dame has shown that the time devoted to this course "pays for itself" by permitting senior engineering courses to proceed at a faster pace and at a higher level than they normally would. When engineering courses must be disrupted to introduce mathematical topics on an "as needed" basis, both the engineering and the mathematics suffer from a lack of continuity. Moreover, sufficient time can seldom be taken to develop the mathematics in depth.

To make the most effective use of mathematics, it is necessary that the engineer understand mathematics in a meaningful manner. Thus, in this text, emphasis is given to explaining how abstract mathematical concepts are used to represent concrete physical phenomena. To achieve breadth, a wide variety of mathematical methods are developed. In addition, examples are chosen to illustrate a broad range of application to such areas as control systems, vibrations, heat transfer, fluid dynamics, electrical and mechanical circuits, dynamics, structures, electromagnetics, field theory, analogies, etc.

The first four chapters are primarily concerned with the application of ordinary differential equations to engineering. Both classical and transform methods are presented so that by comparison the reader can gain better appreciation of the relative merits of each method. Chapter 1 begins with a development of the Fourier series, which is extended to the Fourier transformation, and then to the Laplace transformation. In Chapter 2, basic properties and characteristics of Laplace transformations are developed. The Laplace transformation is given a rather thorough treatment because other transformation methods, which are presented later in the book, follow the same pattern. Chapter 3 is primarily concerned with the solution of linear differential equations, and Chapter 4 considers a variety of engineering applications.

Basic principles of the theory of complex variables are explained in Chapters 5 and 6. These principles are required for an understanding of the more advanced methods which are presented in later chapters.

Partial differential equations are covered in Chapters 7 and 8. Because the wave equation, heat equation, Laplace's equation, and Poisson's equation have extensive engineering applications, they are given special consideration. In addition to the Laplace transform method, it is also shown that the sine and cosine transformations provide a powerful method of solution. A basis for comparison between classical and transformation methods is provided by the method of separation of variables. This section is concluded with a general discussion of elliptic, parabolic, and hyperbolic partial differential equations.

Chapters 9 and 10 are concerned with field theory and mapping. This section begins with a discussion of vector analysis, which is essential to the understanding of vector fields. It is then shown how basic principles of complex variables and vector analysis are combined to develop vector-field theory. Mapping has extensive applications, since many problems which appear complex may be mapped, or transformed, to relatively simple situations.

Differential equations with variable coefficients are discussed in Chapter 11. Special consideration is given to Bessel's and Legendre's differential equations. It is shown that the Hankel transformation is well suited to the solution of partial differential equations in cylindrical coordinates, and, similarly, Legendre's transformation is well suited to spherical coordinates. Again, both classical and transformation methods are described in detail. This chapter concludes with a general discussion of orthogonal functions.

In the first part of Chapter 12, the application of vector-matrix theory to the solution of engineering problems is explained. Special consideration is given to systems of algebraic and systems of differential equations. The last part of Chapter 12 is concerned with computer techniques and methods of numerical analysis.

This text is the outgrowth of notes developed by the author for his classes at the University of Notre Dame. Appreciation is expressed for the encouragement and suggestions afforded by the students who used the original notes. Although it is not possible to acknowledge the contribution of each student individually, particular recognition is due Robert Gilmore, Russ Jandresivits, and Joe Weaver. Special thanks are due Jim Ackert for his valuable assistance in the preparation of this text and for his many fine suggestions.

The author also wishes to express his gratitude for the encouragement of his colleagues at the University of Notre Dame, especially to Dr. Edward W. Jerger, Head of the Department of Mechanical Engineering. In addition appreciation is expressed to Professors Arthur G. Hansen and Joseph E. Shigley of the University of Michigan for their very fine recommendations. Thanks are also due Mrs. Ella Levee, who typed the original notes and manuscript from which this text developed.

Deepest gratitude is expressed to the author's wife Therese, a constant source of inspiration.

FRANCIS H. RAVEN

CONTENTS

Contents

chapter I

The Laplace transformation

A transformation is an operation which converts a mathematical expression to a different but equivalent form. For example, the equation for a curve may be transformed from rectangular coordinates to polar coordinates, or vice versa. Also, the location of the curve may be referred to a new set of coordinate axes by a transformation of coordinates. A transformation is ordinarily employed to simplify the solution of a problem. Thus, the well-known transformation logarithms reduce multiplication and division to the simpler processes of addition and subtraction.

The Laplace transformation is a transformation which enables one to solve differential equations by the use of algebraic methods. It may be regarded as an extension of the Fourier series. This viewpoint yields a good understanding of its physical significance and meaning.

1.1 FOURIER SERIES

A periodic function of period T is shown in Fig. 1.1. This periodic function $f(t)$ may be expressed as a sum of sine and cosine terms,

$$f(t) = K + \sum_{n=1}^{\infty} \left(A_n \cos \frac{2\pi nt}{T} + B_n \sin \frac{2\pi nt}{T} \right) \tag{1.1}$$

where K, A_n, and B_n are constants. Equation (1.1) is the Fourier series expansion for a periodic function. As time t goes from $-T/2$ to $T/2$, the term t/T goes from $-\frac{1}{2}$ to $\frac{1}{2}$. Thus, as illustrated in

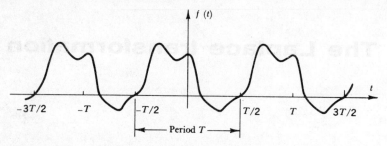

Fig. 1.1. Periodic function.

Fig. 1.2, as time t proceeds through one complete period, then $\cos(2\pi nt/T)$ goes through n complete cycles. Similarly, as t proceeds through a period, then $\sin(2\pi nt/T)$ also completes n cycles.

Equation (1.1) may be expressed in the form

$$f(t) = K + \sum_{n=1}^{\infty} (A_n \cos n\omega_0 t + B_n \sin n\omega_0 t) \tag{1.2}$$

where $\omega_0 = 2\pi/T$ is the angular frequency.

The constant K is evaluated by integrating each term in Eq. (1.2) over a complete period. Each cosine and sine term in the infinite summation vanishes when integrated over a period; hence

$$\int_{-T/2}^{T/2} f(t)\,dt = \int_{-T/2}^{T/2} K\,dt = KT$$

or

$$K = \frac{1}{T} \int_{-T/2}^{T/2} f(t)\,dt \tag{1.3}$$

The term K (the area of the function divided by the length of the period) is the average value of $f(t)$.

An expression for evaluating A_n is obtained by first multiplying each term in Eq. (1.2) by $\cos m\omega_0 t$, where m is a specific integer. Upon integrating each term over a period and noting that

$$\int_{-T/2}^{T/2} \cos n\omega_0 t \cos m\omega_0 t\,dt = \begin{cases} 0 & \text{for } n \neq m \\ \dfrac{T}{2} & \text{for } n = m \end{cases}$$

and

$$\int_{-T/2}^{T/2} \sin n\omega_0 t \cos m\omega_0 t\,dt = 0$$

then it follows that

$$A_n = \frac{2}{T} \int_{-T/2}^{T/2} f(t) \cos n\omega_0 t\,dt \qquad n = 1, 2, 3, \ldots \tag{1.4}$$

The constant A_0 is now defined as the value of Eq. (1.4) when $n = 0$. That is,

$$A_0 = \frac{2}{T} \int_{-T/2}^{T/2} f(t)\, dt$$

Comparison with Eq. (1.3) reveals the interesting result that the average value K is equal to $A_0/2$; that is,

$$K = \frac{A_0}{2}$$

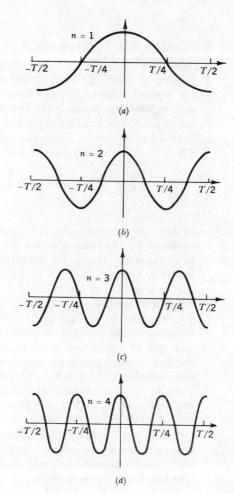

Fig. 1.2. Plot of $\cos(n\omega_0 t/T)$ for values of $n = 1, 2, 3,$ and 4.

3

An expression for evaluating B_n is obtained by multiplying each term in Eq. (1.2) by $\sin m\omega_0 t$ and then integrating each term over a period. Because

$$\int_{-T/2}^{T/2} \sin n\omega_0 t \sin m\omega_0 t \, dt = \begin{cases} 0 & n \neq m \\ \dfrac{T}{2} & n = m \end{cases}$$

and

$$\int_{-T/2}^{T/2} \cos n\omega_0 t \sin m\omega_0 t \, dt = 0$$

then

$$B_n = \frac{2}{T} \int_{-T/2}^{T/2} f(t) \sin n\omega_0 t \, dt \qquad n = 1, 2, 3, \ldots \tag{1.5}$$

Raising or lowering the horizontal axis in Fig. 1.1 merely changes the average value of $f(t)$, and hence K. The values of A_n and B_n are unaffected by changing the location of the horizontal axis. This fact is proved by letting C represent the amount the horizontal axis is raised or lowered. Thus, the equation for the new function is $f(t) + C$. The substitution of $f(t) + C$ for $f(t)$ in Eq. (1.4) gives

$$A_n = \frac{2}{T} \int_{-T/2}^{T/2} f(t) \cos n\omega_0 t \, dt + \frac{2C}{T} \int_{-T/2}^{T/2} \cos n\omega_0 t \, dt \tag{1.6}$$
$$n = 1, 2, \ldots$$

In Fig. 1.2 is shown a plot of $\cos n\omega_0 t$ for $n = 1, 2, 3$, and 4. It is apparent for any value of n that the integral of $\cos n\omega_0 t$ from $-T/2$ to $T/2$ is zero. Hence, the last term in Eq. (1.6) vanishes, and thus Eq. (1.6) becomes identical to Eq. (1.4). In a similar manner, it can be proved that the value of B_n is unaffected by changing the location of the horizontal axis.

Moving the vertical axis to the left or right is equivalent to changing the beginning of the period from $-T/2$ to some other value. The end of the period is, of course, changed accordingly. This shifting changes the upper and lower limits of integration in the equations for K, A_n, and B_n. This does not affect K, for the average value of $f(t)$ is unaffected by shifting the vertical axis. However, the values of A_n and B_n are changed (unless the vertical axis is moved a full period T).

It is not necessary that a function be continuous in order to possess a Fourier expansion. Any periodic function which satisfies the "Dirichlet conditions" may be represented by a Fourier series. The two Dirichlet conditions are:

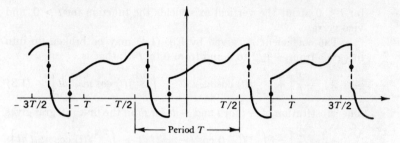

Fig. 1.3. Periodic function which is piecewise continuous.

1. The function must be bounded.
2. In any one period the function must have at most a finite number of discontinuities and a finite number of maxima or minima.

The preceding conditions are such that almost any conceivable periodic function encountered in engineering possesses a Fourier series expansion. Functions which satisfy the Dirichlet conditions are said to be *piecewise continuous*.

In Fig. 1.3 is shown a discontinuous function which is periodic. The Dirichlet conditions are satisfied, and thus the Fourier series expansion converges to the function at all points where $f(t)$ is continuous. At points where the function is discontinuous, the expansion converges to the average values of the right- and left-hand limits (i.e., at the dots in Fig. 1.3).

1.2 EVEN AND ODD FUNCTIONS

These functions are periodic functions which have a special type of symmetry. Because of this symmetry, certain simplifications can be made in the general equations for these periodic functions.

Even functions

An even function is one for which

$$f(t) = f(-t) \tag{1.7}$$

The cosine curve, which is a well-known example of an even function, is shown in Fig. 1.4a. Other even functions are illustrated in Fig. 1.4b to d. An even function is symmetrical about the vertical ($t = 0$) axis. That is, reflecting the part of the function

5

for $t < 0$ about the vertical axis yields the function for $t > 0$, and vice versa.

The coefficient A_n given by Eq. (1.4) may be broken up into integrals from $-T/2$ to 0 and from 0 to $T/2$; that is,

$$A_n = \frac{2}{T} \left[\int_{-T/2}^{0} f(t) \cos n\omega_0 t \, dt + \int_{0}^{T/2} f(t) \cos n\omega_0 t \, dt \right] \quad (1.8)$$

The substitution of $-t$ for t and $-dt$ for dt in the first integral gives

$$A_n = \frac{2}{T} \left[- \int_{T/2}^{0} f(-t) \cos (-n\omega_0 t) \, dt + \int_{0}^{T/2} f(t) \cos n\omega_0 t \, dt \right]$$

Inverting the limits of integration for the first integral changes the sign. Also, noting that $\cos (-n\omega_0 t) = \cos n\omega_0 t$ gives

$$A_n = \frac{2}{T} \left[\int_{0}^{T/2} f(-t) \cos n\omega_0 t \, dt + \int_{0}^{T/2} f(t) \cos n\omega_0 t \, dt \right] \quad (1.9)$$

Because $f(-t) = f(t)$ for an even function, Eq. (1.9) becomes

$$A_n = \frac{4}{T} \int_{0}^{T/2} f(t) \cos n\omega_0 t \, dt \qquad n = 1, 2, 3, \ldots \quad (1.10)$$

Comparison of Eqs. (1.4) and (1.10) shows that, because of the symmetry of an even function, it is necessary only to integrate over the half range from 0 to $T/2$ and then double the value of the constant (that is, $2/T$ becomes $4/T$) in order to evaluate A_n for an even function. By proceeding in a similar manner, the equation for K [Eq. (1.3)] may be written as follows for an even function:

$$K = \frac{2}{T} \int_{0}^{T/2} f(t) \, dt \quad (1.11)$$

The Fourier coefficient B_n may be written in the form

$$B_n = \frac{2}{T} \left[\int_{-T/2}^{0} f(t) \sin n\omega_0 t \, dt + \int_{0}^{T/2} f(t) \sin n\omega_0 t \, dt \right] \quad (1.12)$$

The substitution of $-t$ for t in the first integral gives

$$B_n = \frac{2}{T} \left[- \int_{T/2}^{0} f(-t) \sin (-n\omega_0 t) \, dt + \int_{0}^{T/2} f(t) \sin n\omega_0 t \, dt \right]$$

Inverting the limits of integration in the first integral and noting that $\sin (-n\omega_0 t) = -\sin n\omega_0 t$ gives

$$B_n = \frac{2}{T} \left[- \int_{0}^{T/2} f(-t) \sin n\omega_0 t \, dt + \int_{0}^{T/2} f(t) \sin n\omega_0 t \, dt \right] \quad (1.13)$$

Because $f(-t) = f(t)$ for an even function,

$$B_n = 0 \tag{1.14}$$

The substitution of $B_n = 0$ into Eq. (1.2) gives

$$f(t) = K + \sum_{n=1}^{\infty} A_n \cos n\omega_0 t \tag{1.15}$$

This is the Fourier series expansion for an even function.

Odd functions

Examples of odd functions are illustrated in Fig. 1.5. The sine curve shown in Fig. 1.5a is probably the best-known example of an odd function. The symmetry of an odd function may be observed by reflecting the function about the vertical ($t = 0$) axis and then reflecting it about the horizontal axis. For an odd function

$$f(t) = -f(-t) \tag{1.16}$$

Replacing $-f(-t)$ by $f(t)$ in Eq. (1.13) shows that for an odd function

$$B_n = \frac{4}{T} \int_0^{T/2} f(t) \sin n\omega_0 t \, dt \tag{1.17}$$

Similarly replacing $f(-t)$ by $-f(t)$ in Eq. (1.9) gives

$$A_n = 0 \tag{1.18}$$

Hence, there are no cosine terms in the Fourier series for an odd function. Because an odd function is symmetrical about the horizontal axis, the area over a period is zero. This means that $K = 0$. Thus, for an odd function

$$f(t) = \sum_{n=1}^{\infty} B_n \sin n\omega_0 t \tag{1.19}$$

Comparison of Figs. 1.4a and 1.5a shows that the only difference between the cosine (even function) and the sine (odd function) is the location of the vertical axis. The even functions shown in Fig. 1.4b and c may be changed to odd functions by changing the location of both the horizontal and the vertical axes as shown in Fig. 1.5b and c. Many functions are such that they may be regarded as either even or odd depending on the location of the coordinates. However, the even function shown in Fig. 1.4d cannot be changed to an odd function. Also, the odd function of Fig. 1.5d cannot be changed to an even function.

Mathematics of engineering systems

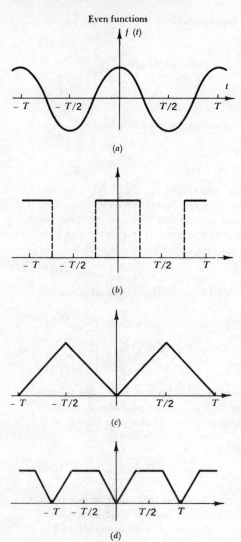

Even functions

$f(t)$

(a)

(b)

(c)

(d)

Fig. 1.4. Even functions.

Odd functions

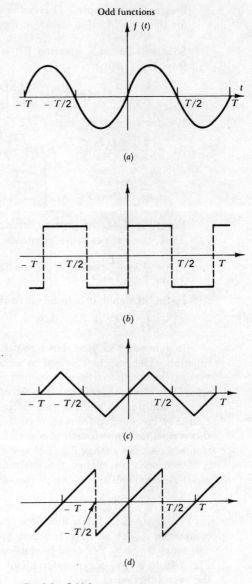

(a)

(b)

(c)

(d)

Fig. 1.5. Odd functions.

Illustrative example   Determine the Fourier series expansion for the even function shown in Fig. 1.4c.

Solution   For this function $f(t) = t$ for $0 \leq t \leq T/2$; hence, from Eq. (1.10),

$$A_n = \frac{4}{T} \int_0^{T/2} t \cos n\omega_0 t \, dt = \frac{4}{n\omega_0 t} \left[\frac{\cos n\omega_0 t}{n\omega_0} + t \sin n\omega_0 t \right]_0^{T/2}$$

Because $\omega_0 = 2\pi/T$, then

$$A_n = \frac{2}{n\pi} \left[\frac{\cos (2\pi n t/T)}{(2\pi n/T)} + t \sin \frac{2\pi n t}{T} \right]_0^{T/2}$$

$$= \frac{2}{n\pi} \frac{(\cos \pi n) - 1}{2\pi n/T} = \begin{cases} \dfrac{-2T}{(n\pi)^2} & n = 1, 3, 5, \ldots \\ 0 & n = 2, 4, 6, \ldots \end{cases}$$

From Fig. 1.4c, the average value of the function is $K = T/4$. Thus, the Fourier series expansion is

$$f(t) = \frac{T}{4} - \frac{2T}{\pi^2} \cos \omega_0 t - \frac{2T}{9\pi^2} \cos 3\omega_0 t - \frac{2T}{25\pi^2} \cos 5\omega_0 t - \cdots$$

$$(1.20)$$

In this example, it is to be noted that

$$A_2 = A_4 = A_6 = \cdots = 0 \tag{1.21}$$

In general, it may be shown that Eq. (1.21) holds for any even function which can be changed to an odd function by shifting the vertical axis one-fourth of a period (that is, $T/4$). This is so regardless of whether or not it is necessary to shift the horizontal axis in addition to shifting the vertical axis. For example, Fig. 1.4a may be changed from an even to an odd function by shifting the vertical axis one-fourth of a period with no change in the horizontal axis. To change Fig. 1.4b and c from even to odd functions, it is necessary to change the horizontal axis in addition to the quarter-period shifting of the vertical axis. In all three cases

$$A_2 = A_4 = A_6 = \cdots = 0$$

To verify Eq. (1.21), first note that Eq. (1.10) shows that, in computing A_n for an even function, only the values of $f(t)$ in the half range $0 \leq t \leq T/2$ need be considered. The even functions of Fig. 1.4a to c have a symmetry about the $T/4$ vertical axis which the even function of Fig. 1.4d lacks. This additional symmetry is that, in the half range $0 \leq t \leq T/2$, these functions may be regarded as odd functions about a vertical axis through $T/4$. Thus, for the half range the equation for the functions of Fig. 1.4a to c may be written in the form

$$f(t) = C + [\text{odd function}]$$

where C is the amount the horizontal axis must be raised or lowered. Substitution of the preceding expression for $f(t)$ into Eq. (1.10) gives

$$A_n = \frac{4C}{T} \int_0^{T/2} \cos n\omega_0 t \, dt + \frac{4}{T} \int_0^{T/2} [\text{odd function}] \cos n\omega_0 t \, dt \tag{1.22}$$

In Fig. 1.2, it is to be noted that for any value of n the area of the cosine functions from 0 to $T/2$ is zero; hence Eq. (1.22) becomes

$$A_n = \frac{4}{T} \int_0^{T/2} [\text{odd function}] \cos n\omega_0 t \, dt \tag{1.23}$$

For $n = 2, 4, 6, \ldots$, Fig. 1.2 shows that $\cos n\omega_0 t$ is an even function with respect to the $T/4$ vertical axis in the half range from $0 \le t \le T/2$. The product of an odd function and an even function is an odd function. This is analogous to the fact that the sum of an odd number and an even number is an odd number. The product of even or odd functions is even or odd in the same manner as the sum of even or odd numbers is even or odd. Thus, for $n = 2, 4, 6, \ldots$, the integrand of Eq. (1.23) is an odd function (i.e., the product of an odd function and an even function). Because the integral of an odd function over its interval is zero, A_n as given by Eq. (1.23) vanishes for $n = 2, 4, 6, \ldots$.

Figure 1.2 shows that, for $n = 1, 3, 5, \ldots$, $\cos n\omega_0 t$ is an odd function about $T/4$ in the half range $0 \le t \le T/2$. Because the product of two odd functions is even, the integrand of Eq. (1.23) is even. The integral of an even function is not necessarily zero, and hence A_n does not necessarily vanish for $n = 1, 3, 5, \ldots$.

Illustrative example Determine the Fourier series expansion for the odd function shown in Fig. 1.5c.

Solution For this function $f(t) = t$ for $0 \le t \le T/4$ and $f(t) = T/2 - t$ for $T/4 \le t \le T/2$; thus

$$B_n = \frac{4}{T} \int_0^{T/4} t \sin n\omega_0 t \, dt + \frac{4}{T} \int_{T/4}^{T/2} \left(\frac{T}{2} - t \right) \sin n\omega_0 t \, dt$$

$$= \frac{2T}{n^2\pi^2} \sin \frac{\pi n}{2} = \begin{cases} \dfrac{2T}{(n\pi)^2} & n = 1, 5, 9, \ldots \\[2mm] \dfrac{-2T}{(n\pi)^2} & n = 3, 7, 11, \ldots \\[2mm] 0 & n = 2, 4, 6, \ldots \end{cases}$$

Thus the Fourier series is

$$f(t) = \frac{2T}{\pi^2} \sin \omega_0 t - \frac{2T}{9\pi^2} \sin 3\omega_0 t + \frac{2T}{25\pi^2} \sin 5\omega_0 t - \cdots$$

$$(1.24)$$

In the preceding series, $B_2 = B_4 = B_6 = \cdots = 0$. By an analysis similar to that just discussed for even functions, it may be shown that

$$B_2 = B_4 = B_6 = \cdots = 0 \qquad (1.25)$$

for any odd function which can be changed to an even function by shifting the vertical axis one-fourth of a period (that is, $T/4$).

Figures 1.4c and 1.5c are identical plots but have different coordinate systems. Thus, Eq. (1.24) can be obtained directly from Eq. (1.20) by a transformation of coordinates. In going from Fig. 1.4c to 1.5c, the substitution of $t + T/4$ for t in Eq. (1.20) accounts for shifting the vertical axis forward a distance $T/4$. Similarly, the substitution of $f(t) + T/4$ for $f(t)$ in Eq. (1.20) accounts for raising the horizontal axis a distance $T/4$. Hence, Eq. (1.20) becomes

$$f(t) = -\frac{2T}{\pi^2} \cos \omega_0 \left(t + \frac{T}{4} \right) - \frac{2T}{9\pi^2} \cos 3\omega_0 \left(t + \frac{T}{4} \right)$$
$$- \frac{2T}{25\pi^2} \cos 5\omega_0 \left(t + \frac{T}{4} \right) - \cdots$$

Because $\omega_0 = 2\pi/T$ and thus $T/4 = \pi/2\omega_0$, the preceding equation becomes

$$f(t) = -\frac{2T}{\pi^2} \cos \left(\omega_0 t + \frac{\pi}{2} \right) - \frac{2T}{9\pi^2} \cos \left(3\omega_0 t + \frac{3\pi}{2} \right)$$
$$- \frac{2T}{25\pi^2} \cos \left(5\omega_0 t + \frac{5\pi}{2} \right) - \cdots \qquad (1.26)$$

By noting that $\sin \alpha = -\cos (\alpha + \pi/2) = -\cos (\alpha + 5\pi/2) = \cdots$ and that $\sin \alpha = \cos (\alpha + 3\pi/2) = \cos (\alpha + 7\pi/2) = \cdots$, Eq. (1.26) becomes identical to Eq. (1.24).

In Fig. 1.6 are shown the original function $f(t)$ of Fig. 1.5c and the first two terms of the series expansion given by Eq. (1.24). The first term, $(2T/\pi^2) \sin \omega_0 t$, is called the *fundamental component*. This term yields the most significant contribution to the function. The addition of each successive term in the series results in a closer approximation. In general, the first few terms of the Fourier series for a function yield a very good approximation to the function.

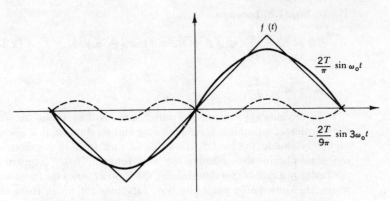

Fig. 1.6. Fourier series approximation for a triangular wave.

Fourier series expansions have numerous applications to engineering problems. For example, one of the most powerful methods of analyzing nonlinear control systems is the describing-function technique. The basis for the describing-function method is the representation of periodic functions as a sum of sine terms. Frequently the periodic function is odd, in which case sine terms automatically result. In the general case, it is possible to represent a periodic function by sine terms and phase angles. To prove this, let the constants A_n and B_n of Eq. (1.2) be the length of the sides of a right triangle as shown in Fig. 1.7. From this triangle, it follows that $A_n = \sqrt{A_n^2 + B_n^2} \sin \phi_n$ and $B_n = \sqrt{A_n^2 + B_n^2} \cos \phi_n$. Thus, the summation quantity in Eq. (1.2) may be written in the form

$$
\begin{aligned}
A_n \cos n\omega_0 t &+ B_n \sin n\omega_0 t \\
&= \sqrt{A_n^2 + B_n^2} \, (\cos n\omega_0 t \sin \phi_n + \sin n\omega_0 t \cos \phi_n) \\
&= \sqrt{A_n^2 + B_n^2} \sin (n\omega_0 t + \phi_n)
\end{aligned}
$$

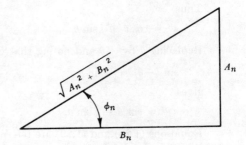

Fig. 1.7. Triangular representation for coefficients A_n and B_n.

Hence Eq. (1.2) becomes

$$f(t) = K + \sum_{n=1}^{\infty} \sqrt{A_n{}^2 + B_n{}^2} \sin (n\omega_0 t + \phi_n) \tag{1.27}$$

where

$$\phi_n = \tan^{-1} \frac{A_n}{B_n}$$

An obvious application of Fourier series is that, if one obtains the solution of a problem for $A_n \cos n\omega_0 t$ and $B_n \sin n\omega_0 t$, it is necessary only to add the contribution due to each term in the Fourier series to obtain the solution for any function $f(t)$. Numerical methods are available for determining the Fourier series for functions which are known only pointwise (i.e., functions for which there are no equations but merely points as in the case of experimentally determined data).

A more immediate application of Fourier series is that the basic Laplace transformation relationships are obtained by expressing the series in exponential form and then extending this form to non-periodic functions.

1.3 EXPONENTIAL FORM OF THE FOURIER SERIES

The series expansions for $\cos \theta$, $\sin \theta$, and $e^{j\theta}$ are

$$\cos \theta = 1 - \frac{\theta^2}{2!} + \frac{\theta^4}{4!} - \cdots$$

$$\sin \theta = \theta - \frac{\theta^3}{3!} + \frac{\theta^5}{5!} - \cdots$$

$$e^{j\theta} = 1 + j\theta + \frac{(j\theta)^2}{2!} + \frac{(j\theta)^3}{3!} + \frac{(j\theta)^4}{4!} + \frac{(j\theta)^5}{5!} + \cdots$$

$$= \left(1 - \frac{\theta^2}{2!} + \frac{\theta^4}{4!} - \cdots\right) + j\left(\theta - \frac{\theta^3}{3!} + \frac{\theta^5}{5!} - \cdots\right)$$

Thus,

$$e^{j\theta} = \cos \theta + j \sin \theta \tag{1.28}$$

Replacing θ by $-\theta$ and noting that $\cos (-\theta) = \cos \theta$ and

$$\sin (-\theta) = -\sin \theta$$

gives

$$e^{-j\theta} = \cos \theta - j \sin \theta \tag{1.29}$$

Equations (1.28) and (1.29) are referred to as *Euler's equations.*

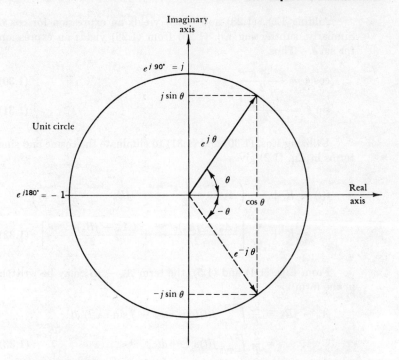

Fig. 1.8. Unit circle.

The term $e^{j\theta}$ may be regarded as a unit vector which is rotated counterclockwise an angle θ from the real axis, as is illustrated in Fig. 1.8. The horizontal (real) component of this unit vector is $\cos\theta$, and the vertical (imaginary) component of this vector is $j\sin\theta$. Thus, expressing this vector in terms of its two components gives $e^{j\theta} = \cos\theta + j\sin\theta$. The magnitude of $e^{j\theta}$ is always unity,

$$|e^{j\theta}| = \sqrt{\cos^2\theta + \sin^2\theta} = 1$$

When θ is 180°, then Fig. 1.8 shows that $e^{j180°} = -1$. Taking the square root gives $e^{j90°} = \sqrt{-1}$. Because $e^{j90°}$ is a unit vector along the j axis (that is, $e^{j90°} = j$), the well-known fact that $j = \sqrt{-1}$ is established.

In a similar manner, the term $e^{-j\theta}$ is a unit vector which is rotated an angle $-\theta$ from the real axis, as shown in Fig. 1.8. The real component of this vector is $\cos\theta$, and the imaginary component is $-j\sin\theta$, and thus $e^{-j\theta} = \cos\theta - j\sin\theta$.

15

Adding Eqs. (1.28) and (1.29) yields an expression for $\cos\theta$. Similarly, subtracting Eq. (1.29) from (1.28) yields an expression for $\sin\theta$. Thus

$$\cos\theta = \frac{e^{j\theta} + e^{-j\theta}}{2} \tag{1.30}$$

$$\sin\theta = \frac{e^{j\theta} - e^{-j\theta}}{2j} \tag{1.31}$$

Utilizing Eqs. (1.30) and (1.31) to eliminate the cosine and sine terms in Eq. (1.2) gives

$$
\begin{aligned}
f(t) &= K + \sum_{n=1}^{\infty} \left(A_n \frac{e^{jn\omega_0 t} + e^{-jn\omega_0 t}}{2} - jB_n \frac{e^{jn\omega_0 t} - e^{-jn\omega_0 t}}{2} \right) \\
&= K + \sum_{n=1}^{\infty} \left[\frac{(A_n - jB_n)e^{jn\omega_0 t}}{2} + \frac{(A_n + jB_n)e^{-jn\omega_0 t}}{2} \right]
\end{aligned} \tag{1.32}
$$

From Eqs. (1.4) and (1.5), the term $A_n - jB_n$ may be written in the form

$$
\begin{aligned}
A_n - jB_n &= \frac{2}{T} \int_{-T/2}^{T/2} f(t)(\cos n\omega_0 t - j\sin n\omega_0 t)\,dt \\
&= \frac{2}{T} \int_{-T/2}^{T/2} f(t)e^{-jn\omega_0 t}\,dt
\end{aligned} \tag{1.33}
$$

Similarly

$$
\begin{aligned}
A_n + jB_n &= \frac{2}{T} \int_{-T/2}^{T/2} f(t)(\cos n\omega_0 t + j\sin n\omega_0 t)\,dt \\
&= \frac{2}{T} \int_{-T/2}^{T/2} f(t)e^{jn\omega_0 t}\,dt
\end{aligned} \tag{1.34}
$$

Substitution of the results of Eqs. (1.33) and (1.34) into Eq. (1.32) gives

$$
\begin{aligned}
f(t) = K &+ \sum_{n=1}^{\infty} \frac{e^{jn\omega_0 t}}{T} \int_{-T/2}^{T/2} f(t)e^{-jn\omega_0 t}\,dt \\
&+ \sum_{n=1}^{\infty} \frac{e^{-jn\omega_0 t}}{T} \int_{-T/2}^{T/2} f(t)e^{jn\omega_0 t}\,dt
\end{aligned} \tag{1.35}
$$

The last summation in Eq. (1.35) may be expressed in the form

$$\sum_{n=1}^{\infty} \frac{e^{-jn\omega_0 t}}{T} \int_{-T/2}^{T/2} f(t)e^{jn\omega_0 t}\,dt = \sum_{n=-1}^{-\infty} \frac{e^{jn\omega_0 t}}{T} \int_{-T/2}^{T/2} f(t)e^{-jn\omega_0 t}\,dt$$

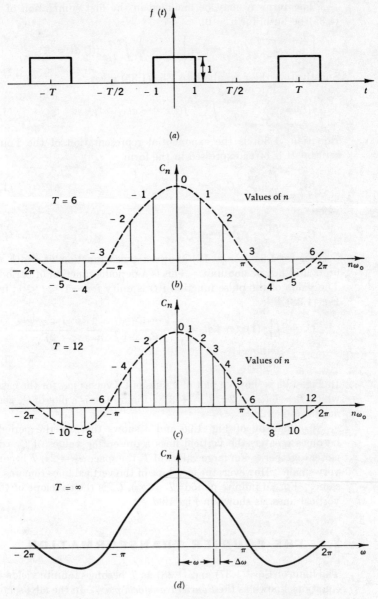

Fig. 1.9. Periodic pulse function. (a) The function; (b) C_n for $T = 6$; (c) C_n for $T = 12$; (d) C_n for $T = \infty$.

17

The term K may be included in the first summation of Eq. (1.35) because, for $n = 0$,

$$\left[\frac{e^{jn\omega_0 t}}{T} \int_{-T/2}^{T/2} f(t)e^{-jn\omega_0 t}\, dt \right]_{n=0} = \frac{1}{T} \int_{-T/2}^{T/2} f(t)\, dt = K$$

Substituting these results into Eq. (1.35) gives

$$f(t) = \sum_{n=-\infty}^{\infty} \frac{e^{jn\omega_0 t}}{T} \int_{-T/2}^{T/2} f(t)e^{-jn\omega_0 t}\, dt \tag{1.36}$$

Equation (1.36) is the exponential representation of the Fourier series. It is often expressed in the form

$$f(t) = \frac{1}{T} \sum_{n=-\infty}^{\infty} C_n e^{jn\omega_0 t} \tag{1.37}$$

where

$$C_n = \int_{-T/2}^{T/2} f(t)e^{-jn\omega_0 t}\, dt \tag{1.38}$$

Figure 1.9a (see p. 17) shows a pulse function of width two units and height one unit. This is a periodic function of period T. The value of the pulse function $f(t)$ is unity for $-1 < t < 1$; from Eq. (1.38)

$$C_n = \int_{-1}^{1} (1)e^{-jn\omega_0 t}\, dt = -\frac{e^{-jn\omega_0 t}}{jn\omega_0}\bigg|_{-1}^{1} = \frac{2}{n\omega_0} \frac{e^{jn\omega_0} - e^{-jn\omega_0}}{2j}$$

$$= 2\frac{\sin n\omega_0}{n\omega_0} \tag{1.39}$$

In Fig. 1.9b is shown a plot of values of C_n versus $n\omega_0$ for the case in which $T = 6$ ($\omega_0 = 2\pi/T = \pi/3$). Figure 1.9c is a plot of C_n versus $n\omega_0$ for $T = 12$.

Comparison of Fig. 1.9b and c shows that, as the period T becomes greater, the vertical lines representing values of C_n come closer together. For large values of T, the space $\omega_0 = 2\pi/T$ becomes very small. However, the envelope of the vertical lines remains the same. For an infinite period ($T = \infty$), C_n is the envelope of these vertical lines, as shown in Fig. 1.9d.

1.4 THE FOURIER TRANSFORMATION

The limit of Eqs. (1.37) and (1.38) as T becomes infinite yields two equations known as the *Fourier transform pair*. In the next section, it is shown that a slight modification of these Fourier transform expressions yields the basic Laplace transform relationships.

As n changes by one unit, the function $2\pi n/T$ changes by $2\pi/T$ units. As T becomes very large, the change $2\pi/T$ becomes very small and $2\pi n/T$ approaches a continuous function. Thus, for large T it is more appropriate to designate the change $\omega_0 = 2\pi/T$ by $\Delta\omega$. That is,

$$\omega_0 = \Delta\omega \qquad \text{for large } T$$

Similarly for large T, then $n\omega_0 = 2\pi n/T$ approaches the continuous function ω. That is,

$$n\omega_0 = \omega \qquad \text{for large } T$$

Taking the limit of Eq. (1.38) as T becomes very large gives

$$C_n = \int_{-\infty}^{\infty} f(t)e^{-j\omega t}\, dt = F(j\omega) \tag{1.40}$$

The evaluation of t at the upper and lower limits of integration (that is, $+\infty$ and $-\infty$) eliminates the term t from the preceding expression for C_n. Thus, the resulting expression for C_n is a function of $j\omega$ [that is, $C_n = F(j\omega)$].

Similarly, for very large values of T, Eq. (1.37) becomes

$$f(t) = \frac{\omega_0}{2\pi} \sum_{n=-\infty}^{\infty} C_n e^{jn\omega_0 t} = \frac{1}{2\pi} \sum_{n=-\infty}^{\infty} F(j\omega)e^{j\omega t}\, \Delta\omega$$

In the limit, the preceding summation may be written in the integral form

$$f(t) = \frac{1}{2\pi} \int_{-\infty}^{\infty} F(j\omega)e^{j\omega t}\, d\omega \tag{1.41}$$

Because the term $F(j\omega)$ is a function of the variable ω (j is a constant), it suffices to designate $F(j\omega)$ as simply $F(\omega)$.

Equations (1.40) and (1.41) comprise the Fourier transform pair. In particular, Eq. (1.40) is the direct Fourier transform and Eq. (1.41) is the inverse Fourier transform. With Eq. (1.40) a nonperiodic function $f(t)$ is transformed to a function of ω [that is, $F(\omega)$]. The inverse operation of finding the function $f(t)$ corresponding to a given $F(\omega)$ is accomplished by Eq. (1.41). There is a unique Fourier transform $F(\omega)$ corresponding to each function $f(t)$. Thus knowing $F(\omega)$ is equivalent to knowing $f(t)$, and vice versa.

Equations (1.40) and (1.41) show that $f(t)$ and $F(\omega)$ are two representations for the same function. The representation in the time domain is $f(t)$, and the representation in the frequency domain is $F(\omega)$.

Illustrative example   Determine the direct Fourier transform for the function

$$f(t) = \begin{cases} e^{at} & t \geq 0 \\ 0 & t < 0 \end{cases}$$

Solution   Application of Eq. (1.40) gives

$$F(\omega) = \int_0^\infty e^{at} e^{-j\omega t}\, dt = \frac{e^{at} e^{-j\omega t}}{a - j\omega}\Big|_0^\infty$$

For $a < 0$, the preceding expression becomes

$$F(\omega) = 0 - \frac{1}{a - j\omega} = \frac{1}{j\omega - a} \qquad a < 0 \tag{1.42}$$

For $a > 0$, then e^{at} becomes infinite when evaluated at $t = \infty$, and thus $F(\omega)$ is divergent.   For $a = 0$, $F(\omega)$ is indeterminate.

A function $f(t)$ possesses a Fourier transform $F(\omega)$ as given by Eq. (1.40), if the integral exists.   This requires that $f(t)$ be piecewise continuous and that

$$|F(\omega)| \leq \int_{-\infty}^\infty |f(t)|\, |e^{-j\omega t}|\, dt = \int_{-\infty}^\infty |f(t)|\, dt < \infty \tag{1.43}$$

Because $e^{-j\omega t}$ is a unit vector, its magnitude is 1.

Equation (1.43) shows that the integral of $|f(t)|$ from $-\infty$ to $+\infty$ must be finite.   Application of this requirement to the preceding example gives

$$\int_{-\infty}^\infty |f(t)|\, dt = \int_0^\infty e^{at}\, dt = \frac{e^{at}}{a}\Big|_0^\infty = \begin{cases} \dfrac{-1}{a} & a < 0 \\ \infty & a \geq 0 \end{cases}$$

Thus, this function possesses a transform only when $a < 0$.

Many functions do not satisfy Eq. (1.43) and thus do not possess Fourier transforms.   This, of course, limits the usefulness of the Fourier transformation.   However, in the next section, it is shown that multiplication of $f(t)$ by a converging factor results in a new transform (the Laplace transform) which is applicable to most functions encountered in engineering work.

1.5   THE LAPLACE TRANSFORMATION

Multiplication of $f(t)$ in Eq. (1.40) by $e^{-\sigma t}$ gives

$$\int_{-\infty}^\infty f(t) e^{-(\sigma + j\omega)t}\, dt = F(\sigma + j\omega) \tag{1.44}$$

where σ is a real number. Equation (1.44) is called the *two-sided Laplace transformation.* Note that, after performing the indicated integration, the term t vanishes when evaluated at the upper and lower limits of integration. Hence, the resulting expression in Eq. (1.44) is a function of $\sigma + j\omega$ and is thus designated $F(\sigma + j\omega)$.

Because the lower limit of integration is minus infinity $(t = -\infty)$, a complete knowledge of the past history of the phenomena under investigation is required. For most engineering work, one is interested in determining the behavior of the system after some arbitrary time $t = 0$. The initial conditions to which the system is subjected at $t = 0$ are the result of the past history of the system for $t < 0$. Thus, the effect of the operation of the system for negative time is accounted for by the initial conditions at $t = 0$. This suggests multiplying $f(t)$ in Eq. (1.40) by the converging factor

$$e^{-\sigma t} \quad t \geq 0$$
$$0 \quad t < 0$$

Thus

$$F(\sigma + j\omega) = \int_0^\infty f(t)e^{-(\sigma+j\omega)t}\,dt \tag{1.45}$$

It is customary to use the symbol s rather than $\sigma + j\omega$, in which case Eq. (1.45) becomes the direct Laplace transformation. This is commonly referred to simply as the *Laplace transformation.*

$$F(s) = \int_0^\infty f(t)e^{-st}\,dt = \mathcal{L}[f(t)] \tag{1.46}$$

The symbol $\mathcal{L}$ is read "Laplace transform of" so that Eq. (1.46) is the general expression for obtaining the Laplace transformation of a time function $f(t)$. With this expression, time functions such as those appearing in differential equations are transformed to functions of another variable s, whence the solution of the differential equation is accomplished by algebraic means. Practically any time function $f(t)$ encountered in engineering work is Laplace-transformable. A function $f(t)$ is Laplace-transformable if:

1. The function is piecewise continuous (i.e., satisfies the Dirichlet conditions).
2. There is a value of σ such that

$$\int_0^\infty |f(t)|e^{-\sigma t}\,dt < \infty \tag{1.47}$$

Condition 1 follows from the fact that the Laplace transformation is an extension of the Fourier series, which in turn required that the Dirichlet conditions be satisfied. Condition 2 follows from the fact that the transform integral, Eq. (1.46), must exist. An integral which has an infinite range of integration is called an *improper integral*. From mathematics, it is known that an improper integral exists if the integrand is piecewise continuous and if the integral is uniformly convergent. An improper integral is uniformly convergent if the magnitude of the integral approaches a finite value as the upper limit becomes infinite. Thus, Eq. (1.46) is uniformly convergent if

$$\left| \int_0^\infty f(t)e^{-st}\, dt \right| \le \int_0^\infty |f(t)|\, |e^{-st}|\, dt = \int_0^\infty |f(t)|e^{-\sigma t}\, dt < \infty$$

(1.48)

where

$$|e^{-st}| = |e^{-\sigma t}e^{-j\omega t}| = e^{-\sigma t}$$

(1.49)

Hence, condition 2 ensures that the integral is uniformly convergent.

Condition 2 is often expressed in a different form. This alternate form is obtained by first rewriting Eq. (1.47) as follows:

$$\int_0^\infty |f(t)|e^{-\sigma t}\, dt = \int_0^T |f(t)|e^{-\sigma t}\, dt + \int_T^\infty |f(t)|e^{-\sigma t}\, dt < \infty$$

(1.50)

where T may be very large but not infinite.

Because $f(t)$ is a bounded function (condition 1), the integral from 0 to T exists. The integral from T to ∞ is called the *remainder*. If this remainder can be made arbitrarily small by making T sufficiently large, the integral given is said to be *uniformly convergent*. That is,

$$\int_T^\infty |f(t)|e^{-\sigma t}\, dt \approx 0 \qquad t > T$$

(1.51)

The preceding equation is satisfied if there is a σ such that

$$|f(t)| \le M e^{(\sigma - \epsilon)t} \qquad t > T$$

(1.52)

where $\epsilon > 0$ is a very small number and M is any finite constant. Substitution of Eq. (1.52) into Eq. (1.51) shows that, if $f(t)$ satisfies Eq. (1.52), then Eq. (1.51) is automatically satisfied.

$$\int_T^\infty |f(t)|e^{-\sigma t}\, dt \le M \int_T^\infty e^{-\epsilon t}\, dt = \left(\frac{M}{\epsilon}\right)e^{-\epsilon T} \approx 0$$

For a given M and ϵ, it is possible to make T sufficiently large so that the preceding result is arbitrarily small.

Equation (1.52) shows that condition 2 is satisfied if for $t > T$ the function $f(t)$ does not grow more rapidly than the exponential $e^{(\sigma - \epsilon)t}$. Such a function $f(t)$ is said to be *of exponential order*. Because σ may be chosen as large as desired (but not infinite), essentially all functions encountered in engineering practice satisfy Eq. (1.52) and are thus of exponential order. In summary, a function $f(t)$ is Laplace-transformable if it is (1) piecewise continuous and (2) of exponential order.

The fact that $f(t)$ must be of exponential order can also be ascertained by application of the Weierstrass test for uniform convergence of an integral. This test states that the integral

$$\int_a^\infty f(x,t)\,dt = F(x)$$

is uniformly convergent if there is a function $G(t)$ such that $|f(x,t)| \leq G(t)$ and such that $\int_a^\infty G(t)\,dt$ exists. In the case of the Laplace transform x is replaced by s, a by T, and furthermore $f(s,t) = f(t)e^{-st}$. The function $G(t)$ is $Me^{-\epsilon t}$; that is,

$$|f(t)e^{-st}| = |f(t)|e^{-\sigma t} \leq Me^{-\epsilon t}$$

Thus, the Weierstrass test verifies that, to be Laplace-transformable, $f(t)$ must be of exponential order.

Let us now investigate to see whether or not the function $f(t) = t^n$ is Laplace-transformable. For large values of n this function grows very rapidly as t approaches infinity. To be of exponential order, there must be a σ such that Eq. (1.52) is satisfied; i.e.,

$$\lim_{t \to \infty} t^n \leq Me^{(\sigma - \epsilon)t}$$

or

$$\lim_{t \to \infty} \frac{t^n}{Me^{(\sigma - \epsilon)t}} \leq 1$$

Application of L'Hospital's rule gives

$$\lim_{t \to \infty} \frac{d^n/dt^n(t^n)}{d^n/dt^n(Me^{(\sigma - \epsilon)t})} = \lim_{t \to \infty} \frac{n!}{(\sigma - \epsilon)^n Me^{(\sigma - \epsilon)t}} = 0 \qquad \sigma > \epsilon$$

$$(1.53)$$

Hence, t^n is of exponential order because, for any $\sigma > \epsilon > 0$, the preceding expression is satisfied and thus Eq. (1.52) is also.

Illustrative example Let it be desired to obtain the Laplace transform for the function

$$f(t) = e^{at}$$

Solution

$$\mathcal{L}(e^{at}) = \int_0^\infty e^{at}e^{-st} \, dt = \frac{e^{(a-\sigma-j\omega)t}}{a - \sigma - j\omega}\bigg|_0^\infty$$

$$= \frac{1}{s - a} \qquad \sigma > a \qquad (1.54)$$

The arbitrary parameter σ may be chosen as large as possible; hence, the Laplace transform for e^{at} exists for any finite value of a whether positive or negative. (Note that the Fourier transform for this function existed for negative values of a only.)

In using the Laplace transform method to solve differential equations, it suffices to know that there is a σ or a range of values of σ such that the transform exists.

Inverse Laplace transform

The inverse Fourier transform for obtaining the time function $f(t)$ corresponding to $F(j\omega)$ is given by Eq. (1.41). The inverse Laplace transform for the time function $f(t)$ corresponding to

$$F(s) = F(\sigma + j\omega)$$

is obtained by substituting $\sigma + j\omega$ for $j\omega$ in Eq. (1.41). The limits of integration in Eq. (1.41) are at $\omega = \pm\infty$. Because $s = \sigma + j\omega$, the new limits are at $s = \sigma \pm j\infty$. Similarly in replacing $j\omega$ by s, then $jd\omega = ds$. Thus, $d\omega$ is replaced by ds/j. The resulting equation for obtaining the inverse Laplace transform is

$$f(t) = \frac{1}{2\pi j} \int_{\sigma-j\infty}^{\sigma+j\infty} F(s)e^{st} \, ds \qquad (1.55)$$

The inverse Laplace transform is an integral expression for converting a function $F(s)$ back to the original time function $f(t)$ from which it came. Symbolically, this is

$$\mathcal{L}^{-1}[F(s)] = f(t) \qquad (1.56)$$

where $\mathcal{L}^{-1}$ is read "inverse transform of."

Because

$$\mathcal{L}(e^{at}) = \frac{1}{s - a}$$

then

$$\mathcal{L}^{-1}\left(\frac{1}{s-a}\right) = e^{at} \tag{1.57}$$

The function e^{at} and its transform $1/(s-a)$ are said to form a *transform pair*. By a transform pair is meant a function $f(t)$ and its corresponding transform $F(s)$.

$$\mathcal{L}[f(t)] = F(s)$$
$$\mathcal{L}^{-1}[F(s)] = \mathcal{L}^{-1}\{\mathcal{L}[f(t)]\} = f(t) \tag{1.58}$$

For each time function $f(t)$, there is a unique transform $F(s)$. Because transform pairs are unique, there is no need to apply both Eqs. (1.46) and (1.55). If one applies Eq. (1.46) to obtain the transform $F(s)$ for a certain time function $f(t)$, application of Eq. (1.55) to obtain the inverse of $F(s)$ merely yields the original time function $f(t)$.

Determination of the Laplace transform of various functions by using Eq. (1.46) can become exceedingly tedious and difficult because of the integration involved. In the next chapter, several techniques are developed for obtaining Laplace transformations by other means. These methods can save considerable time and effort. Many other important properties of Laplace transformations are also presented in Chap. 2. These basic characteristics help one to obtain a better understanding of Laplace transformations and thus use them more effectively.

PROBLEMS

1.1 Show that

$$\int_{-T/2}^{T/2} \cos n\omega_0 t \cos m\omega_0 t \, dt = \begin{cases} 0 & \text{for } n \neq m \\ \dfrac{T}{2} & \text{for } n = m \end{cases}$$

$$\int_{-T/2}^{T/2} \sin n\omega_0 t \cos m\omega_0 t \, dt = 0$$

$$\int_{-T/2}^{T/2} \sin n\omega_0 t \sin m\omega_0 t \, dt = \begin{cases} 0 & \text{for } n \neq m \\ \dfrac{T}{2} & \text{for } n = m \end{cases}$$

1.2 For the even function shown in Fig. 1.4b

$$f(t) = \begin{cases} 0 & -\dfrac{T}{2} < t < -\dfrac{T}{4} \\ 2 & -\dfrac{T}{4} < t < +\dfrac{T}{4} \\ 0 & \dfrac{T}{4} < t < +\dfrac{T}{2} \end{cases}$$

Determine the trigonometric form of the Fourier series for this function.

1.3 For the even function shown in Fig. 1.4d,

$$f(t) = \begin{cases} t & 0 \le t \le \dfrac{T}{4} \\[2mm] \dfrac{T}{4} & \dfrac{T}{4} \le t \le \dfrac{T}{2} \end{cases}$$

Determine the trigonometric form of the Fourier series for this function.

1.4 For the odd function shown in Fig. 1.5b,

$$f(t) = \begin{cases} -1 & -\dfrac{T}{2} < t < 0 \\[2mm] +1 & 0 < t < \dfrac{T}{2} \end{cases}$$

Determine the trigonometric form for the Fourier series for this function. Verify this answer by using the result of Prob. 1.2, and then transform the coordinates.

1.5 For the odd function shown in Fig. 1.5d,

$$f(t) = t \qquad \dfrac{-T}{2} < t < \dfrac{T}{2}$$

Determine the trigonometric form for the Fourier series for this function.

1.6 The derivative of the Fourier series for a function yields the series for the derivative of the function. The derivative of the function shown in Fig. 1.4c is the function shown in Fig. 1.5b. Thus, obtain the Fourier series for Fig. 1.5b by differentiating the series for Fig. 1.4c [that is, Eq. (1.20)].

1.7 Integration of the Fourier series for a function yields the series for the integral of the function. The equation for the function shown

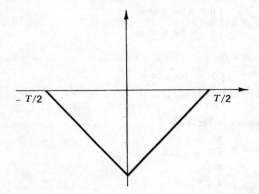

Fig. P1.7

in Fig. 1.5b is given in Prob. 1.4. Integrating this function from $-T/2$ to $+T/2$ yields the new function shown in Fig. P1.7, that is,

$$f(t) = \begin{cases} -t - \dfrac{T}{2} & \dfrac{-T}{2} < t \leq 0 \\ t - \dfrac{T}{2} & 0 \leq t < \dfrac{T}{2} \end{cases}$$

Determine the Fourier series for this new function. Verify this result by integrating the result of Prob. 1.4.

1.8 The equations for A_n, B_n, and C_n are

$$A_n = \frac{2}{T} \int_{-T/2}^{T/2} f(t) \cos n\omega_0 t \, dt$$

$$B_n = \frac{2}{T} \int_{-T/2}^{T/2} f(t) \sin n\omega_0 t \, dt$$

$$C_n = \int_{-T/2}^{T/2} f(t) e^{-jn\omega_0 t} \, dt$$

Utilize Euler's equation that $e^{-j\theta} = \cos \theta - j \sin \theta$ to show that

$$C_n = \frac{T}{2} (A_n - jB_n)$$

Thus, $(T/2)A_n$ is the real part of C_n, and $-(T/2)B_n$ is the imaginary part of C_n (note also that $C_0 = (T/2)A_0 = TK = \int_{-T/2}^{T/2} f(t) \, dt$ is the area of the function for one period).

1.9 (a) Determine the trigonometric form of the Fourier series for the function shown in Fig. 1.9a of the text.
(b) The equation for C_n for the exponential form of this series is given by Eq. (1.39). Check that the answer for part a and Eq. (1.39) satisfy the relationship $C_n = (T/2)(A_n - jB_n)$.

1.10 Determine the exponential form of the Fourier series for the function whose trigonometric form was obtained in:

(a) Prob. 1.2
(b) Prob. 1.3
(c) Prob. 1.4
(d) Prob. 1.5

Check that the relationship $C_n = (T/2)(A_n - jB_n)$ is satisfied in each case.

1.11 Figure P1.11 shows a periodic function $f(t)$ and its derivative $f'(t)$.

(a) Determine the trigonometric form of the Fourier series representation for the derivative $f'(t)$. That is, specify K, A_n, and B_n.
(b) Integrate the series representation for $f'(t)$ to obtain the series for $f(t)$.

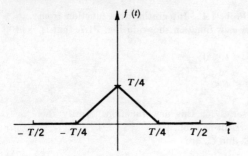

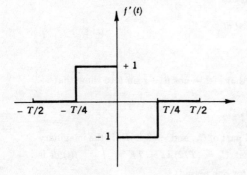

Fig. P1.11

1.12 Determine the exponential form of the Fourier series for the function shown in:

(a) Fig. 1.4c
(b) Fig. 1.5c

1.13 A unit impulse is the limit as ϵ approaches zero of a function of width ϵ and height $1/\epsilon$. Thus, for a unit impulse, the width is zero and

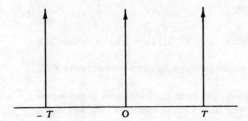

Fig. P1.13

height infinite, but area $\epsilon(1/\epsilon) = 1$ is 1. For the train of unit impulses shown in Fig. P1.13:

(a) Determine the trigonometric form of the Fourier series.
(b) Determine the exponential form of the Fourier series.
(c) Check that the answers for parts a and b satisfy the relationship $C_n = (T/2)(A_n - jB_n)$.

Hint: Because the duration of a unit impulse is so short, the integral of the product of a unit impulse $u_1(t)$ and a function $f(t)$ is

$$\int_0^\epsilon u_1(t)f(t)\ dt = f(0)\int_0^\epsilon u_1(t)\ dt = f(0)$$

The function may be factored from the integral because it remains essentially constant during the short interval of the integral. The last integral is the area of the impulse which is 1, and $f(0)$ is the value of $f(t)$ at $t = 0$ when it is being multiplied by the impulse.

1.14 For two functions each of which is Laplace-transformable, determine whether or not:

(a) The sum is transformable.
(b) The product is transformable.

1.15 For each of the functions given below, determine the range of values of σ for which these functions are Laplace-transformable:

(a) $\sin \omega t$ (b) $\sinh \omega t$
(c) $C = \text{const}$ (d) te^{at}
(e) $e^{at}\sin \omega t$ (f) $C + t + e^{at}$

1.16 Which of the functions given in Prob. 1.15 are Fourier-transformable?

REFERENCES

1 Zygmund, A.: "Trigonometric Series," Cambridge University Press, New York, 1959.
2 Carslaw, H. S.: "Introduction to the Theory of Fourier Series and Integrals," 3d ed., The Macmillan Company, New York, 1930.
3 Churchill, R. V.: "Fourier Series and Boundary Value Problems," 2d ed., McGraw-Hill Book Company, New York, 1963.
4 Titchmarsh, E. C.: Introduction to the Theory of Fourier Integrals," Oxford University Press, New York, 1948.
5 Byerly, W. E.: "Fourier Series and Spherical Harmonics," Ginn and Company, Boston, 1893.
6 LePage, W. R.: "Complex Variables and the Laplace Transform for Engineers," McGraw-Hill Book Company, New York, 1961.

chapter 2

Properties of the
Laplace transformation

Many properties and theorems pertaining to Laplace transformations are developed in this chapter. It is necessary to be familiar with these basic characteristics in order to make effective use of Laplace transformations in solving engineering problems.

2.1 TRANSFORMS OF FUNCTIONS

The function shown in Fig. 2.1a is called a *unit step function*. This function, $f(t) = 1$, is designated as $u(t)$. The application of Eq. (1.46) to obtain the Laplace transform of this function gives

$$\mathcal{L}[u(t)] = \int_0^\infty (1)e^{-st}\,dt = -\left.\frac{e^{-st}}{s}\right|_0^\infty = \frac{1}{s} \qquad \sigma > 0 \qquad (2.1)$$

The function $f(t) = t$ shown in Fig. 2.1b is called a *ramp function*. Its transform is

$$\mathcal{L}(t) = \int_0^\infty te^{-st}\,dt = -\left.\left(\frac{t}{s} + \frac{1}{s^2}\right)e^{-st}\right|_0^\infty = \frac{1}{s^2} \qquad \sigma > 0 \qquad (2.2)$$

In Fig. 2.1c is shown the parabolic function $f(t) = t^2$. Its transform is

$$\mathcal{L}(t^2) = \int_0^\infty t^2 e^{-st}\,dt = -\left.\left(\frac{t^2}{s} - \frac{2t}{s^2} - \frac{2}{s^3}\right)e^{-st}\right|_0^\infty = \frac{2}{s^3} \qquad \sigma > 0 \qquad (2.3)$$

From Eq. (1.46), if a is a constant, then

$$\mathcal{L}[af(t)] = a\int_0^\infty f(t)e^{-st}\,dt = aF(s) \qquad (2.4)$$

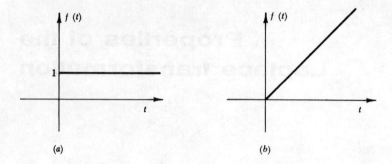

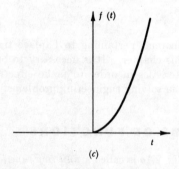

Fig. 2.1. (a) Step function; (b) ramp function; (c) parabolic function.

Similarly, from Eq. (1.46), it follows that

$$\mathcal{L}[af_1(t) + bf_2(t)] = aF_1(s) + bF_2(s) \tag{2.5}$$

A linear transformation is one in which the transform of the sum of two functions is the sum of the transform of each individual function. Thus, Eq. (2.5) shows that the Laplace transformation is a linear transformation.

The transform of the function

$$f(t) = \cosh at = \tfrac{1}{2}(e^{at} + e^{-at})$$

is determined as follows:

$$\mathcal{L}\left(\frac{e^{at} + e^{-at}}{2}\right) = \tfrac{1}{2}\mathcal{L}(e^{at}) + \tfrac{1}{2}\mathcal{L}(e^{-at})$$

$$= \frac{1}{2}\frac{1}{s-a} + \frac{1}{2}\frac{1}{s+a} = \frac{s}{s^2 - a^2} \tag{2.6}$$

where the transform of e^{at} is given by Eq. (1.54). The transform of e^{-at} is obtained by substituting $-a$ for a.

2.2 TRANSFORMS OF DERIVATIVES AND INTEGRALS

To solve differential equations, it is necessary to obtain the transform of the differential terms which appear in the differential equation. An integrodifferential equation is one which contains integral terms as well as differentials.

The transform of the derivative with respect to time of a function is

$$\mathcal{L}\left[\frac{d}{dt}f(t)\right] = \int_0^\infty \left[\frac{d}{dt}f(t)\right]e^{-st}\,dt$$

By letting $u = e^{-st}$ and $v = f(t)$, integration by parts gives

$$\int_0^\infty \left[\frac{df(t)}{dt}\right]e^{-st}\,dt = f(t)e^{-st}\Big|_0^\infty + s\int_0^\infty f(t)e^{-st}\,dt \tag{2.7}$$

To be Laplace-transformable, $f(t)$ must be of exponential order. From Eqs. (1.49) and (1.52), it follows that for large values of t

$$|f(t)e^{-st}| = |f(t)|e^{-\sigma t} \leq Me^{-\epsilon t}$$

Hence

$$\lim_{t\to\infty} |f(t)e^{-st}| \leq \lim_{t\to\infty} Me^{-\epsilon t} \approx 0$$

Thus, the term $f(t)e^{-st}$ in Eq. (2.7) vanishes at the upper limit, $t = \infty$. At the lower limit, $t = 0$, the term $f(t)e^{-st}$ reduces to $f(0)$. Therefore, Eq. (2.7) may be written in the form

$$\mathcal{L}[f'(t)] = sF(s) - f(0) \tag{2.8}$$

The transform of higher derivatives is similarly found by continuing this process of integration by parts; hence,

$$\mathcal{L}[f''(t)] = s^2F(s) - sf(0) - f'(0)$$
$$\mathcal{L}[f'''(t)] = s^3F(s) - s^2f(0) - sf'(0) - f''(0) \tag{2.9}$$
$$\mathcal{L}[f^n(t)] = s^nF(s) - s^{n-1}f(0) - s^{n-2}f'(0) - \cdots - f^{n-1}(0)$$

where

$$f'(t) = \frac{df(t)}{dt} \qquad f''(t) = \frac{d^2f(t)}{dt^2} \qquad f^n(t) = \frac{d^nf(t)}{dt^n}$$

and $f'(0)$ is the initial value of $f'(t)$ evaluated at $t = 0$, etc.

Equation (2.8) may be used to obtain the transform of a function $f(t)$ when the transform of its derivative $f'(t)$ is known.

Illustrative example Let it be desired to determine the transform of the function $f(t) = t^2$. It is known that

$$\mathfrak{L}[f'(t)] = \mathfrak{L}(2t) = \frac{2}{s^2}$$

Solution First solve Eq. (2.8) for $F(s)$.

$$F(s) = \frac{\mathfrak{L}[f'(t)] + f(0)}{s} \tag{2.10}$$

Because $f(t) = t^2$, $F(s) = \mathfrak{L}[f(t)] = \mathfrak{L}(t^2)$. The initial condition is $f(0) = f(t) \big|_{t=0} = 0$. Substituting these results into the preceding expression gives

$$F(s) = \mathfrak{L}(t^2) = \frac{2}{s^3} \tag{2.11}$$

In a similar manner, Eq. (2.9) enables one to obtain the transform of a function when the transform of the nth derivative is known.

Illustrative example Determine the transform of the function $f(t) = t^n$.

Solution Successive differentiation gives $f'(t) = nt^{n-1}$,

$$f''(t) = n(n - 1)t^{n-2}, \ldots, f^n(t) = n!$$

Because $n!$ is a constant, a plot of $n!$ for positive values of time would be a step function whose height is $n!$; hence

$$\mathfrak{L}[f^n(t)] = \mathfrak{L}(n!) = \frac{n!}{s}$$

Solving Eq. (2.9) for $F(s)$ gives

$$F(s) = \frac{\mathfrak{L}[f^n(t)] + s^{n-1}f(0) + \cdots + f^{n-1}(0)}{s^n}$$

Because $F(s) = \mathfrak{L}[f(t)] = \mathfrak{L}(t^n)$ and all the initial conditions are zero [that is, $f(0) = f'(0) = \cdots = f^{n-1}(0) = 0$],

$$F(s) = \mathfrak{L}(t^n) = \frac{n!/s}{s^n} = \frac{n!}{s^{n+1}} \tag{2.12}$$

Many functions may be represented as power series; i.e.,

$$f(t) = \sum_{n=0}^{\infty} a_n t^n$$

For functions which are expressed as power series, the transform is

$$F(s) = \sum_{n=0}^{\infty} \frac{a_n n!}{s^{n+1}} \qquad (2.13)$$

For functions that have a derivative that is a constant times the original function, Eq. (2.9) may be employed directly to obtain the transform. That is, it is not necessary to know the transform of the nth derivative of the function.

Illustrative example   Determine the transform of $f(t) = \sin at$.

Solution   Successive differentiation of $f(t)$ gives

$$f'(t) = a \cos at$$
$$f''(t) = -a^2 \sin at = -a^2 f(t)$$

Note that $f''(t)$ is a constant $-a^2$ times the original function. Because $f''(t) = -a^2 f(t)$, then

$$\mathcal{L}[f''(t)] = -a^2 \mathcal{L}[f(t)] = -a^2 F(s)$$

Solving the equation for the transform of the second derivative of a function [Eq. (2.9)] for $F(s)$ gives

$$F(s) = \frac{\mathcal{L}[f''(t)] + sf(0) + f'(0)}{s^2} \qquad (2.14)$$

The initial conditions are

$$f(0) = \sin at \Big|_{t=0} = 0 \quad \text{and} \quad f'(0) = a \cos at \Big|_{t=0} = a$$

Thus, the preceding equation for $F(s)$ becomes

$$F(s) = \frac{-a^2 F(s) + a}{s^2}$$

Solving for $F(s) = \mathcal{L}(\sin at)$ gives

$$F(s) = \mathcal{L}(\sin at) = \frac{a}{s^2 + a^2} \qquad (2.15)$$

In general, the transform of a function may be ascertained by application of one of the general expressions for obtaining the transform of a derivative [Eq. (2.8) or Eqs. (2.9)] when:

1. The transform of the nth derivative of the function is known [for example, $(d^n/dt^n)(t^n) = n!$, and $\mathcal{L}(n!) = n!/s$].
2. A derivative of the function is a constant times the function [for example, $(d^2/dt^2) \sin \omega t = -\omega^2 \sin \omega t$ or $(d/dt)e^{at} = ae^{at}$].

Laplace transforms provide a very convenient method for evaluating definite integrals of the type

$$\int_0^\infty f(t)e^{-at}\, dt \tag{2.16}$$

For example, Eq. (2.12) states that

$$\int_0^\infty t^n e^{-st}\, dt = \frac{n!}{s^{n+1}}$$

This integral exists for any $s = \sigma + j\omega$ in which $\sigma > 0$. For the case in which $\omega = 0$ and σ is replaced by a,

$$\int_0^\infty t^n e^{-at}\, dt = \frac{n!}{a^{n+1}} \qquad a > 0 \tag{2.17}$$

This technique is applicable for any function $f(t)$ whose Laplace transform is known.

Transform of definite integrals

The Laplace transform of a definite integral is obtained as follows:

$$\mathcal{L}\left[\int_0^t f(t)\, dt\right] = \int_0^\infty \left[\int_0^t f(t)\, dt\right] e^{-st}\, dt \tag{2.18}$$

The preceding expression may be integrated by parts by letting

$$u = \int_0^t f(t)\, dt \text{ and } dv = e^{-st}\, dt$$

Hence

$$\mathcal{L}\left[\int_0^t f(t)\, dt\right] = -\frac{1}{s}e^{-st}\int_0^t f(t)\, dt\,\Big|_0^\infty + \frac{1}{s}\int_0^\infty f(t)e^{-st}\, dt \tag{2.19}$$

In order that the integral be Laplace-transformable, it must be of exponential order. Therefore, the first term on the right-hand side of the preceding expression vanishes at the upper limit. This term also vanishes at the lower limit because at $t = 0$ the integration is from zero to zero. Thus, Eq. (2.19) becomes

$$\mathcal{L}\left[\int_0^t f(t)\, dt\right] = \frac{1}{s} F(s) \tag{2.20}$$

where $F(s) = \mathcal{L}[f(t)]$.

Illustrative example It is known that the Laplace transform of t is $1/s^2$. Use Eq. (2.20) to determine the transform of t^2.

Solution Because $t^2 = \int_0^t 2t\, dt$, then

$$\mathcal{L}[t^2] = \mathcal{L}\left(2 \int_0^t t\, dt\right) = \frac{2}{s} F(s) = \frac{2}{s^3}$$

Transform of indefinite integrals

An indefinite integral has the form

$$y(t) = \int f(t)\, dt = \int_0^t f(t)\, dt + C \tag{2.21}$$

The constant of integration C is obtained by evaluating the preceding expression at time $t = 0$; hence

$$y(0) = 0 + C = C \tag{2.22}$$

Thus, C is seen to be the initial value of $y(t)$ at $t = 0$. Substitution of $C = y(0)$ in Eq. (2.21) gives

$$y(t) = \int_0^t f(t)\, dt + y(0)$$

The term $\int_0^t f(t)\, dt$ is the change in $y(t)$ from the initial value $y(0)$ to time t. Because $y(0)$ is a constant,

$$\mathcal{L}[y(t)] = \mathcal{L}\left[\int f(t)\, dt\right] = \mathcal{L}\left[\int_0^t f(t)\, dt + y(0)\right]$$
$$= \frac{1}{s}[F(s) + y(0)] \tag{2.23}$$

where $F(s) = \mathcal{L}[f(t)]$.

Illustrative example Determine the transform of $y(t) = t^2 + 3$ by using the relationship $\mathcal{L}(t) = 1/s^2$.

Solution For this function $y(t) = \int_0^t 2t\,dt + 3$, in which $f(t) = t$ and $y(0) = 3$; thus

$$\mathcal{L}(t^2 + 3) = \mathcal{L}\left(2 \int_0^t t\,dt + 3\right) = \frac{2F(s)}{s} + \frac{3}{s} = \frac{2}{s^3} + \frac{3}{s}$$

Illustrative example Use the result of Eq. (2.15) that

$$\mathcal{L}(\sin at) = \frac{a}{s^2 + a^2}$$

to determine the transform of $\cos at$.

Solution First note that

$$y(t) = \int_0^t -a \sin at\,dt + y(0) = \cos at \Big|_0^t + 1 = \cos at$$

Because $f(t) = -a \sin at$, $F(s) = -a^2/(s^2 + a^2)$, and $y(0) = 1$, application of Eq. (2.23) gives

$$\mathcal{L}(\cos at) = \mathcal{L}\left(-a \int_0^t \sin at\,dt + 1\right)$$

$$= \frac{-a^2}{s(s^2 + a^2)} + \frac{1}{s} = \frac{s}{s^2 + a^2}$$

The equations for the transforms of definite and indefinite integrals [Eqs. (2.20) and (2.23)] enable one to determine the transform of the integral of a function when the transform of the function $f(t)$ is known.

2.3 TRANSFORM PROPERTIES

General properties of Laplace transforms are developed in this section.

Change of scale

A change of scale is effected in a time function by the substitution of at for t. If the transform of $f(t)$ is $F(s)$, the transform of $f(at)$ is

$$\mathcal{L}[f(at)] = \frac{1}{a} F\left(\frac{s}{a}\right) \tag{2.24}$$

This result is verified as follows: Application of Eq. (1.46) to obtain the transform of $f(at)$ gives

$$\mathcal{L}[f(at)] = \int_0^\infty f(at)e^{-st}\,dt$$

By letting $\tau = at$, and thus $d\tau = a\,dt$, then

$$\mathcal{L}[f(at)] = \frac{1}{a}\int_0^\infty f(\tau)e^{-(s/a)\tau}\,d\tau \qquad (2.25)$$

The transform for $F(s/a)$ is obtained by substituting s/a for s in Eq. (1.46); that is,

$$F\left(\frac{s}{a}\right) = \int_0^\infty f(\tau)e^{-(s/a)\tau}\,d\tau$$

The substitution of this result into Eq. (2.25) verifies Eq. (2.24).

Illustrative example The transform of $\cos t$ is

$$\mathcal{L}(\cos t) = \frac{s}{s^2+1}$$

Determine the transform of $\cos at$.

Solution Direct application of Eq. (2.24) gives

$$\mathcal{L}(\cos at) = \frac{1}{a}\frac{s/a}{(s/a)^2+1} = \frac{s}{s^2+a^2}$$

Transform of $tf(t)$

Differentiation of Eq. (1.46) with respect to s yields an expression for obtaining the transform of $tf(t)$; that is,

$$\frac{d}{ds}F(s) = \frac{d}{ds}\int_0^\infty f(t)e^{-st}\,dt = \int_0^\infty \frac{d}{ds}f(t)e^{-st}\,dt$$

$$= -\int_0^\infty tf(t)e^{-st}\,dt = -\mathcal{L}[tf(t)] \qquad (2.26)$$

Solving the preceding for the transform of $tf(t)$ yields the desired result.

$$\mathcal{L}[tf(t)] = -\frac{d}{ds}F(s) \qquad (2.27)$$

This result [Eq. (2.27)] is applicable for any function $f(t)$ which is Laplace-transformable.

In the development of Eq. (2.26), differentiation was taken inside the integral. In general, differentiation may be taken inside the integral sign

$$\frac{d}{ds} \int_a^\infty f(s,t)\, dt = \int_a^\infty \frac{d}{ds} f(s,t)\, dt \tag{2.28}$$

if $f(s,t)$ is continuous with respect to s, if $f(s,t)$ is continuous or piecewise continuous with respect to t, and if $\int_a^\infty (d/ds)f(s,t)\, dt$ exists. To prove Eq. (2.28), let $G(s) = \int_a^\infty f(s,t)\, dt$. Thus, Eq. (2.28) may be written in the form

$$\begin{aligned}
\frac{d}{ds} \int_a^\infty f(s,t)\, dt = \frac{d}{ds} G(s) &= \lim_{\Delta s \to 0} \frac{G(s + \Delta s) - G(s)}{\Delta s} \\
&= \lim_{\Delta s \to 0} \int_a^\infty \frac{[f(s + \Delta s, t) - f(s,t)]\, dt}{\Delta s} \\
&= \int_a^\infty \frac{d}{ds} f(s,t)\, dt
\end{aligned}$$

In order that the preceding limit exist as Δs approaches zero, $f(s,t)$ must be continuous with respect to s. It suffices to have $f(s,t)$ piecewise continuous with respect to t.

For the case of Eq. (2.26), then $f(s,t) = f(t)e^{-st}$. Thus,

$$\int_a^\infty (d/ds)f(t)e^{-st}\, dt = - \int_0^\infty tf(t)e^{-st}\, dt$$

From Eq. (1.48), it follows that this integral exists if $tf(t)$ is of exponential order. If $f(t)$ is of exponential order, $tf(t)$ is also of exponential order. Similarly, if $f(t)$ is piecewise continuous, $f(s,t) = f(t)e^{-st}$ is continuous with respect to s and piecewise continuous with respect to t. Thus, the requirements for taking the differentiation inside the integral are automatically satisfied for any function $f(t)$ which is Laplace-transformable.

By proceeding in a similar manner, it may be shown that, if $f(t)$ is Laplace-transformable,

$$\mathcal{L}[t^n f(t)] = (-1)^n \frac{d^n}{ds^n} F(s) \tag{2.29}$$

For the case in which $n = 1$, then Eq. (2.29) reduces to Eq. (2.27).

As an illustration of the application of this theorem, let it be desired to determine the transform of te^{at}, in which case $f(t) = e^{at}$ and $F(s) = 1/(s - a)$; hence

$$\mathcal{L}(te^{at}) = - \frac{d}{ds}\left(\frac{1}{s - a} \right) = \frac{1}{(s - a)^2} \tag{2.30}$$

Transform of $f(t)/t$

Integration of both sides of Eq. (1.46) with respect to s yields a convenient expression for obtaining the transform of $f(t)/t$; that is,

$$\int_s^\infty F(s)\,ds = \int_s^\infty \left[\int_0^\infty f(t)e^{-st}\,dt \right] ds = \int_0^\infty \left[\int_s^\infty f(t)e^{-st}\,ds \right] dt$$

$$= \int_0^\infty \left[-\frac{f(t)}{t} e^{-st} \right]_s^\infty dt = \int_0^\infty \frac{f(t)}{t} e^{-st}\,dt = \mathcal{L}\left[\frac{f(t)}{t} \right]$$

$$(2.31)$$

Hence

$$\mathcal{L}\left[\frac{f(t)}{t} \right] = \int_s^\infty F(s)\,ds \qquad (2.32)$$

This theorem [Eq. (2.32)] is applicable if $f(t)$ is Laplace-transformable and if $\lim_{t \to 0} f(t)/t$ exists.

In the development of Eq. (2.31), it is necessary to interchange the order of integration. In general, the order of integration may be interchanged,

$$\int_a^b \int_c^\infty f(x,t)\,dt\,dx = \int_c^\infty \int_a^b f(x,t)\,dx\,dt \qquad (2.33)$$

if the improper integral

$$\int_c^\infty f(x,t)\,dt = F(x) \qquad (2.34)$$

exists. An improper integral exists if the integrand is piecewise continuous and if the integral is uniformly convergent (see Sec. 1.5). Substitution of Eq. (2.34) into the left-hand side of Eq. (2.33) shows that the order of integration may be interchanged. That is,

$$\int_a^b F(x)\,dx = \sum_{k=0}^n F(a + k\,\Delta x)\,\Delta x$$

$$= F(a)\,\Delta x + F(a + \Delta x)\,\Delta x + \cdots$$

$$= \int_c^\infty f(a,t)\,\Delta x\,dt + \int_c^\infty f(a + \Delta x,t)\,\Delta x\,dt + \cdots$$

$$= \int_c^\infty \sum_{k=0}^n f(a + k\,\Delta x,t)\,\Delta x\,dt$$

$$= \int_c^\infty \int_a^b f(x,t)\,dx\,dt \qquad (2.35)$$

When the upper limit b is infinite, then it is also necessary to show that the resultant integral in Eq. (2.35) exists.

To interchange the order of integration as indicated in Eq. (2.31), it is necessary that

$$\int_0^\infty f(t)e^{-st}\,dt$$

exist. In addition, it is necessary to show that the resultant integral

$$\int_0^\infty \int_s^\infty f(t)e^{-st}\,ds\,dt = \int_0^\infty \frac{f(t)}{t}\,e^{-st}\,dt$$

exists. The first of these integrals exists if $f(t)$ is Laplace-transformable, and the second exists if $f(t)/t$ is Laplace-transformable.

In order that $f(t)/t$ be Laplace-transformable, it is necessary that $f(t)/t$ be piecewise continuous and of exponential order. If $f(t)$ is of exponential order, $f(t)/t$ is also of exponential order. If $f(t)$ is piecewise continuous, $f(t)/t$ is piecewise continuous provided that $\lim_{t\to 0} f(t)/t$ exists. Thus, Eq. (2.32) is applicable if $f(t)$ is Laplace-transformable and if $\lim_{t\to 0} f(t)/t$ exists.

To illustrate this theorem, let it be desired to determine the transform of $(\sin at)/t$, whence $f(t) = \sin at$ and $F(s) = a/(s^2 + a^2)$.

$$\mathfrak{L}\left(\frac{\sin at}{t}\right) = \int_s^\infty \frac{a}{s^2 + a^2}\,ds = \tan^{-1}\frac{s}{a}\bigg|_s^\infty = \frac{\pi}{2} - \tan^{-1}\frac{s}{a}$$

$$= \cot^{-1}\frac{s}{a} \tag{2.36}$$

As another illustration, let it be desired to determine the function $f(t)$ whose transform is $1/(s - a)^2$.

$$\mathfrak{L}\left[\frac{f(t)}{t}\right] = \int_s^\infty \frac{1}{(s - a)^2}\,ds = \frac{-1}{s - a}\bigg|_s^\infty = \frac{1}{s - a}$$

$$\frac{f(t)}{t} = \mathfrak{L}^{-1}\mathfrak{L}\left[\frac{f(t)}{t}\right] = \mathfrak{L}^{-1}\left(\frac{1}{s - a}\right) = e^{at}$$

Hence

$$f(t) = te^{at} \tag{2.37}$$

2.4 TRANSFORM THEOREMS

Interesting theorems pertaining to the Laplace transformation itself are developed in this section.

First shifting theorem

The substitution of $s - a$ for s in Eq. (1.46) gives

$$F(s - a) = \int_0^\infty f(t)e^{-(s-a)t}\,dt = \int_0^\infty e^{at}f(t)e^{-st}\,dt = \mathfrak{L}[e^{at}f(t)]$$

Hence

$$\mathcal{L}[e^{at}f(t)] = F(s - a) \tag{2.38}$$

The transform of $e^{at}f(t)$ is obtained by substituting $s - a$ for s in the transform of $f(t)$. This substitution of $s - a$ for s shifts the transform of the original function by an amount a. Thus, this is referred to as the *first shifting theorem*.

To illustrate the application of this theorem, let it be desired to obtain the transform of $t^n e^{at}$. The transform of t^n is given by Eq. (2.12); hence the substitution of $s - a$ for s gives

$$\mathcal{L}(t^n e^{at}) = \frac{n!}{(s - a)^{n+1}} \tag{2.39}$$

As another example, the transform of $\sin kt$ is $k/(s^2 + k^2)$. Thus

$$\mathcal{L}(e^{at} \sin kt) = \frac{k}{(s - a)^2 + k^2} \tag{2.40}$$

Second shifting theorem

In Fig. 2.2 is shown a function $f(t)$ whose transform is $F(s)$. The function $f(t - a)$ is delayed a time a. The transform of this delayed function is

$$\mathcal{L}[f(t - a)] = \int_0^\infty f(t - a)e^{-st}\,dt = \int_a^\infty f(t - a)e^{-st}\,dt \tag{2.41}$$

The lower limit of integration is changed from 0 to a because the function $f(t - a)$ has no value for $t < a$. The substitution of

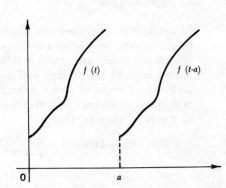

Fig. 2.2. Delayed function.

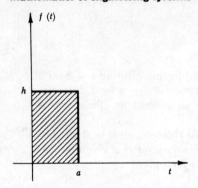

Fig. 2.3.   Pulse function.

$\tau = t - a$ and $d\tau = dt$ into Eq. (2.41) gives

$$\mathcal{L}[f(t - a)] = \int_0^\infty f(\tau)e^{-s(\tau+a)}\,d\tau = e^{-as}\int_0^\infty f(\tau)e^{-s\tau}\,d\tau$$

$$= e^{-as}F(s) \tag{2.42}$$

The lower limit of integration in Eq. (2.41) is at $t = a$; hence the corresponding limit in Eq. (2.42) is at $\tau = t - a = 0$. The transform of a delayed function $f(t - a)$ is obtained by multiplying the transform of the original function by e^{-as}. In the second shifting theorem [Eq. (2.42)], the time function is shifted by a, whereas, in the first shifting theorem, the transform is shifted by a.

In Fig. 2.3 is shown a pulse function of height h and duration a. This function may be regarded as a step function which begins at time $t = 0$, minus an equal step function which begins at time $t = a$. Hence the transform of the pulse is

$$\frac{h}{s} - \frac{h}{s}e^{-as} = \frac{h}{s}(1 - e^{-as}) \tag{2.43}$$

For the case in which $h = k/a$, the area of the pulse is

$$\text{Area} = ha = \frac{k}{a}a = k$$

In the limit as a becomes zero, the width also becomes zero and the height infinite, but the area is always k. This resulting function of infinite height and zero width is called an *impulse function of area k*. A unit impulse has an area of 1 and is designated by the symbol $u_1(t)$. An impulse of area k is designated as $ku_1(t)$. Its transform is obtained by letting $h = k/a$ in Eq. (2.43) and then taking the limit as a approaches zero; thus

$$\mathcal{L}[ku_1(t)] = k\mathcal{L}[u_1(t)] = \lim_{a \to 0} \frac{k}{as}(1 - e^{-as})$$

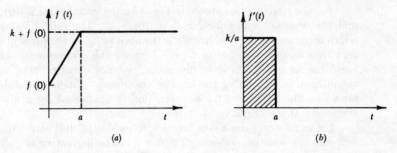

Fig. 2.4. Function whose derivative is a pulse. (a) The function; (b) the derivative.

Because $e^{-as} \approx 1$ as a becomes zero, the preceding limit is indeterminate. Application of L'Hospital's rule gives

$$k\mathcal{L}[u_1(t)] = \lim_{a \to 0} \frac{(d/da)[k(1 - e^{-as})]}{(d/da)(as)} = \lim_{a \to 0} \frac{kse^{-as}}{s} = k \qquad (2.44)$$

Hence the transform of an impulse is equal to its area k.

In Fig. 2.4a is shown a function $f(t)$. The initial value of this function is $f(0)$. The derivative of this function $f'(t) = df(t)/dt$ is the pulse function shown in Fig. 2.4b. By taking the limit as a approaches zero (slope k/a becomes infinite), the function $f(t)$ in Fig. 2.4a approaches the step function shown in Fig. 2.5a. Similarly, the limit of the derivative shown in Fig. 2.4b (as a approaches zero and thus k/a becomes infinite) is the impulse of area k shown in Fig. 2.5b. As is illustrated in Fig. 2.5a and b, the derivative of a step change is an impulse whose area k is equal to the change k in the step.

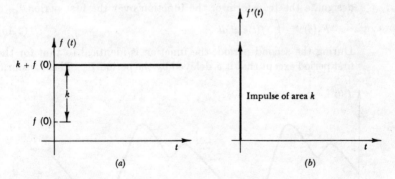

Fig. 2.5. Function whose derivative is an impulse. (a) The function (step function); (b) the derivative.

For the step function shown in Fig. 2.5a, the initial value is $f(0)$, and the constant value for $t > 0$ is $f(0) + k$. The initial value which accounts for the past history is the limit of $f(t)$ as t approaches zero from negative values of time. Because the discontinuity at $t = 0$ has no width, then the discontinuity contributes nothing to the integral in computing the Laplace transform. Similarly it follows that the transform for any function is unaffected by a discontinuity at the origin.

Let us now consider a step function $f(t)$ whose initital value $f(0)$ is 2, and the constant value for $t > 0$ is 5. The derivative of this function is an impulse whose area is equal to the step change $5 - 2 = 3$. The transform of the derivative is $\mathcal{L}[f'(t)] = 3$. Thus, Eq. (2.8) may be solved for $F(s)$ to determine the transform of the original step function.

$$F(s) = \frac{\mathcal{L}[f'(t)] + f(0)}{s} = \frac{3 + 2}{s} = \frac{5}{s}$$

A constant is the particular step function in which there is no step change from the initial value $f(0)$ to the constant value for $t > 0$. Thus, for the constant $f(t) = 5$ the initial value is $f(0) = 5$. Because the derivative is zero, application of the preceding expression yields for the transform

$$F(s) = \frac{0 + 5}{s} = \frac{5}{s}$$

This show that the transform is unaffected by the discontinuity.

Periodic functions

In Fig. 2.6 is shown a periodic function of period T. Let $F_1(s)$ designate the transform of the function over the first period.

$$F_1(s) = \int_0^T f(t)e^{-st}\, dt \tag{2.45}$$

During the second period, the function is identical to that for the first period except that it is delayed by the period T. The transform

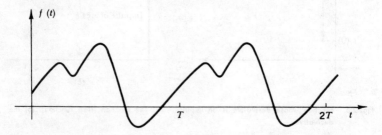

Fig. 2.6. Periodic function.

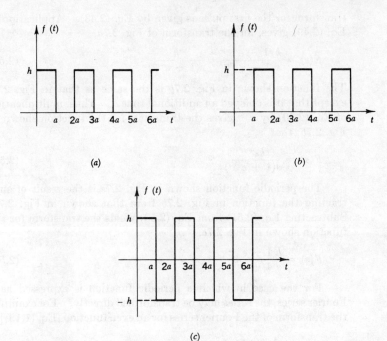

Fig. 2.7. Periodic functions. (a) Series of pulse functions; (b) series of delayed pulse functions; (c) alternating positive and negative pulse functions.

for the second period is $F_1(s)e^{-Ts}$. For the third period, the delay is $2T$, etc. Adding the transform for each period gives

$$F(s) = F_1(s)(1 + e^{-Ts} + e^{-2Ts} + \cdots) \qquad (2.46)$$

Let S equal the sum of the terms in the parentheses.

$$S = 1 + e^{-Ts} + e^{-2Ts} + e^{-3Ts} + \cdots \qquad (2.47)$$

Multiplying the preceding summation by e^{-Ts} gives

$$e^{-Ts}S = e^{-Ts} + e^{-2Ts} + e^{-3Ts} + \cdots \qquad (2.48)$$

Subtracting Eq. (2.48) from Eq. (2.47) gives

$$S - e^{-Ts}S = 1 \qquad \text{or} \qquad S = \frac{1}{1 - e^{-Ts}}$$

Thus, the transform $F(s)$ of a periodic function is

$$F(s) = F_1(s)\frac{1}{1 - e^{-Ts}} \qquad (2.49)$$

where $F_1(s)$ is the transform of the function over the first period. For the periodic function shown in Fig. 2.7a, the period is $2a$. The transform for the first pulse is given by Eq. (2.43). Application of Eq. (2.49) gives, for the transform of Fig. 2.7a,

$$F(s) = \frac{h}{s} \frac{1 - e^{-as}}{1 - e^{-2as}} = \frac{h}{s(1 + e^{-as})} \tag{2.50}$$

The function shown in Fig. 2.7b is the same as that in Fig. 2.7a except that it is delayed an additional time a. Thus, multiplication of Eq. (2.50) by e^{-as} gives the transform for the function shown in Fig. 2.7b; that is,

$$F(s) = \frac{he^{-as}}{s(1 + e^{-as})} \tag{2.51}$$

The periodic function shown in Fig. 2.7c is the result of subtracting the function in Fig. 2.7b from that shown in Fig. 2.7a. Subtracting Eq. (2.51) from Eq. (2.50) yields the transform for the function shown in Fig. 2.7c.

$$F(s) = \frac{h}{s} \frac{1 - e^{-as}}{1 + e^{-as}} \tag{2.52}$$

For the case in which a periodic function is expressed as a Fourier series, the series may be transformed directly. For example, the transform of the Fourier series for an even function [Eq. (1.15)] is

$$F(s) = \frac{A_0}{s} + \sum_{n=1}^{\infty} \frac{sA_n}{s^2 + (n\omega_0)^2}$$

Similarly the transform of the Fourier series for an odd function [Eq. (1.19)] is

$$F(s) = \sum_{n=1}^{\infty} \frac{n\omega_0 B_n}{s^2 + (n\omega_0)^2}$$

Adding the two preceding results gives the transform for a general periodic function.

Final-value theorem

This theorem provides a means for determining directly from the transform $F(s)$ the final value of the time function $f(t)$ at $t = \infty$. The final-value theorem is applicable only to functions which have a limit at $t = \infty$; otherwise $f(\infty)$ is meaningless. Examples of functions which do not have a limit at $t = \infty$ are $\sin \omega t$, which oscillates indefinitely, and functions such as e^{at} (where $a > 0$), which become infinite at $t = \infty$. To develop this theorem, first write the equation for the transform of the derivative of a function in the form

$$\mathcal{L}[f'(t)] = \int_0^\infty f'(t)e^{-st}\,dt = sF(s) - f(0) \tag{2.53}$$

Evaluating the preceding expression at $s = 0$ gives

$$\int_0^\infty f'(t)\,dt = [sF(s)]_{s=0} - f(0) \tag{2.54}$$

There is no e^{-st} term in the first integral because, for $s = 0$, $e^{-st} = 1$. The left-hand side of the preceding expression may be written in the form

$$\int_0^\infty f'(t)\,dt = [f(t)]_0^\infty = f(\infty) - f(0) \tag{2.55}$$

where $f(\infty)$ is the final value of $f(t)$ at $t = \infty$. Comparison of Eqs. (2.54) and (2.55) shows that

$$f(\infty) = [sF(s)]_{s=0} \tag{2.56}$$

Equation (2.56) is the mathematical statement of the final-value theorem.

To illustrate the application of this theorem, let it be desired to obtain the final value at $t = \infty$ of the function $f(t) = t^n e^{at}$ (where $a < 0$, for otherwise the function is infinite at $t = \infty$). The transform $F(s)$ of $t^n e^{at}$ is given by Eq. (2.39); hence

$$t^n e^{at}\Big|_{t=\infty} = [sF(s)]_{s=0} = \frac{sn!}{(s-a)^{n+1}}\Bigg]_{s=0} = 0 \qquad a < 0 \tag{2.57}$$

Note that the transform [Eq. (2.39)] is valid for any finite value of a, whereas the final-value theorem is applicable for negative values of a only.

Initial-value theorem

The initial-value theorem enables one to determine the value of the function $f(t)$ at time $t = 0+$ directly from the transform $F(s)$. To derive this theorem, first write the equation for the transform of a derivative in the form

$$\mathcal{L}[f'(t)] = \int_0^\infty f'(t)e^{-st}\,dt = \int_0^{0+} f'(t)\,dt + \int_{0+}^\infty f'(t)e^{-st}\,dt \tag{2.58}$$

There is no e^{-st} term in the integration from 0 to 0+ because, for $t \approx 0$, $e^{-st} \approx 1$. By taking the limit of the preceding expression as

s approaches infinity, the last integral vanishes because $e^{-\infty} \approx 0$. Hence

$$\lim_{s \to \infty} \mathcal{L}[f'(t)] = \int_0^{0+} f'(t) \, dt = f(0+) - f(0)$$

The limit of Eq. (2.8) as s approaches infinity is

$$\lim_{s \to \infty} \mathcal{L}[f'(t)] = \lim_{s \to \infty} [sF(s)] - f(0)$$

Equating the right-hand sides of the two preceding expressions and solving for $f(0+)$ gives

$$f(0+) = [sF(s)]_{s=\infty} \tag{2.59}$$

It is to be noted that the initial-value theorem yields, not the initial value $f(0)$, but rather the value at some time slightly greater than zero, i.e., at time $t = 0+$. The distinction between $f(0)$ and $f(0+)$ is explained in the following discussion of initial conditions.

Initial conditions

In the study of engineering systems, it is desired to know how the system responds to various excitations, or forcing functions. The time when the excitation begins is designated as the time $t = 0$ for the system. The state of a system just before this excitation occurs (i.e., at $t = 0-$) is called the *initial state*. This initial state is specified by the given set of initial conditions for the system (these initial conditions account for the past history of the system). The initial condition $f(0-)$ means the limit of the value of $f(t)$ as t approaches zero from a negative direction. For the sake of simplicity, the initial condition is designated as $f(0)$ rather than $f(0-)$ throughout this text. The fact that the initial-condition terms in the Laplace transform are to be evaluated at $t = 0-$ rather than $t = 0+$ is illustrated by the following example.

For the circuit shown in Fig. 2.8a, the switch is open for $t < 0$ so that $i(0) = 0$. The switch is then closed for $t > 0$. As is shown in Fig. 2.8b, the voltage e applied across the circuit is zero for $t < 0$ (switch opened) and then $e = E_0$ for $t > 0$ (switch closed). The differential equation for this circuit is

$$Ri + \frac{1}{C} \int i \, dt = e \tag{2.60}$$

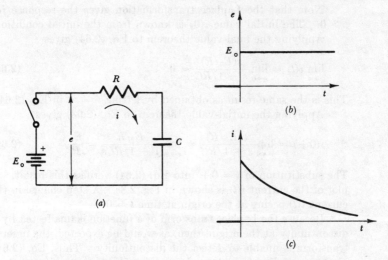

Fig. 2.8. Series RC circuit. (a) The circuit; (b) the voltage e; (c) the current i.

where the voltage drop e across the resistor and capacitor is

$$e = \begin{cases} 0 & t < 0 \\ E_0 & t > 0 \end{cases}$$

Differentiating Eq. (2.60) gives

$$RDi + \frac{1}{C} i = De \tag{2.61}$$

where $D = d/dt$ is the differential operator.

The resulting transform is

$$R[sI(s) - i(0)] + \frac{1}{C} I(s) = sE(s) - e(0) \tag{2.62}$$

The transform of the applied voltage e shown in Fig. 2.8b is $E(s) = E_0/s$. Solving Eq. (2.62) for $I(s)$ and noting that $e(0) = i(0) = 0$, then

$$I(s) = \frac{E_0}{Rs + 1/C} = \frac{E_0/R}{s + 1/RC} \tag{2.63}$$

Inverting the preceding expression gives the equation for the current for $t > 0$; that is,

$$i(t) = \frac{E_0}{R} e^{-t/RC} \qquad t > 0 \tag{2.64}$$

51

Note that the Laplace transformation gives the response for $t > 0$. The initial value $i(0)$ is known from the initial condition. Applying the final-value theorem to Eq. (2.64) gives

$$\lim_{t \to \infty} i(t) = \lim_{s \to 0} \frac{s(E_0/R)}{s + 1/RC} = 0 \tag{2.65}$$

This is the same result as obtained by letting $t = \infty$ in Eq. (2.64). Applying the initial-value theorem to Eq. (2.63) gives

$$i(0+) = \lim_{s \to \infty} \frac{s(E_0/R)}{s + 1/RC} = \lim_{s \to \infty} \frac{E_0/R}{1 + 1/RCs} = \frac{E_0}{R} \tag{2.66}$$

The substitution of $t = 0+$ into Eq. (2.64) verifies this result. A plot of the current $i(t)$ is shown in Fig. 2.8c. A step change in the current $i(t)$ occurs at the origin at time $t = 0$.

Because the Laplace transform of a function is unaffected by a discontinuity at the origin, then, as would be expected, the inverse transform is unable to detect the discontinuity. Thus, Eq. (2.64) gives the solution for values of time after the discontinuity (that is, $t > 0$). Because the initial state of the system must be known, then substitution of $t = 0+$ into the solution and comparison with the initial value reveals whether or not the solution has a discontinuity.

A major advantage of the Laplace transformation method is that the given initial conditions at time $t = 0-$ are employed even though there is a discontinuity at the origin. In the classical method the constants of integration must be evaluated in a region where the response is continuous. Thus, to solve this problem by classical methods, it is necessary to know the state of the system at time $t = 0+$. It should be pointed out that the state of the system at time $t = 0+$ is not the initial state. It is possible to calculate the state of the system at time $t = 0+$ by integrating the original differential equation between the limits $t = 0-$ and $t = 0+$. Thus, integration of Eq. (2.61) gives

$$R \int_{0-}^{0+} Di(t)\, dt + \frac{1}{C} \int_{0-}^{0+} i(t)\, dt = \int_{0-}^{0+} De(t)\, dt$$

$$Ri(t) \Big|_{0-}^{0+} + 0 = e(t) \Big|_{0-}^{0+}$$

$$R[i(0+) - i(0-)] = e(0+) - e(0-) = E_0$$

$$i(0+) = \frac{E_0}{R}$$

Because $e(t)$ is the given input as shown in Fig. 2.8b, it is known that $e(0+) = E_0$ and $e(0-) = 0$. For a second-order differential equation it is necessary to integrate each term twice and then evaluate the result between $0-$ and $0+$ in order to evaluate $y(0+)$. Next each term is integrated once to determine $y'(0+)$. In a similar manner, an nth-order differential equation requires integrating each term n times, and then $n-1$ times, $n-2$ times, etc., in order completely to specify the state of the system at time $t = 0+$. Another disadvantage of the classical method is that, since the behavior of the system is analyzed only after time $t = 0+$, impulses which occur at the origin ($t = 0$) must be ignored. In contrast, the Laplace transform method provides a direct, straightforward solution for these engineering problems.

A list of common Laplace transforms is given in Table 2.1, and a summary of Laplace transform properties and theorems is given in Table 2.2.

2.5 CLASSICAL METHOD

A better understanding of the transformation method of solving differential equations is obtained by first considering the classical method of solution.

In the classical method the solution consists of two parts. The first is called the *complementary solution*, and the second is called the *particular solution*. To illustrate this method, consider the example problem of the preceding section. The right-hand side of Eq. (2.61) is De, where e is the step function shown in Fig. 2.8b. The deriva-

Table 2.1 Laplace transforms

$f(t)$	$F(s)$
$u(t)$ (step)	$1/s$
$u_1(t)$ (impulse)	1
t (ramp)	$1/s^2$
t^n	$n!/s^{n+1}$
$t^n e^{at}$	$n!/(s-a)^{n+1}$
$af(t)$	$aF(s)$
e^{at}	$1/(s-a)$
e^{-at}	$1/(s+a)$
$\sin at$	$a/(s^2+a^2)$
$\cos at$	$s/(s^2+a^2)$
$\sinh at$	$a/(s^2-a^2)$
$\cosh at$	$s/(s^2-a^2)$

Table 2.2

Transform of Derivatives

$$\mathcal{L}[f'(t)] = sF(s) - f(0)$$
$$\mathcal{L}[f^n(t)] = s^nF(s) - s^{n-1}f(0) - \cdots - f^{n-1}(0)$$

Transform of Definite Integral

$$\mathcal{L}\left[\int_0^t f(t)\, dt\right] = F(s)/s$$

Transform of Indefinite Integral

$$\mathcal{L}[y(t)] = \mathcal{L}[\smallint f(t)\, dt] = \frac{F(s) + y(0)}{s}$$

Change of Scale

$$\mathcal{L}[f(at)] = (1/a)F(s/a)$$

Multiplication by t^n

$$\mathcal{L}[t^nf(t)] = (-1)^n(d^n/ds^n)F(s)$$

Division by t

$$\mathcal{L}[f(t)/t] = \int_s^\infty F(s)\, ds$$

First Shifting Theorem

$$\mathcal{L}[e^{at}f(t)] = F(s - a)$$

Second Shifting Theorem

$$\mathcal{L}[f(t - a)] = e^{-as}F(s)$$

Final-value Theorem

$$f(\infty) = [sF(s)]_{s=0}$$

Initial-value Theorem

$$f(0+) = [sF(s)]_{s=\infty}$$

---

tive of a step is an impulse whose area is equal to the step change. Thus, $De = E_0 u_1(t)$ is an impulse of area E_0. Equation (2.61) may be written in the form

$$\left(D + \frac{1}{RC}\right)i = \frac{E_0}{R}u_1(t) \tag{2.67}$$

The complementary solution of a differential equation is the solution which satisfies the differential equation when the right-hand side is zero. Thus, the complementary solution satisfies the equation

$$\left(D + \frac{1}{RC}\right)i = 0 \tag{2.68}$$

Substitution of $i = K_1 e^{-t/RC}$ in the preceding expression gives

$$\frac{d}{dt}(K_1 e^{-t/RC}) + \frac{K_1}{RC}e^{-t/RC} = -\frac{K_1}{RC}e^{-t/RC} + \frac{K_1}{RC}e^{-t/RC} = 0$$

Because $i = K_1 e^{-t/RC}$ satisfies Eq. (2.68), this is the complementary solution. In the classical method, it is necessary to work in regions where the response is continuous. In this case, that means after time $t = 0+$. The right-hand side of Eq. (2.67) is an impulse at $t = 0$, which is thereby omitted. Thus, the original differential equation [Eq. (2.67)] becomes the same as Eq. (2.68). In this case, the complementary solution is also the solution of the original equation (i.e., the particular solution is zero when the right-hand side of the original equation is zero). It remains now to evaluate the constant of integration K_1. In the preceding section, it was shown that $i(0+) = E_0/R$; thus

$$i(0+) = K_1 e^{-t/RC}\Big|_{t=0+} = K_1 = \frac{E_0}{R}$$

Hence, the solution is

$$i(t) = K_1 e^{-t/RC} = \frac{E_0}{R}e^{-t/RC}$$

As another example, let it be desired to solve the following differential equation for the case in which all the initial conditions are zero.

$$(D^2 + 5D + 6)y = u_1(t) + u(t) \tag{2.69}$$

The complementary solution satisfies the differential equation

$$(D^2 + 5D + 6)y = (D + 2)(D + 3)y = 0 \tag{2.70}$$

The complementary solution y_c is

$$y_c = K_1 e^{-2t} + K_2 e^{-3t} \tag{2.71}$$

Substitution of Eq. (2.71) into Eq. (2.70) verifies the fact that the complementary solution satisfies Eq. (2.70). By neglecting the impulse $u_1(t)$ in Eq. (2.69), the particular solution is the value of y such that $(D^2 + 5D + 6)y = 1$ for $t > 0$. The particular solution y_p is simply the constant

$$y_p = \tfrac{1}{6} \tag{2.72}$$

The general solution is the sum of the complementary plus the particular solution.

$$y = y_c + y_p = K_1 e^{-2t} + K_2 e^{-3t} + \tfrac{1}{6} \tag{2.73}$$

It remains now to evaluate the constants of integration K_1 and K_2. Because this is a second-order differential equation, there are two constants of integration. To determine the state of the system at time $t = 0+$, first integrate each term of Eq. (2.69) twice and then evaluate between the limits of $0-$ to $0+$. Thus

$$[y(0+) - y(0-)] + 5[\int y(t)\, dt]_{0-}^{0+} + 6[\int \int y(t)\, dt\, dt]_{0-}^{0+}$$
$$= [\int \int u_1(t)\, dt\, dt]_{0-}^{0+} + [\int \int u(t)\, dt\, dt]_{0-}^{0+}$$

All the preceding integrals vanish when evaluated at the limits; thus $y(0+) - y(0-) = 0$ or $y(0+) = y(0-)$.

Because the initial condition $y(0-)$ is zero, $y(0+)$ is zero also. Integrating each term of Eq. (2.69) once gives

$$y'(0+) - y'(0-) + 5[y(0+) - y(0-)] + 6[\int y(t)\, dt]_{0-}^{0+}$$
$$= [\int u_1(t)\, dt]_{0-}^{0+} + [\int u(t)\, dt]_{0-}^{0+}$$

The integral of $u_1(t)$ is the area of a unit impulse which is unity. Thus, $y'(0+) - y'(0-) = 1$. Because the initial conditions are zero, $y'(0+) = 1$.

Differentiating Eq. (2.73) to obtain an equation for $y'(t)$ gives

$$y'(t) = -2K_1 e^{-2t} - 3K_2 e^{-3t} \tag{2.74}$$

Evaluating Eqs. (2.73) and (2.74) at $t = 0+$ and utilizing the information that $y(0+) = 0$ and $y'(0+) = 1$ gives two equations for evaluating the constants K_1 and K_2. Thus, it is found that $K_1 = \tfrac{1}{2}$ and $K_2 = -\tfrac{2}{3}$. Substitution of these values into the general solution [Eq. (2.73)] gives for the final result

$$y = \tfrac{1}{6} + \tfrac{1}{2} e^{-2t} - \tfrac{2}{3} e^{-3t} \tag{2.75}$$

In general, an nth-order differential equation may be expressed in the form

$$(D - r_1)(D - r_2) \cdots (D - r_n)y = f(t) \qquad (2.76)$$

where $r_1, r_2, \ldots, r_n$ are the n distinct roots of the differential operator. The complementary solution has the form

$$y_c(t) = K_1 e^{r_1 t} + K_2 e^{r_2 t} + \cdots + K_n e^{r_n t} \qquad (2.77)$$

These n constants must eventually be determined by solving n boundary-condition equations simultaneously.

The complementary solution assumes a slightly different form when the operator contains a root r which is repeated q times; that is,

$$(D - r)^q (D - r_1) \cdots (D - r_n)y = f(t) \qquad (2.78)$$

For repeated roots, the complementary solution has the form

$$y_c(t) = (C_0 + C_1 t + C_2 t^2 + \cdots + C_{q-1} t^{q-1})e^{rt} \\ + K_1 e^{r_1 t} + \cdots + K_n e^{r_n t} \qquad (2.79)$$

Equation (2.79) shows that the number of unknown constants is equal to the order of the differential equation.

The particular solution y_p is the function that when operated on by the differential operator yields the right-hand side of the differential equation. The particular solution usually has the same form as the right-hand side of the differential equation plus the sum of all the derivatives of this form. For example, in Eq. (2.69) the right-hand side is considered to be $u(t) = 1$, which is a constant for $t > 0$. Hence, the form of the particular solution is a constant, say $y_p = A$. Substitution of $y_p = A$ into Eq. (2.69) gives $(D^2 + 5D + 6)A = 6A = 1$ or $A = \frac{1}{6}$. As a more general case, suppose that the right-hand side of Eq. (2.69) is t^2. The assumed particular solution has the form of t^2 plus its derivatives; that is, $y_p = At^2 + Bt + C$. The constants A, B, and C are obtained by first substituting this value of y_p into the original equation. Thus

$$(D^2 + 5D + 6)(At^2 + Bt + C) = 6At^2 + (10A + 6B)t \\ + (2A + 5B + 6C) = t^2 \qquad (2.80)$$

Equating coefficients of like terms gives $6A = 1$, $10A + 6B = 0$, and $2A + 5B + 6C = 0$, whence the constants are found to be $A = \frac{1}{6}$, $B = -\frac{5}{18}$, and $C = \frac{19}{108}$. It is important to notice that the particular solution is determined entirely from the original differential equation and thus does not depend upon the initial conditions. For engineering work, the right-hand side of the differential equation (i.e., excitation) is often a sinusoidal of the form $\sin \omega t$. For this case, the particular solution has the form $A \sin \omega t + B \cos \omega t$.

To solve Eq. (2.69) by Laplace transforms, first transform each term.

$$s^2 Y(s) - sy(0) - y'(0) + 5[sY(s) - y(0)] + 6Y(s)$$
$$= 1 + \frac{1}{s} = \frac{s+1}{s}$$

Because all the initial conditions are zero, solving for $Y(s)$ gives

$$Y(s) = \frac{s+1}{s(s+2)(s+3)} = \frac{\frac{1}{6}}{s} + \frac{\frac{1}{2}}{s+2} - \frac{\frac{2}{3}}{s+3} \qquad (2.81)$$

The right-hand side of the preceding expression is the result of a partial fraction expansion. The subject of partial fractions is covered in the next chapter. Inverting Eq. (2.81) verifies the result of Eq. (2.75).

A significant advantage of the transform method is that the initial conditions are substituted directly into the transformed equation, whence they are automatically incorporated into the solution. Also, in the transform method, one always utilizes directly the given initial state of the system at time $t = 0-$ regardless of whether or not a discontinuity occurs in an initial condition in going from time $t = 0-$ to $t = 0+$. The classical method is restricted to continuous intervals so that it is necessary to begin with the state of the system at $t = 0+$ when such a discontinuity occurs. Another feature of the transform method is that the entire solution is obtained in one operation without need of determining separately the complementary and the particular solutions. With the classical method various situations or cases arise which require specialized techniques for obtaining the general solution. An example of such a case is when a term in the particular solution (for example, Ae^{-4t}) has the same form as a term in the complementary solution (for example, Ke^{-4t}). Such special situations do not arise in solving problems by Laplace transforms. With transforms the same procedure is always followed. That is,

1. Transform the differential equation.
2. Insert the values of the initial conditions.
3. Solve this equation for the transformed variable [that is, $Y(s)$].
4. Express $Y(s)$ as a sum of simple fractions.
5. Invert to obtain the desired result [that is, $y(t)$].

In the application of the Laplace transform method, most, if not all, references begin the solution of problems at time $t = 0+$ rather than at the initial state. This has many of the same disadvantages as classical methods, such as requiring the additional step of ascertaining beforehand the state of the system at $t = 0+$, neglecting impulses, etc. The method described in this text in which the given initial state is employed as the initial conditions is a more direct and a more powerful method.

Additional characteristics and properties of Laplace transformations are brought out in the following chapters when the application of transformations to the solution of engineering problems is explained in greater detail. It is shown that this method gives tremendous insight into the actual behavior of engineering systems, as well as saving considerable time and effort in carrying out the solution.

PROBLEMS

2.1 Use the equation for the transform of the second derivative [i.e., Eq. (2.9)] to determine the Laplace transform of each of the following functions:

(a) $\cos kt$
(b) $\sinh kt$
(c) $\sin (kt + 4)$

2.2 Make use of identities to determine the Laplace transform of each of the following functions [for example, $\sin^2 \alpha = (1 - \cos 2\alpha)/2$]:

(a) $\sin^2 kt$
(b) $\cos^2 kt$
(c) $\sin kt \cos kt$

2.3 Determine the Laplace transform of each of the following functions by transforming each term in the series expansion for the functions (for example, $e^t = 1 + t + t^2/2! + \cdots$):

(a) e^t
(b) $\sin t$

2.4 Determine the Laplace transform of each of the following functions. Start with the fact that $\mathcal{L}(1) = 1/s$, and use the theorem pertaining to the transform of the integral of a function:

59

    (a) $y = t + a$
    (b) $y = t^2 + at + b$

2.5   Derive a general expression for the transform of:

    (a)   The definite integral $\int_0^t \int_0^t f(t) \, dt \, dt$
    (b)   The indefinite integral $y(t) = \int\int f(t) \, dt \, dt$

2.6   Obtain the transform of each of the following functions by application of the theorem for obtaining the transform of $tf(t)$:

    (a) $t \sin t$
    (b) $t \cos t$
    (c) $t \sinh t$

2.7   Obtain the transform of each of the following functions by application of the theorem for obtaining the transform of $f(t)/t$:

    (a) $(\sin t)/t$
    (b) $(1 - e^{-t})/t$
    (c) $(1 - \cos t)/t$

2.8   Apply the first shifting theorem to the results obtained in Prob. 2.6 to obtain the transform of the following functions:

    (a) $te^{at} \sin t$
    (b) $te^{at} \cos t$

2.9   Evaluate the following definite integrals for $a > 0$:

    (a) $\int_0^\infty e^{-at} \sin kt \, dt$
    (b) $\int_0^\infty e^{-at} \sinh kt \, dt$
    (c) $\int_0^\infty \dfrac{e^{-at} \sin t}{t} \, dt$

2.10   Determine the Laplace transform for the train of impulse functions shown in:

    (a)   Fig. P2.10a
    (b)   Fig. P2.10b

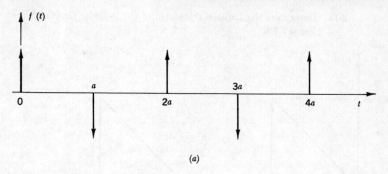

Area of each impulse $= h$

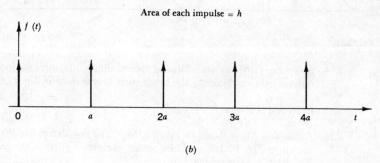

(b)

Fig. P2.10

2.11 The function shown in Fig. P2.10a is the derivative of the function shown in Fig. 2.7a. Determine the transform of Fig. P2.10a by applying the theorem pertaining to the transform of the derivative of a function to Fig. 2.7a.

2.12 Figure P2.12 is the integral of the periodic function shown in Fig. 2.7c. Use the theorem pertaining to the transform of the integral of a function to determine the transform $F(s)$ for the function shown in Fig. P2.12.

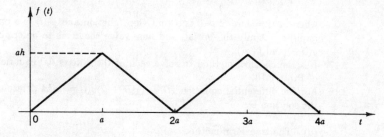

Fig. P2.12

2.13   Determine the Laplace transform of the periodic function shown in Fig. P2.13.

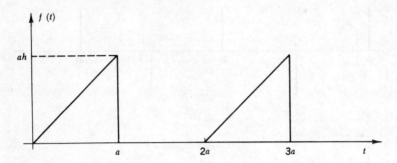

Fig. P2.13

2.14   Solve Eq. (2.60) by transforming the indefinite integral [i.e., do not differentiate to eliminate the integration as was done in Eq. (2.61)]

$$Ri + \frac{1}{C} \int i \, dt = e$$

2.15   A function $f(t)$ is shown in Fig. P2.15a. The first derivative $f'(t)$ is shown in Fig. P2.15b, and the second derivative $f''(t)$ is shown in Fig. P2.15c. Determine:

   (a)   The transform of $f''(t)$
   (b)   The transform of $f(t)$
   (c)   The transform of the periodic function shown in Fig. P2.15d

2.16   Use Laplace transforms to solve each of the following differential equations:

   (a) $y' + ay = u_1(t)$ $y(0) = 0$
   (b) $y' + ay = 0$ $y(0) = 1$
   (c) $y' + ay = u(t)$ $y(0) = 0$

where $u(t) = 1$ for $t > 0$ is a unit step function and $u_1(t)$ is a unit impulse. Apply the initial- and final-value theorems to determine $y(0+)$ and $y(\infty)$.

2.17   Use the classical method to solve each of the differential equations of Prob. 2.16.

2.18   For the differential equation $(D + 1)(D + 2)y(t) = u_1(t)$ determine the solution by

   (a)   The transform method
   (b)   The classical method

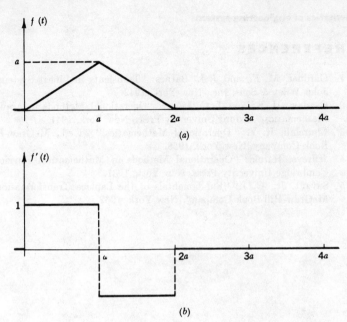

(a)

(b)

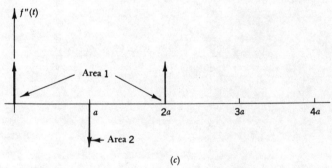

(c)

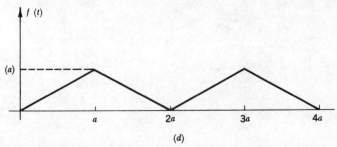

(d)

Fig. P2.15

63

REFERENCES

1 Gardner, M. F., and J. L. Barnes: "Transients in Linear Systems," John Wiley & Sons, Inc., New York, 1942.
2 Carslaw, H. S., and J. C. Jaeger: "Operational Methods in Applied Mathematics," Oxford University Press, New York, 1941.
3 Churchill, R. V.: "Operational Mathematics," 2d ed., McGraw-Hill Book Company, New York, 1958.
4 Jeffreys, Harold: "Operational Methods in Mathematical Physics," Cambridge University Press, New York, 1931.
5 Savant, Jr., C. J.: "Fundamentals of the Laplace Transformation," McGraw-Hill Book Company, New York, 1962.

chapter 3

Linear differential equations

Linear differential equations describe the operation of numerous engineering phenomena. In this chapter, it is shown how the Laplace transformation method is applied to typical engineering problems. This method yields significant information, because Laplace transformations bridge the gap between the abstract mathematical representation (i.e., the differential equation) and the actual physical situation.

3.1 DIFFERENTIAL EQUATIONS

In Fig. 3.1 is shown a mass-spring-damper combination. The differential equation of operation for this system is

$$M \frac{d^2y}{dt^2} + B \frac{dy}{dt} + Ky = f(t)$$

Dividing through by M and letting

$$D = \frac{d}{dt} \quad \text{and} \quad D^2 = \frac{d^2}{dt^2}$$

gives

$$D^2y + \frac{B}{M} Dy + \frac{K}{M} y = \frac{1}{M} f(t) \tag{3.1}$$

where $f(t)$ is the applied external force, which varies as a function of time, M is the mass, B is the coefficient of viscous damping, and K

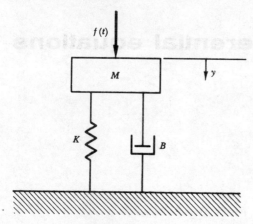

Fig. 3.1. Mass-spring-damper system.

is the spring constant. Transforming each term of Eq. (3.1) gives

$$[s^2 Y(s) - sy(0) - y'(0)] + \frac{B}{M}[sY(s) - y(0)] + \frac{K}{M} Y(s)$$
$$= \frac{1}{M} F(s)$$

Thus, the transform for $Y(s)$ is

$$Y(s) = \frac{(1/M)F(s)}{s^2 + (B/M)s + K/M} + \frac{(s + B/M)y(0) + y'(0)}{s^2 + (B/M)s + K/M}$$

$$(3.2)$$

The inverse of the first term on the right-hand side of the preceding equation yields the portion of the solution due to the forcing function $f(t)$, while the inverse of the second term yields the response due to the initial conditions $y(0)$ and $y'(0)$.

In general a differential equation may be written in the form

$$(D^n + a_{n-1}D^{n-1} + \cdots + a_1 D + a_0)y(t)$$
$$= (b_m D^m + b_{m-1}D^{m-1} + \cdots + b_1 D + b_0)f(t) \quad (3.3)$$

where $D^r = d^r/dt^r, r = 1, 2, \ldots, n$. Note that the coefficient of the D^n term can always be made equal to unity. This was done in Eq. (3.1) by dividing through by M. By using the notation

$$L_n(D) = D^n + a_{n-1}D^{n-1} + \cdots + a_1 D + a_0 \quad (3.4)$$

and

$$L_m(D) = b_m D^m + b_{m-1} D^{m-1} + \cdots + b_1 D + b_0 \tag{3.5}$$

then Eq. (3.3) may be written in the form

$$y(t) = \frac{L_m(D)f(t)}{L_n(D)} \tag{3.6}$$

The general transformed expression for $Y(s)$ is

$$Y(s) = \frac{L_m(s)F(s)}{L_n(s)} + \frac{I(s)}{L_n(s)} \tag{3.7}$$

where $L_n(s)$ and $L_m(s)$ are obtained by substituting s for D in Eqs. (3.4) and (3.5), $F(s)$ is the transform of the excitation $f(t)$, and $I(s)$ represents the sum of all the initial-condition terms.

Comparison of Eqs. (3.6) and (3.7) shows that, when all the initial conditions are zero, the transformed equation [Eq. (3.7)] is obtained by substituting s for D, $F(s)$ for $f(t)$, and $Y(s)$ for $y(t)$ in the original differential equation [Eq. (3.6)].

To illustrate the procedure for determining the inverse, let it be desired to determine the response $y(t)$ of Eq. (3.1) to a step change in $f(t)$, as shown in Fig. 3.2a. All the initial conditions are zero, and the particular system parameters are $M = 1$, $B = 3$, and $K = 2$. As illustrated in Fig. 3.2, initial values are indicated by an φ.

Because $f(t) = F_0$, a constant for $t > 0$, then $F(s) = F_0/s$. Hence, from Eq. (3.2) it follows that the transformed equation for $Y(s)$ is

$$Y(s) = \frac{(1/M)(F_0/s)}{s^2 + (B/M)s + K/M} = \frac{F_0}{s(s^2 + 3s + 2)} \tag{3.8}$$

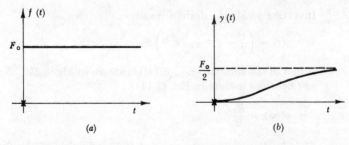

Fig. 3.2. Response of mass-spring-damper system to a step input. (a) The step input; (b) the response.

This may be written in the form

$$\frac{F_0}{s(s + 1)(s + 2)} = \frac{K_1}{s} + \frac{K_2}{s + 1} + \frac{K_3}{s + 2} \tag{3.9}$$

Multiplying through both sides of Eq. (3.9) by s and taking the limit as s approaches zero makes all the terms on the right-hand side vanish except K_1. That is,

$$\lim_{s \to 0} s \frac{F_0}{s(s + 1)(s + 2)} = \lim_{s \to 0} \left(K_1 + \frac{sK_2}{s + 1} + \frac{sK_3}{s + 2} \right) = K_1$$

or

$$K_1 = \frac{F_0}{(s + 1)(s + 2)} \bigg|_{s = 0} = \frac{F_0}{2} \tag{3.10}$$

To evaluate K_2, multiply Eq. (3.9) by $s + 1$, and then take the limit as s approaches -1; hence

$$\lim_{s \to -1} (s + 1) \frac{F_0}{s(s + 1)(s + 2)} = \lim_{s \to -1} \left[\frac{(s + 1)K_1}{s} + K_2 \right. $$
$$\left. + \frac{(s + 1)K_3}{s + 2} \right] = K_2$$

or

$$K_2 = \frac{F_0}{s(s + 2)} \bigg|_{s = -1} = -F_0 \tag{3.11}$$

Similarly, the constant K_3 is

$$K_3 = \frac{F_0}{s(s + 1)} \bigg|_{s = -2} = \frac{F_0}{2} \tag{3.12}$$

Thus, Eq. (3.9) becomes

$$Y(s) = \frac{F_0}{s(s + 1)(s + 2)} = \frac{F_0}{2s} - \frac{F_0}{s + 1} + \frac{F_0}{2(s + 2)} \tag{3.13}$$

Inverting yields the desired result,

$$y(t) = \left(\frac{1}{2} - e^{-t} + \frac{1}{2} e^{-2t} \right) F_0 \tag{3.14}$$

A plot of the solution [Eq. (3.14)] is shown in Fig. 3.2b. Note that, as t becomes infinite in Eq. (3.14),

$$y(\infty) = \frac{F_0}{2} \tag{3.15}$$

This final value may be verified by application of the final-value theorem to Eq. (3.8). When the input is a constant [$f(t) = F_0$], an alternative method for obtaining the final value is to let all derivatives in the original differential equation equal zero. Because the derivative of a constant is zero, powers of D operating on the constant input and constant final value vanish. Thus, Eq. (3.1) becomes $y(\infty) = f(\infty)/K = F_0/K = F_0/2$.

As illustrated in the preceding example, a partial fraction expansion enables one to express the transform of a differential equation as a sum of simple fractions. By inverting each of these simple fractions the inverse is readily obtained.

3.2 PARTIAL FRACTION EXPANSION

In this section, a general systematic procedure for obtaining partial fraction expansions is developed. Consider the function

$$\frac{A(s)}{B(s)} = \frac{A(s)}{(s - r_1)(s - r_2) \cdots (s - r_n)(s - r)^q} \tag{3.16}$$

where $A(s)$ is the numerator polynomial and $B(s)$ is the denominator polynomial. As indicated by Eq. (3.16), $B(s)$ has n distinct linear factors $(s - r_1) \cdots (s - r_n)$ and a repeated factor $s - r$, where q is the number of times that this factor is repeated.

Note that, when $s = r_1$, $B(s) = 0$. Hence, $r_1, r_2, \ldots, r_n$ and r are all roots of the equation $B(s) = 0$.

Distinct roots

Equation (3.16) may be written in the form

$$\frac{A(s)}{B(s)} = \frac{K_1}{s - r_1} + \frac{G_1(s)}{H_1(s)} \tag{3.17}$$

where $G_1(s)/H_1(s)$ is the sum of the other fractions in the expansion.

The common denominator for Eq. (3.17) is $B(s) = (s - r_1)H_1(s)$ or $H_1(s) = B(s)/(s - r_1)$. Because $s - r_1$ is distinct, $H_1(s)$ does not contain this factor. Multiplying Eq. (3.17) by $s - r_1$ and taking the limit as s approaches the root r_1 gives

$$\lim_{s \to r_1} (s - r_1) \frac{A(s)}{B(s)} = K_1 + \lim_{s \to r_1} (s - r_1) \frac{G_1(s)}{H_1(s)}$$

The last term in the preceding expression vanishes; thus

$$K_1 = \lim_{s \to r_1} (s - r_1) \frac{A(s)}{B(s)} \tag{3.18}$$

To illustrate the preceding, note that Eq. (3.8) may be written in the form

$$\frac{F_0}{s(s+1)(s+2)} = \frac{K_1}{s} + \frac{G_1(s)}{H_1(s)} = \frac{F_0/2}{s} - \frac{(s+3)F_0/2}{(s+1)(s+2)}$$

where $K_1 = F_0/2$, $H_1(s) = (s+1)(s+2)$, and $G_1(s) = -(s+3)F_0/2$.
Next, Eq. (3.17) may be expanded in the form

$$\frac{A(s)}{B(s)} = \frac{K_1}{s - r_1} + \frac{K_2}{s - r_2} + \frac{G_2(s)}{H_2(s)} \tag{3.19}$$

where $G_2(s)/H_2(s)$ is the sum of the remaining fractions. $H_2(s)$ does not contain an $s - r_1$ or $s - r_2$ factor. Multiplying Eq. (3.19) by $s - r_2$ and taking the limit as s approaches r_2 gives

$$\lim_{s \to r_2} (s - r_2) \frac{A(s)}{B(s)} = K_2 + \lim_{s \to r_2} (s - r_2)\left[\frac{K_1}{s - r_1} + \frac{G_2(s)}{H_2(s)}\right]$$

Because the last term vanishes when the limit is taken,

$$K_2 = \lim_{s \to r_2} (s - r_2) \frac{A(s)}{B(s)} \tag{3.20}$$

Correspondingly, Eq. (3.8) may be written in the form

$$\frac{F_0}{s(s+1)(s+2)} = \frac{F_0/2}{s} + \frac{K_2}{s+1} + \frac{G_2(s)}{H_2(s)} = \frac{F_0/2}{s} - \frac{F_0}{s+1} + \frac{F_0/2}{s+2}$$

where $K_2 = -F_0$, $H_2(s) = s + 2$, and $G_2(s) = F_0/2$. Following this procedure for each of the linear factors gives

$$\frac{A(s)}{B(s)} = \frac{K_1}{s - r_1} + \cdots + \frac{K_i}{s - r_i} + \cdots + \frac{K_n}{s - r_n} + \frac{G_r(s)}{(s - r)^q} \tag{3.21}$$

The constant K_i corresponding to the $s - r_i$ factor is

$$K_i = \lim_{s \to r_i} (s - r_i) \frac{A(s)}{B(s)} \qquad i = 1, 2, \ldots, n \tag{3.22}$$

Taking the inverse transform of $K_i/(s - r_i)$ gives

$$\mathcal{L}^{-1}\left(\frac{K_i}{s - r_i}\right) = K_i e^{r_i t} \tag{3.23}$$

Hence, a distinct root yields an exponential term in the response expression.

Another method of evaluating K_i is obtained by writing Eq. (3.22) in the form

$$K_i = \lim_{s \to r_i} \frac{A(s)}{B(s)/(s - r_i)}$$

Because $B(s)$ contains an $s - r_i$ factor,

$$\lim_{s \to r_i} \frac{B(s)}{s - r_i} = \frac{0}{0}$$

Application of L'Hospital's rule gives

$$\lim_{s \to r_i} \frac{(d/ds)B(s)}{(d/ds)(s - r_i)} = \lim_{s \to r_i} \frac{(d/ds)B(s)}{1} = \lim_{s \to r_i} B'(s) \qquad (3.24)$$

Hence

$$K_i = \lim_{s \to r_i} \frac{A(s)}{B'(s)} \qquad (3.25)$$

Application of Eq. (3.25) to obtain the partial fraction expansion of Eq. (3.8) gives

$$\frac{A(s)}{B(s)} = \frac{F_0}{s(s + 1)(s + 2)} = \frac{F_0}{s^3 + 3s^2 + 2s}$$

$$K_1 = \lim_{s \to 0} \frac{A(s)}{B'(s)} = \frac{F_0}{3s^2 + 6s + 2}\bigg|_{s=0} = \frac{F_0}{2}$$

$$K_2 = \lim_{s \to -1} \frac{A(s)}{B'(s)} = \frac{F_0}{3s^2 + 6s + 2}\bigg|_{s=-1} = -F_0$$

$$K_3 = \lim_{s \to -2} \frac{A(s)}{B'(s)} = \frac{F_0}{3s^2 + 6s + 2}\bigg|_{s=-2} = \frac{F_0}{2}$$

Thus, the results of Eqs. (3.10) to (3.12) are verified.

Repeated roots

For a repeated factor of the form $(s - r)^q$,

$$\frac{A(s)}{B(s)} = \frac{K_1}{s - r_1} + \cdots + \frac{K_n}{s - r_n} + \frac{G_r(s)}{(s - r)^q}$$

$$= \frac{G(s)}{H(s)} + \frac{C_q}{(s - r)^q} + \frac{C_{q-1}}{(s - r)^{q-1}} + \cdots$$

$$+ \frac{C_2}{(s - r)^2} + \frac{C_1}{s - r} \qquad (3.26)$$

where $C_1, C_2, \ldots, C_q$ are constants and $G(s)/H(s)$ is the sum of the first n fractions in the expansion. Multiplication of Eq. (3.26) by $(s - r)^q$ gives

$$(s - r)^q \frac{A(s)}{B(s)} = (s - r)^q \frac{G(s)}{H(s)}$$
$$+ [C_q + (s - r)C_{q-1} + (s - r)^2 C_{q-2} + \cdots] \quad (3.27)$$

Because $H(s)$ does not contain an $s - r$ factor, the limit of Eq. (3.27) as s approaches r is

$$C_q = \lim_{s \to r} (s - r)^q \frac{A(s)}{B(s)} \quad (3.28)$$

To evaluate C_{q-1}, first differentiate Eq. (3.27) with respect to s,

$$\frac{d}{ds} \left[(s - r)^q \frac{A(s)}{B(s)} \right] = \frac{d}{ds} \left[(s - r)^q \frac{G(s)}{H(s)} \right]$$
$$+ [C_{q-1} + 2(s - r)C_{q-2} + 3(s - r)^2 C_{q-3} + \cdots] \quad (3.29)$$

The limit of the preceding expression as s approaches r is

$$C_{q-1} = \lim_{s \to r} \frac{d}{ds} \left[(s - r)^q \frac{A(s)}{B(s)} \right] \quad (3.30)$$

The next constant, C_{q-2}, is obtained by differentiating Eq. (3.29) with respect to s and then taking the limit as s approaches r. Thus,

$$C_{q-2} = \frac{1}{2} \frac{d^2}{ds^2} \left[(s - r)^q \frac{A(s)}{B(s)} \right]_{s=r}$$

By proceeding in a similar manner, the general expression for the constant C_{q-i} is found to be

$$C_{q-i} = \frac{1}{i!} \frac{d^i}{ds^i} \left[(s - r)^q \frac{A(s)}{B(s)} \right]_{s=r} \qquad i = 0, 1, \ldots, q - 1$$
$$(3.31)$$

The portion of the response due to a repeated root is obtained by taking the inverse transformation of the part of the partial fraction expansion due to the repeated root. Inverting this portion of Eq. (3.26) gives

$$\left[\frac{C_q t^{q-1}}{(q-1)!} + \frac{C_{q-1} t^{q-2}}{(q-2)!} + \cdots + \frac{C_3 t^2}{2!} + \frac{C_2 t}{1!} + C_1 \right] e^{rt} \quad (3.32)$$

When the repeated root r is negative, the corresponding response term [Eq. (3.32)] approaches zero as time becomes infinite.

To illustrate the application of this procedure for repeated roots, let $K = 2.25$ rather than 2 in Eq. (3.8). Thus, Eq. (3.8) becomes

$$Y(s) = \frac{A(s)}{B(s)} = \frac{F_0}{s(s^2 + 3s + 2.25)} = \frac{F_0}{s(s + 1.5)^2}$$

$$= \frac{K_1}{s} + \frac{C_2}{(s + 1.5)^2} + \frac{C_1}{s + 1.5} \quad (3.33)$$

The constant K_1 is evaluated in the usual manner for distinct roots.

$$K_1 = \lim_{s \to 0} \frac{F_0}{(s + 1.5)^2} = \frac{F_0}{2.25}$$

The constants C_2 and C_1 are evaluated as follows:

$$C_2 = \lim_{s \to -1.5} \frac{F_0}{s} = -\frac{F_0}{1.5}$$

$$C_1 = \lim_{s \to -1.5} \frac{d}{ds}\left(\frac{F_0}{s}\right) = -\frac{F_0}{s^2}\Big|_{s=-1.5} = -\frac{F_0}{2.25}$$

Hence, the partial fraction expansion is

$$Y(s) = \frac{F_0}{s(s + 1.5)^2} = \frac{F_0}{2.25s} - \frac{F_0}{1.5(s + 1.5)^2} - \frac{F_0}{2.25(s + 1.5)}$$

Taking the inverse transform of the preceding expression gives the response

$$y(t) = [1 - (1.5t + 1)e^{-1.5t}]\frac{F_0}{2.25}$$

This same method for handling repeated roots is applicable regardless of the number of repeated roots in $B(s)$. For the sake of simplicity, only the one $(s - r)^q$ was considered in this development.

The initial-value theorem provides an alternative method for evaluating the partial-fraction-expansion constants for repeated roots. An advantage of the initial-value theorem is that it eliminates the differentiation. This fact is now illustrated by evaluating the constant C_1 in the preceding example. Inverting Eq. (3.33) gives the general form of the response; i.e.,

$$y(t) = K_1 + (C_2 t + C_1)e^{-1.5t}$$

Because $K_1 = F_0/2.25$, then for $t = 0+$ the preceding expression becomes

$$y(0+) = \frac{F_0}{2.25} + C_1$$

Application of the initial-value theorem to Eq. (3.33) gives

$$y(0+) = \lim_{s \to \infty} \frac{F_0}{(s + 1.5)^2} = 0$$

From the two preceding equations, it follows that $C_1 = -F_0/2.25$. When the repeated root occurs three times rather than twice, the initial-value theorem must be applied to $y(t)$ and $y'(t)$ to obtain two equations for evaluating the two partial-fraction-expansion constants that involve differentiation.

Illustrative example The differential equation of operation for a system is

$$(D + 1)^3 y = (D^3 + 3D^2 + 3D + 1)y = (D + 2)f(t)$$

The input $f(t)$ is the function shown in Fig. 3.3. The initial value is $f(0) = 5$, and then $f(t) = 10$ for $t > 0$. Determine the response $y(t)$ for the case in which $y(0) = 1$, $y'(0) = 2$, and $y''(0) = 6$.

Solution Transforming each term in the differential equation gives

$$[s^3 Y(s) - s^2 y(0) - s y'(0) - y''(0)] + 3[s^2 Y(s) - s y(0) - y'(0)]$$
$$+ 3[s Y(s) - y(0)] + Y(s) = [s F(s) - f(0)] + 2F(s)$$

Solving for $Y(s)$,

$$Y(s) = \frac{(s + 2)F(s) - f(0)}{(s + 1)^3}$$
$$+ \frac{(s^2 + 3s + 3)y(0) + (s + 3)y'(0) + y''(0)}{(s + 1)^3}$$

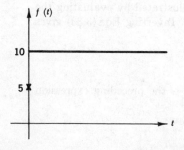

Fig. 3.3. Step function.

Substitution of $F(s) = 10/s$ and the values of the initial conditions into the preceding expression gives

$$Y(s) = \frac{s^3 + 5s^2 + 20s + 20}{s(s+1)^3}$$

The partial fraction expansion has the form

$$Y(s) = \frac{K_1}{s} + \frac{C_3}{(s+1)^3} + \frac{C_2}{(s+1)^2} + \frac{C_1}{(s+1)}$$

Application of Eq. (3.22) to evaluate K_1 and then application of Eq. (3.31) to evaluate C_3, C_2, and C_1 gives

$$K_1 = \lim_{s \to 0} \frac{s^3 + 5s^2 + 20s + 20}{(s+1)^3} = 20$$

$$C_3 = \lim_{s \to -1} \frac{s^3 + 5s^2 + 20s + 20}{s} = -4$$

$$C_2 = \lim_{s \to -1} \frac{d}{ds} \left(\frac{s^3 + 5s^2 + 20s + 20}{s} \right)$$

$$= \lim_{s \to -1} \frac{d}{ds} \left(s^2 + 5s + 20 + \frac{20}{s} \right)$$

$$= \left[2s + 5 - \frac{20}{s^2} \right]_{s=-1} = -17$$

$$C_1 = \lim_{s \to -1} \frac{1}{2!} \frac{d^2}{ds^2} \left(\frac{s^3 + 5s^2 + 20s + 20}{s} \right)$$

$$= \lim_{s \to -1} \frac{1}{2!} \frac{d}{ds} \left(2s + 5 - \frac{20}{s^2} \right)$$

$$= \frac{1}{2} \left[2 + \frac{40}{s^3} \right]_{s=-1} = -19$$

Thus, the partial fraction expansion is

$$Y(s) = \frac{20}{s} - \frac{4}{(s+1)^3} - \frac{17}{(s+1)^2} - \frac{19}{s+1}$$

Inverting gives the desired response

$$y(t) = 20 - (2t^2 + 17t + 19)e^{-t}$$

The initial-value theorem provides an alternative method for evaluating the constants C_1 and C_2 (this eliminates the differentiation). After evaluating K_1 and C_3, the partial fraction expansion has the form

$$Y(s) = \frac{20}{s} - \frac{4}{(s+1)^3} + \frac{C_2}{(s+1)^2} + \frac{C_1}{s+1}$$

Inverting gives

$$y(t) = 20 - 2t^2e^{-t} + C_2te^{-t} + C_1e^{-t}$$

At time $t = 0+$, then

$$y(0+) = 20 + C_1$$

Application of the initial-value theorem to determine $y(0+)$ gives

$$y(0+) = \lim_{s \to \infty} sY(s) = \frac{s^3 + 5s^2 + 20s + 20}{(s + 1)^3}\bigg|_{s = \infty} \approx \frac{s^3}{s^3} = 1$$

For very large values of s (as $s \to \infty$), then $s^3 \gg s^2 \gg s$. Thus, the limit is determined by the highest order of s in the numerator and in the denominator. Equating the two preceding expressions for $y(0+)$ verifies the result that $C_1 = -19$.

To evaluate C_2, first differentiate the expression for $y(t)$. Thus

$$y'(t) = -2(2t - t^2)e^{-t} + C_2(1 - t)e^{-t} - C_1e^{-t}$$

At time $t = 0+$, then

$$y'(0+) = C_2 - C_1$$

The transform of $y'(t)$ is

$$\mathcal{L}[y'(t)] = sY(s) - y(0) = \frac{s^3 + 5s^2 + 20s + 20}{(s + 1)^3} - 1$$

$$= \frac{2s^2 + 17s + 19}{(s + 1)^3}$$

To determine the value of a function at $t = 0+$ by application of the initial-value theorem, multiply the transform of the function by s, and then take the limit as s approaches infinity. Thus,

$$y'(0+) = \frac{s(2s^2 + 17s + 19)}{(s + 1)^3}\bigg|_{s = \infty} \approx \frac{2s^3}{s^3} = 2$$

Equating the two expressions for $y'(0+)$ gives $C_2 - C_1 = 2$, or $C_2 = -19 + 2 = -17$.

3.3 COMPLEX CONJUGATE ROOTS

Complex roots always occur as complex conjugate pairs. Thus, if there is a complex root $a + jb$, there will also be the conjugate root

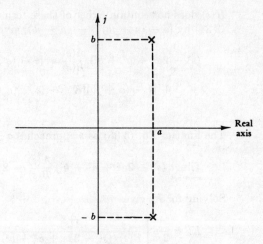

Fig. 3.4. Plot of complex conjugate roots.

$a - jb$. A pair of complex conjugate roots is shown in Fig. 3.4. The real part of each root is a. The imaginary parts, $\pm jb$, are equal but opposite in sign.

When $B(s)$ contains complex conjugate roots, then $B(s)$ has a factor of the form

$$[s - (a + jb)][s - (a - jb)] = s^2 - 2as + a^2 + b^2$$

It is to be noted that when complex conjugates are multiplied together, the resulting quadratic has real coefficients. The equation for an actual system must have real coefficients. Hence, complex conjugate roots must occur in pairs, for otherwise the system equation would have imaginary coefficients.

Each root of a complex conjugate pair is distinct, and thus the response due to each root may be obtained by the technique described for distinct roots. It will now be shown that the response terms due to each complex conjugate root combine to yield an exponential sinusoid.

For complex conjugate roots $A(s)/B(s)$ may be written in the form

$$\frac{A(s)}{B(s)} = \frac{K_c}{s - (a + jb)} + \frac{K_{-c}}{s - (a - jb)} + \frac{G(s)}{H(s)} \tag{3.34}$$

where $G(s)/H(s)$ represents the remaining terms in the expansion. Because $s - (a + jb)$ and $s - (a - jb)$ are distinct factors, then

77

$H(s)$ does not contain either of these terms. Multiplication of Eq. (3.34) by $[s - (a + jb)][s - (a - jb)]$ gives

$$(s^2 - 2as + a^2 + b^2) \frac{A(s)}{B(s)} = [s - (a - jb)]K_c$$

$$+ [s - (a + jb)]K_{-c} + [s - (a + jb)][s - (a - jb)] \frac{G(s)}{H(s)}$$

(3.35)

The limit of Eq. (3.35) as s approaches $a + jb$ is

$$\lim_{s \to a+jb} (s^2 - 2as + a^2 + b^2) \frac{A(s)}{B(s)} = 2jbK_c$$

Solving for K_c gives

$$K_c = \frac{1}{2jb} \left[\frac{A(s)}{B(s)/(s^2 - 2as + a^2 + b^2)} \right]_{s=a+jb}$$

$$= \frac{1}{2jb} \frac{A(a + jb)}{B(a + jb)}$$

(3.36)

where $A(a + jb)/B(a + jb)$ means the limit of $A(s)/B(s)$, in which the quadratic factor $s^2 - 2as + a^2 + b^2$ has been removed from $B(s)$ before the limit is taken.

The constant K_{-c} results by taking the limit of Eq. (3.35) as s approaches $a - jb$. Because $H(s)$ does not contain an $s - (a - jb)$ factor, then

$$\lim_{s \to a-jb} (s^2 - 2as + a^2 + b^2) \frac{A(s)}{B(s)} = -2jbK_{-c}$$

Solving for K_{-c} gives

$$K_{-c} = \frac{-1}{2jb} \left[\frac{A(s)}{B(s)/(s^2 - 2as + a^2 + b^2)} \right]_{s=a-jb}$$

$$= \frac{-1}{2jb} \frac{A(a - jb)}{B(a - jb)}$$

(3.37)

where $A(a - jb)/B(a - jb)$ means the limit of $(s)/B(s)$, in which the factor $s^2 - 2as + a^2 + b^2$ has been removed from $B(s)$ before the limit is taken. The constants $A(a + jb)/B(a + jb)$ and $A(a - jb)/B(a - jb)$ are complex conjugates, as is illustrated in Fig. 3.5. The magnitudes of these vectors are the same; i.e.,

$$\left| \frac{A(a + jb)}{B(a + jb)} \right| = \left| \frac{A(a - jb)}{B(a - jb)} \right|$$

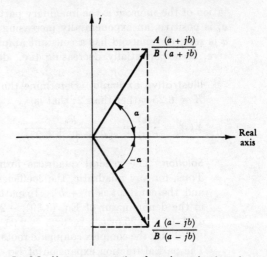

Fig. 3.5. Vector representation of complex conjugate roots.

Hence, K_c and K_{-c} may be expressed in the form

$$K_c = \frac{1}{2jb} \left| \frac{A(a+jb)}{B(a+jb)} \right| e^{j\alpha}$$

$$K_{-c} = \frac{-1}{2jb} \left| \frac{A(a-jb)}{B(a-jb)} \right| e^{-j\alpha} = \frac{-1}{2jb} \left| \frac{A(a+jb)}{B(a+jb)} \right| e^{-j\alpha}$$

where α is the angle of $A(a+jb)/B(a+jb)$.

The time response due to these complex conjugate roots is obtained by taking the inverse of these corresponding terms in Eq. (3.34); hence

$$\mathcal{L}^{-1} \left[\frac{K_c}{s-(a+jb)} + \frac{K_{-c}}{s-(a-jb)} \right] = K_c e^{(a+jb)t} + K_{-c} e^{(a-jb)t}$$

Substituting the corresponding values of K_c and K_{-c} into the preceding equation gives

$$\frac{1}{b} \left| \frac{A(a+jb)}{B(a+jb)} \right| e^{at} \frac{e^{j(bt+\alpha)} - e^{-j(bt+\alpha)}}{2j}$$

or

$$\frac{1}{b} \left| \frac{A(a+jb)}{B(a+jb)} \right| e^{at} \sin(bt+\alpha) \tag{3.38}$$

This response is referred to as an *exponential sinusoid* (i.e., a sine wave multiplied by an exponential factor). The frequency of oscillation of the sinusoid is the imaginary part b. When the exponent, a, is positive, an exponentially increasing sinusoid results. When a is zero, the sinusoid has a constant amplitude. When a is negative, an exponentially decreasing (i.e., damped) sinusoid results.

Illustrative example Determine the inverse of Eq. (3.8) for $K = 6.25$ rather than 2; that is,

$$Y(s) = \frac{A(s)}{B(s)} = \frac{F_0}{s(s^2 + 3s + 6.25)} \tag{3.39}$$

Solution The general quadratic form is $s^2 - 2as + a^2 + b^2$. Thus, for any quadratic, the coefficient of the s term is $-2a$, and the constant is $a^2 + b^2$. In particular, for the quadratic in the denominator of Eq. (3.39), $-2a = 3$ or $a = -1.5$, and $a^2 + b^2 = 6.25$ or $b = \sqrt{6.25 - 2.25} = 2$.

Thus, the complex conjugate roots are $a \pm jb = -1.5 \pm j2$. The partial fraction expansion of Eq. (3.39) is

$$Y(s) = \frac{K_1}{s} + \frac{K_c}{s - (-1.5 + j2)} + \frac{K_{-c}}{s - (-1.5 - j2)}$$

Inverting the preceding expression gives

$$y(t) = K_1 + \frac{1}{b} \left| \frac{A(a + jb)}{B(a + jb)} \right| e^{at} \sin(bt + \alpha) \tag{3.40}$$

where

$$K_1 = \lim_{s \to 0} s \frac{A(s)}{B(s)} = \frac{F_0}{s^2 + 3s + 6.25}\bigg|_{s=0} = \frac{F_0}{6.25}$$

$$\frac{A(a + jb)}{B(a + jb)} = \lim_{s \to a + jb} (s^2 - 2as + a^2 + b^2) \frac{A(s)}{B(s)} = \frac{F_0}{s}\bigg|_{s = -1.5 + j2}$$

$$= \frac{F_0}{-1.5 + j2} = \frac{F_0}{\sqrt{6.25}\underline{/126.9°}} = \frac{F_0}{2.5}\underline{/-126.9°} \tag{3.41}$$

As is illustrated in Fig. 3.6, the vector $-1.5 + j2$ may be expressed in the polar form

$$-1.5 + j2 = \sqrt{1.5^2 + 2^2}\underline{/\tan^{-1}(2/-1.5)} = 2.5\underline{/126.9°}$$

From Eq. (3.41) it follows that

$$\left| \frac{A(a + jb)}{B(a + jb)} \right| = \frac{F_0}{2.5}$$

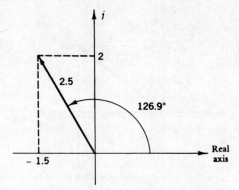

Fig. 3.6. A graphical representation of complex conjugate roots.

and

$$\alpha = \sphericalangle \frac{A(a + jb)}{B(a + jb)} = -126.9°$$

Substituting these values into Eq. (3.40) gives

$$y(t) = \frac{F_0}{6.25} + \frac{1}{2}\frac{F_0}{2.5} e^{-1.5t} \sin (2t - 126.9°)$$

$$= \frac{F_0}{6.25} [1 + 1.25e^{-1.5t} \sin (2t - 126.9°)] \qquad (3.42)$$

The roots of a quadratic are not necessarily complex conjugates. For example, for the quadratic $s^2 + 6s + 8$, it follows that $-2a = 6$, or $a = -3$, and $a^2 + b^2 = 8$, or $b = \sqrt{8 - 9} = \sqrt{-1} = j$. Hence, the roots $a \pm jb$ are $-3 \pm jj = -3 \mp 1 = -4, -2$. These roots are real and distinct. In general, the roots of a quadratic may be ascertained according to Table 3.1.

Table 3.1

b	Roots
Real	Complex conjugate roots
Zero	Two equal roots
Imaginary	Two real roots

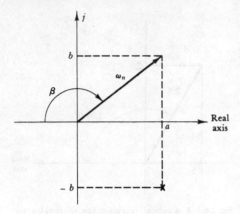

Fig. 3.7. Polar form of complex conjugate roots.

The form $a \pm jb$ is, in effect, the rectangular representation for complex conjugate roots. The polar representation is illustrated in Fig. 3.7. In polar form, the roots $a \pm jb$ are specified by the radius $\omega_n = \sqrt{a^2 + b^2}$ and the angle β. By noting that $a = -\omega_n \cos \beta$, the quadratic form may be expressed as

$$s^2 - 2as + a^2 + b^2 = s^2 + (2\omega_n \cos \beta)s + \omega_n{}^2 \qquad (3.43)$$

In Fig. 3.7, when $\beta = 0$ ($\cos \beta = 1$), equal roots result.

$$s^2 + 2\omega_n s + \omega_n{}^2 = (s + \omega_n)^2 \qquad (3.44)$$

The case of equal roots is the border line between where the roots are complex conjugates and where the roots are real. This represents the transition from the case in which the response is sinusoidal to that in which it is exponential. The ratio of the coefficient of the s term when the roots are complex conjugates (that is, $2\omega_n \cos \beta$) to the value of the s coefficient when the roots are equal (that is, $2\omega_n$) is defined as the damping ratio ζ.

$$\zeta = \frac{2\omega_n \cos \beta}{2\omega_n} = \cos \beta \qquad (3.45)$$

Thus, the angle β determines the damping ratio ζ. In terms of damping ratio, the quadratic form is

$$s^2 + 2\zeta\omega_n s + \omega_n{}^2 \qquad (3.46)$$

For the preceding example, $\omega_n{}^2 = 6.25$ or $\omega_n = 2.5$. Similarly, $2\zeta\omega_n = 3$ or $\zeta = 3/2\omega_n = 0.6$.

The following equations are used to convert from the rectangular form to the polar form:

$$\omega_n = \sqrt{a^2 + b^2}$$

$$\zeta = \frac{-a}{\omega_n} = \frac{-a}{\sqrt{a^2 + b^2}} \tag{3.47}$$

In Chap. 6, it is shown how inverse transforms may also be obtained by the method of residues.

3.4 SYSTEM RESPONSE

When a system has complex conjugate roots, the response is most readily determined by using the rectangular form $a \pm jb$. After having determined the response, more insight into the actual behavior of the system is obtained by converting from the rectangular form to the polar form. This fact is illustrated in the following discussions of step-function response, impulse response, and pulse response.

Step-function response

Let it be desired to determine the response of the mass-spring-damper system of Fig. 3.1 for the case in which $f(t)$ is a step excitation, as shown in Fig. 3.8a. All the initial conditions are zero.

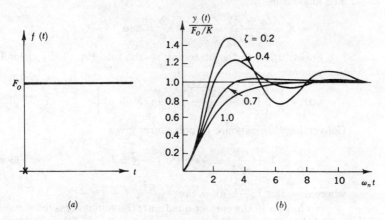

(a)   (b)

Fig. 3.8. Response of a second-order system to a step change. (a) The step input; (b) the response.

Thus, the transformed equation is

$$Y(s) = \frac{F_0/M}{s[s^2 + (B/M)s + (K/M)]} \tag{3.48}$$

The partial fraction expansion is

$$Y(s) = \frac{K_1}{s} + \frac{K_c}{s - (a + jb)} + \frac{K_{-c}}{s - (a - jb)}$$

The corresponding time response is

$$y(t) = K_1 + \frac{1}{b}\left|\frac{A(a + jb)}{B(a + jb)}\right| e^{at} \sin(bt + \alpha) \tag{3.49}$$

The value of K_1 is

$$K_1 = \lim_{s \to 0} \frac{F_0/M}{s^2 + (B/M)s + (K/M)} = \frac{F_0}{K}$$

The term $A(a + jb)/B(a + jb)$ is

$$\frac{A(a + jb)}{B(a + jb)} = \frac{F_0/M}{s}\bigg|_{s=a+jb} = \frac{F_0/M}{a + jb} = \frac{F_0/M}{\sqrt{a^2 + b^2}\,/\tan^{-1}(b/a)}$$

From Eq. (3.48), it follows that $a^2 + b^2 = K/M$, or

$$M = K/(a^2 + b^2)$$

Hence

$$\left|\frac{A(a + jb)}{B(a + jb)}\right| = \frac{F_0}{M\sqrt{a^2 + b^2}} = \frac{\sqrt{a^2 + b^2}\,F_0}{K}$$

The angle α is

$$\alpha = \measuredangle \frac{A(a + jb)}{B(a + jb)} = -\tan^{-1}\frac{b}{a} = \tan^{-1}\frac{-b}{a}$$

Substitution of the preceding results into Eq. (3.49) yields for the response

$$y(t) = \frac{F_0}{K}\left[1 + \frac{\sqrt{a^2 + b^2}}{b} e^{at} \sin(bt + \alpha)\right] \tag{3.50}$$

Converting this response to polar form gives

$$\frac{y(t)}{F_0/K} = 1 + \frac{e^{-\zeta\omega_n t}}{\sqrt{1 - \zeta^2}} \sin(\sqrt{1 - \zeta^2}\,\omega_n t + \alpha) \tag{3.51}$$

where $\alpha = \tan^{-1}(-b/a) = \tan^{-1}(-\sqrt{1 - \zeta^2}/-\zeta)$.

To have α in the correct quadrant, the minus signs must not be canceled in the preceding arc-tangent function. In general, when evaluating α one must not tamper with the signs in the numerator or denominator of the resulting arc-tangent function. The quantity F_0/K is the static deflection of the spring due to the steady force F_0. Thus, $y/(F_0/K)$ is the ratio of the actual deflection to this static deflection.

In Fig. 3.8b is shown the response for various values of ζ. The parameter ζ determines the form of the response. Because the time axis is $\omega_n t$, ω_n determines the speed of response. For example, for $\zeta = 0.4$, the response attains a value of $y/(F_0/K) = 1$ at $\omega_n t = 2$. Hence, $t = 2/\omega_n$, or the larger the value of ω_n, the faster the response.

In terms of the system parameters, the values of ζ and ω_n are

$$\omega_n = \sqrt{\frac{K}{M}}$$

and $\hspace{10cm}$ (3.52)

$$\zeta = \frac{B/M}{2\omega_n} = \frac{B}{2\sqrt{KM}}$$

Maximum overshoot

In Fig. 3.9 is shown the response for the case in which $\zeta = 0.2$. The point at which the maximum overshoot occurs is indicated by time

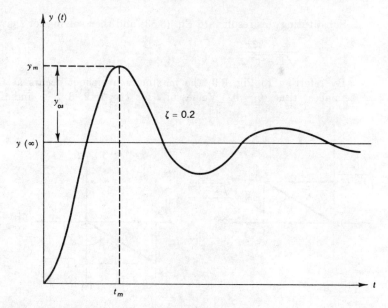

Fig. 3.9. Maximum overshoot.

t_m. This time is obtained by differentiating Eq. (3.51) and then setting the result equal to zero; hence

$$\frac{F_0}{K}\left[\omega_n e^{-\zeta\omega_n t}\cos\left(\sqrt{1-\zeta^2}\,\omega_n t+\alpha\right)\right.$$
$$\left.-\frac{\zeta\omega_n e^{-\zeta\omega_n t}}{\sqrt{1-\zeta^2}}\sin\left(\sqrt{1-\zeta^2}\,\omega_n t+\alpha\right)\right]=0$$

Dividing through by $\cos\left(\sqrt{1-\zeta^2}\,\omega_n t+\alpha\right)$ gives

$$\frac{\sqrt{1-\zeta^2}}{\zeta}=\tan\left(\sqrt{1-\zeta^2}\,\omega_n t+\alpha\right)$$

Taking the arc tangent of the preceding expression gives

$$\tan^{-1}\frac{\sqrt{1-\zeta^2}}{\zeta}=\sqrt{1-\zeta^2}\,\omega_n t+\alpha \tag{3.53}$$

In Fig. 3.10a is shown the angle $\tan^{-1}\left(\sqrt{1-\zeta^2}/\zeta\right)$, and in Fig. 3.10b is shown the angle $\alpha=\tan^{-1}\left(-\sqrt{1-\zeta^2}/-\zeta\right)$. As illustrated in Fig. 3.10a and b, $\tan^{-1}\left(\sqrt{1-\zeta^2}/\zeta\right)$ is displaced from α by 180° (that is, $\pi+k2\pi$). Thus

$$\tan^{-1}\frac{\sqrt{1-\zeta^2}}{\zeta}=\alpha+(\pi+k2\pi)$$

Substituting this result into Eq. (3.53) and then solving for t gives

$$t=\frac{\pi+k2\pi}{\sqrt{1-\zeta^2}\,\omega_n} \tag{3.54}$$

By referring to Fig. 3.9, the maximum overshoot occurs at the smallest time, $k=0$. Values of t for $k=1, 2, 3, \ldots$ indicate

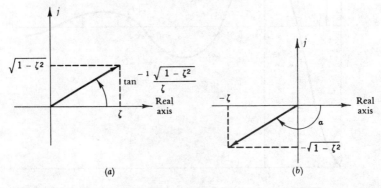

(a) (b)

Fig. 3.10. Representations of (a): $\tan^{-1}\left(\sqrt{1-\zeta^2}/\zeta\right)$; (b): $\tan^{-1}\left(\sqrt{1-\zeta^2}/-\zeta\right)$.

successive places where the derivative is zero. Substitution of $t = \pi/(\sqrt{1 - \zeta^2}\,\omega_n)$ into Eq. (3.51) gives for the maximum value of y

$$y_m = \frac{F_0}{K}\left[1 + \frac{\exp\left(-\zeta\pi/\sqrt{1 - \zeta^2}\right)}{\sqrt{1 - \zeta^2}} \sin\left(\pi + \alpha\right)\right]$$

The angle $\alpha + \pi = \tan^{-1}\left(\sqrt{1 - \zeta^2}/\zeta\right)$ is shown in Fig. 3.10a; hence, $\sin\left(\alpha + \pi\right) = \sqrt{1 - \zeta^2}$. Thus, y_m becomes

$$y_m = \frac{F_0}{K}\left[1 + \exp\left(\frac{-\zeta\pi}{\sqrt{1 - \zeta^2}}\right)\right] \tag{3.55}$$

The final steady-state value for $y(t)$ is found by substituting $t = \infty$ in Eq. (3.51). This gives $y(\infty) = F_0/K$. As shown in Fig. 3.9, the amount of overshoot y_{os} is the maximum value y_m minus the steady-state value $y(\infty)$. Thus,

$$y_{os} = y_m - y(\infty) = \frac{F_0}{K}\exp\left(\frac{-\zeta\pi}{\sqrt{1 - \zeta^2}}\right) \tag{3.56}$$

The overshoot is a function of ζ, which is the parameter that determines the form of the response.

Impulse response

Let it be desired to determine the response to an impulse function as shown in Fig. 3.11a. Let the area of the impulse function be A so that $F(s) = A$. For the case in which all the initial conditions are zero, Eq. (3.2) becomes

$$Y(s) = \frac{A/M}{s^2 + (B/M)s + K/M}$$

The corresponding response is

$$y(t) = \frac{1}{b}\left|\frac{A(a + jb)}{B(a + jb)}\right| e^{at} \sin\left(bt + \alpha\right) \tag{3.57}$$

where

$$\frac{A(a + jb)}{B(a + jb)} = \lim_{s \to a + jb}\frac{A}{M} = \frac{A}{M}$$

Hence, the magnitude is A/M, and the angle α is zero. Substituting these results into Eq. (3.57) and noting that $a^2 + b^2 = K/M$ gives,

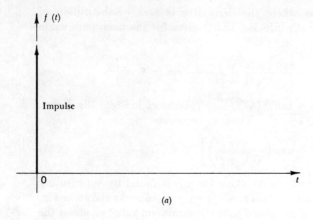

(a)

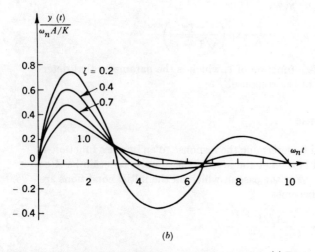

(b)

Fig. 3.11. Response of a second-order system to an impulse. (a) The impulse; (b) the response.

for the response,

$$y(t) = \frac{a^2 + b^2}{b} \frac{A}{K} e^{at} \sin bt$$

Converting to polar form gives

$$\frac{y(t)}{\omega_n A/K} = \frac{1}{\sqrt{1 - \zeta^2}} e^{-\zeta \omega_n t} \sin \sqrt{1 - \zeta^2} \, \omega_n t \qquad (3.58)$$

A plot of the corresponding impulse response is shown in Fig. 3.11b.   As is to be expected, the damping ratio ζ determines the general form of the response, and the natural frequency ω_n determines the speed.

Pulse response

Let it be desired to determine the response of the mass-spring-damper system of Fig. 3.1 to the pulse function shown in Fig. 3.12. Let all the initial conditions be zero.   The transform of this pulse is

$$F(s) = \frac{F_0}{s} (1 - e^{-t_1 s})$$

Substitution of this result into Eq. (3.2) gives

$$Y(s) = \frac{F_0/M}{s[s^2 + (B/M)s + K/M]} - \frac{F_0 e^{-t_1 s}/M}{s[s^2 + (B/M)s + K/M]} \tag{3.59}$$

The operator $e^{-t_1 s}$ means that the second term in Eq. (3.59) does not come into play until $t > t_1$. Hence, for $0 < t \le t_1$, the response is the same as that for a step input as given by Eq. (3.51); that is,

$$y_1(t) = \frac{F_0}{K}\left[1 + \frac{e^{-\zeta \omega_n t}}{\sqrt{1 - \zeta^2}} \sin\left(\sqrt{1 - \zeta^2}\,\omega_n t + \alpha\right)\right]$$
$$0 < t \le t_1 \quad (3.60)$$

For $t > t_1$, the additional contribution due to the second term in Eq. (3.59) must be included.   This additional term is obtained by taking the inverse of the second term without regard to the $e^{-t_1 s}$

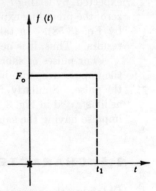

Fig. 3.12. A pulse function.

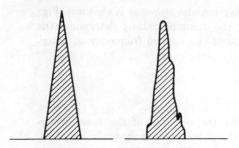

Fig. 3.13. Functions of short duration.

factor and then substituting $t - t_1$ for t to take care of the delayed effect. Because the second term in Eq. (3.59) is the same as the first, with the exception of the $e^{-t_1 s}$ factor, then for $t > t_1$ the response is

$$y(t) = y_1(t) - y_1(t - t_1)$$

where $y_1(t)$ is given by Eq. (3.60) and $y_1(t - t_1)$ is obtained by substituting $t - t_1$ for t in Eq. (3.60). Thus, for $t > t_1$ the response is

$$y_2(t) = \frac{F_0}{K}\left[1 + \frac{e^{-\zeta\omega_n t}}{\sqrt{1 - \zeta^2}}\sin\left(\sqrt{1 - \zeta^2}\,\omega_n t + \alpha\right)\right]$$
$$- \frac{F_0}{K}\left\{1 + \frac{e^{-\zeta\omega_n(t - t_1)}}{\sqrt{1 - \zeta^2}}\sin\left[\sqrt{1 - \zeta^2}\,\omega_n (t - t_1) + \alpha\right]\right\}$$

When $F_0 = A/t_1$, the pulse function has an area A. As would be expected, by letting $F_0 = A/t_1$ and taking the limit as t_1 approaches zero, the preceding expression reduces to the impulse response given by Eq. (3.58). In taking this limit, the indeterminate form $0/0$ results. Thus, it is necessary to apply L'Hospital's rule.

For pulses of short duration, the response is approximated by the impulse response in which the area of the impulse is the area of the pulse. Similarly, for any arbitrary excitation of short duration as illustrated in Fig. 3.13, the response may be approximated by an impulse having the same area.

3.5 CHARACTERISTIC EQUATION

Obtaining the response of a system to a particular input such as a step function or impulse gives an indication of the transient behavior of the system. In practice, the system may be subjected to a wide variety of input excitations. It is quite impractical to determine the response for all possible excitations. Thus, it is desirable to obtain an indication of the transient behavior for any input. The roots of the characteristic equation provide such an indication. Equation (3.7) is the general transform for a differential equation. It may be written in the form

$$Y(s) = \frac{L_m(s)N_{F(s)}}{L_n(s)D_{F(s)}} + \frac{I(s)}{L_n(s)} \tag{3.61}$$

where $F(s) = N_{F(s)}/D_{F(s)}$, in which $N_{F(s)}$ is the numerator of $F(s)$ and $D_{F(s)}$ is the denominator of $F(s)$. If $f(t)$ is a step function of height h, then $N_{F(s)} = h$ and $D_{F(s)} = s$. A partial fraction expansion of the first term on the right-hand side of Eq. (3.61) yields response terms due to the roots of $L_n(s)$ and response terms due to the roots of $D_{F(s)}$. The roots of $D_{F(s)}$ yield terms associated with the particular input. For example, for a step function, the inverse of K_i/s yields a constant K_i. Similarly, if the input is an exponential e^{at}, then the inverse of $K_i/(s - a)$ yields an exponential term $K_i e^{at}$ in the response expression. Thus, $D_{F(s)}$ merely yields response terms appropriate for the particular input.

The function $L_n(s)$ is called the *characteristic function* of the system. The roots of $L_n(s)$ yield response terms associated with the physical system. For example, for the mass-spring-damper system shown in Fig. 3.1, the characteristic function is

$$s^2 + \frac{B}{M}s + \frac{K}{M}$$

If M, B, and K are such that the roots are complex conjugates, a sinusoidal type of term appears in the response expression. If the two roots are real, two exponential terms appear. From Eqs. (3.23), (3.32), and (3.38), it is to be noted that response terms always contain an exponential factor e^{at} or e^{rt}. The value of the exponent is the horizontal distance from the root to the imaginary axis. If any root of the characteristic function lies to the right of the imaginary axis (i.e., is located in the right half plane), then the response always contains an increasing time term. This causes the response to become infinite regardless of the input. Such systems are unstable.

In Fig. 3.14 is shown the type of response terms associated with real roots. A positive real root yields an increasing exponential, while a negative real root yields a decaying exponential. A root at the origin yields a constant term in the response.

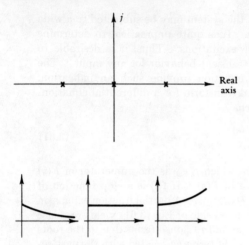

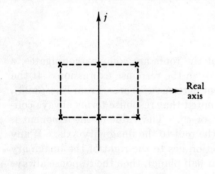

Fig. 3.14.   Response due to real roots.

Figure 3.15 shows the response terms that result from complex conjugate roots. Positive roots yield an increasing sinusoid, while negative roots yield a decreasing sinusoid. Complex conjugate roots on the imaginary axis yield a sinusoidal term of constant amplitude such that the response oscillates indefinitely.

If $L_n(s)$ has a root at the origin and the input has a constant value h such that $F(s) = h/s$, then the denominator has an s^2 term

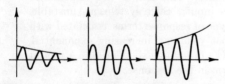

Fig. 3.15.   Response due to complex conjugate roots.

such that a $C_2/s^2 + C_1/s$ term appears in the partial fraction expansion. The resulting response term is of the form

$$(C_2 t + C_1) e^0 = C_2 t + C_1$$

This response increases indefinitely with time, even though the input is constant. Thus, the imaginary axis (including the origin) is to be avoided as a possible location for roots. For stable response all the roots of $L_n(s)$ must be located to the left of the imaginary axis. That is, they must have negative real parts.

A root-locus plot is a plot of the roots of the characteristic equation as one of the parameters is varied. Thus, from a root-locus plot, one can select the value of this parameter to yield the most desirable transient response. There are many fine references ([2] to [5]) which present a thorough explanation of this powerful tool.

Graphical determination of partial-fraction-expansion constants

In Secs. 3.2 and 3.3, numerical methods are presented for evaluating the constants for a partial fraction expansion. These constants may also be evaluated by graphical means. To illustrate the graphical method, let it be desired to determine the expansion for the function

$$Y(s) = \frac{10(s + 4)}{s(s^2 + 4s + 8)} = 10 \frac{s + 4}{s(s + 2 - j2)(s + 2 + j2)} \tag{3.62}$$

The corresponding response expression is

$$y(t) = 10 \left[K_1 + \frac{1}{b} \left| \frac{A(a + jb)}{B(a + jb)} \right| e^{at} \sin(bt + \alpha) \right] \tag{3.63}$$

The constant 10 is factored out, because the graphical method cannot distinguish constant multipliers. In Fig. 3.16a, a specific point $a + jb$ is designated by an ×, and the general point s is designated by a dot. The vector from the × to the dot is $s - (a + jb)$. The values of s which make $Y(s)$ infinite are called the *poles* of $Y(s)$. From Eq. (3.62), the poles of $Y(s)$ occur at $s = 0$, $-2 + j2$, and $-2 - j2$. The poles are plotted as ×'s in Fig. 3.16b. The values of s which make $Y(s)$ zero are called the *zeros* of $Y(s)$. Zeros are plotted as circles, ⊙. The zero for the function of Eq. (3.62) occurs at $s = -4$.

93

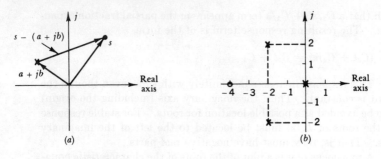

(a)

(b)

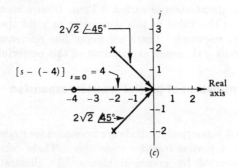

(c)

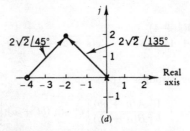

(d)

Fig. 3.16. Graphical determination of partial-fraction-expansion constants.

The value of K_1 is

$$
K_1 = \left. \frac{s + 4}{(s + 2 - j2)(s + 2 + j2)} \right|_{s=0}
$$

$$
= \left. \frac{s - (-4)}{[s - (-2 + j2)][s - (-2 - j2)]} \right|_{s=0}
$$

$$
= \frac{4 \underline{/0°}}{2 \sqrt{2} \underline{/-45°}\, 2 \sqrt{2} \underline{/+45°}} = \frac{1}{2} \tag{3.64}
$$

As shown in Fig. 3.16c, the numerator term $s - (-4)$ is the vector from the zero at (-4) to the point at which s is to be evaluated (i.e., at $s = 0$). In general, each numerator term in a partial fraction expansion may be represented graphically as the vector from the particular zero (circle) to the point at which s is to be evaluated. Similarly, each denominator term is represented by the vector from each $\times$ to the point at which s is to be evaluated.

The term $A(a + jb)/B(a + jb)$ associated with the complex conjugate pair of roots is evaluated graphically from Fig. 3.16d. Thus,

$$\frac{A(a + jb)}{B(a + jb)} = \frac{s + 4}{s}\bigg|_{s = a + jb} = \frac{s - (-4)}{s - (0)}\bigg|_{s = -2 + j2}$$

$$= \frac{2\sqrt{2}\,\underline{/45°}}{2\sqrt{2}\,\underline{/135°}} = 1\,\underline{/-90°} \quad (3.65)$$

Substitution of the preceding results into Eq. (3.63) gives for the response

$$y(t) = 5 + 5e^{-2t}\sin(2t - 90°) \quad (3.66)$$

The graphical method shows that the value of the constants that arise in a partial fraction expansion depends upon the relative location of the various poles and zeros.

3.6 VERTICAL SHIFTING

In the study of engineering systems, it is convenient to take the initial state of a system as being a steady-state operating condition. By *steady state* is meant that both the input $f(t)$ and output $y(t)$ have constant values. Thus, at a steady-state operating condition, derivatives of the input $[Df(t), D^2f(t), \ldots]$ and derivatives of the output $[Dy(t), D^2y(t), \ldots]$ are all zero. A system is initially at a steady-state operating condition when the initial value of each derivative is zero.

Suppose that y_1 is the known solution of the following differential equation in which f_1 is the input.

$$(D^n + a_{n-1}D^{n-1} + \cdots + a_1D + a_0)y_1$$
$$= (b_mD^m + \cdots + b_1D + b_0)f_1 \quad (3.67)$$

In Fig. 3.17a is shown a new input f_2 which is displaced vertically a fixed amount from f_1; that is, $f_2 = f_1 + [f_2(0) - f_1(0)]$. The vertical-shifting theorem states that, if the system is initially at a

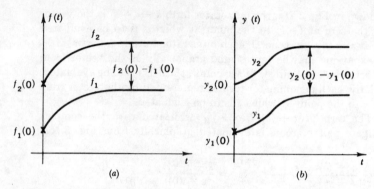

Fig. 3.17. Vertical shifting. (a) Input function f_2 which is vertically displaced from f_1; (b) response function y_2 which is vertically displaced from y_1.

steady-state operating condition (for both the case in which f_1 is the input and the case in which f_2 is the input), then the new response y_2 is

$$y_2 = y_1 + [y_2(0) - y_1(0)] = y_1 + \frac{b_0}{a_0}[f_2(0) - f_1(0)] \qquad (3.68)$$

As is shown in Fig. 3.17b, this resulting response function y_2 is shifted vertically a constant amount $y_2(0) - y_1(0)$ from y_1.

This vertical-shifting theorem is proved by checking to see that y_2 as given by Eq. (3.68) satisfies the original differential equation when f_2 is the input.

$$(D^n + \cdots + a_0)\{y_1 + [y_2(0) - y_1(0)]\}$$
$$= (b_m D^m + \cdots + b_0)\{f_1 + [f_2(0) - f_1(0)]\} \quad (3.69)$$

Because y_1 and f_1 are known to satisfy the differential equation, then subtracting Eq. (3.67) from (3.69) gives

$$(D^n + \cdots + a_0)[y_2(0) - y_1(0)]$$
$$= (b_m D^m + \cdots + b_0)[f_2(0) - f_1(0)]$$

Because $y_2(0) - y_1(0)$ and $f_2(0) - f_1(0)$ are constants, powers of D operating on these constants vanish. Thus, the preceding expression reduces to

$$a_0[y_2(0) - y_1(0)] = b_0[f_2(0) - f_1(0)] \qquad (3.70)$$

It remains now to prove that Eq. (3.70) is a valid expression for any system which is initially at a steady-state operating condition. Figure 3.17a and b shows that derivatives are unaffected by a vertical shifting. That is, at any time t the derivatives of f_2 and y_2 are the same as those for f_1 and y_1 (including the value of the initial derivatives at time $t = 0$). The differential equation relating f_2 and y_2 evaluated at time $t = 0$ and the differential equation relating f_1 and y_1 evaluated at time $t = 0$ are

$$y_2{}^n(0) + \cdots + a_1 y_2'(0) + a_0 y_2(0)$$
$$= b_m f_2{}^m(0) + \cdots + b_0 f_2(0) \quad (3.71)$$
$$y_1{}^n(0) + \cdots + a_1 y_1'(0) + a_0 y_1(0)$$
$$= b_m f_1{}^m(0) + \cdots + b_0 f_1(0) \quad (3.72)$$

Because the values of the initial derivative terms are unaffected by a vertical shift, when Eq. (3.72) is subtracted from Eq. (3.71) all the derivative terms cancel. Thus, Eq. (3.70) is automatically satisfied. In this proof of the vertical-shifting theorem, it is not necessary that the initial derivative terms [see Eqs. (3.71) and (3.72)] be zero, but rather that the derivatives of f_2 and y_2 at time $t = 0$ be the same as those for f_1 and y_1. Thus, a more general statement of the vertical-shifting theorem is that, if:

1. The input is $f_2 = f_1 + [f_2(0) - f_1(0)]$ and
2. The initial derivative terms of y_2 and f_2 are the same as those for y_1 and f_1

then the solution y_2 is given by Eq. (3.68).

In obtaining the response for a system which is initially at a steady-state operating condition, a further simplification occurs when y_1 is taken as the response for the case in which all the initial conditions are zero. Thus, Eq. (3.68) reduces to

$$y_2 = y_1 + y_2(0) = y_1 + \frac{b_0}{a_0} f_2(0) \quad (3.73)$$

Because y_1 is the case in which all the initial conditions are zero, the transform of y_1 is obtained by merely substituting s for D in the differential equation. Hence, the response y_1 is readily obtained. Equation (3.73) shows that, for systems which are initially at a steady-state operating condition, the response y_2 is obtained by merely adding the initial value $y_2(0)$ to the response for the case in which all the initial conditions are zero.

Illustrative example The solution of Eq. (3.1) for the case in which $f(t)$ is the function shown in Fig. 3.2a with $M = 1$, $K = 2$, and $B = 3$ is given by Eq. (3.14). Let it be desired to determine the solution for the case in which $f(t)$ is raised four units as shown in Fig. 3.18a. In both cases, the system is

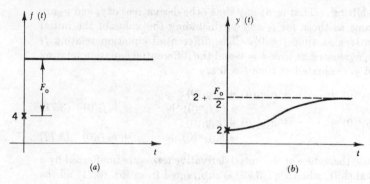

Fig. 3.18. Response to vertically shifted step input.

initially at a steady-state operating condition (i.e., all the initial derivative terms are zero).

Solution From Eq. (3.1)

$$a_0 = \frac{K}{M} = 2 \qquad b_0 = \frac{1}{M} = 1$$

Hence, from Eq. (3.70)

$$y_2(0) - y_1(0) = \frac{b_0}{a_0} [f_2(0) - f_1(0)] = \frac{1}{2} (4) = 2$$

From Eq. (3.68), the new solution is the initial solution as given by Eq. (3.14) plus the change $y_2(0) - y_1(0) = 2$; that is,

$$y(t) = \left(\frac{1}{2} - e^{-t} + \frac{1}{2} e^{-2t} \right) F_0 + 2 \tag{3.74}$$

Note that Fig. 3.18b is identical to Fig. 3.2b except that the former is raised by the change in the initial value

$$y_2(0) - y_1(0) = 2$$

To verify Eq. (3.74) by the transformation method, first obtain the transform of $f(t)$ shown in Fig. 3.18a; that is,

$$F(s) = \frac{F_0 + 4}{s}$$

Next write Eq. (3.1) for time $t = 0$.

$$\frac{1}{M} f(0) = y''(0) + \frac{B}{M} y'(0) + \frac{K}{M} y(0)$$

Because the derivative terms are unchanged, $y''(0) = y'(0) = 0$ as originally given. Hence

$$y(0) = \frac{1}{K} f(0) = 2$$

From Eq. (3.2), the transformed equation is

$$
\begin{aligned}
Y(s) &= \frac{F_0 + 4}{s(s^2 + 3s + 2)} + \frac{2(s + 3)}{s^2 + 3s + 2} \\
&= \frac{F_0}{s(s^2 + 3s + 2)} + \frac{2s^2 + 6s + 4}{s(s^2 + 3s + 2)} \\
&= \frac{F_0}{s(s^2 + 3s + 2)} + \frac{2}{s}
\end{aligned}
\tag{3.75}
$$

The first term on the right-hand side of the preceding expression is identical to the right-hand side of Eq. (3.8); hence, the inverse yields the same result as given by Eq. (3.74). Application of the vertical-shifting theorem is thus seen to save considerable computational effort.

3.7 RESPONSE TO AN ARBITRARY INPUT

Suppose that the forcing function $f(t)$ is some general function such as that indicated by the solid curve of $f(t)$ versus t in Fig. 3.19a. Such a function could be approximated by a series of pulses. However, better accuracy is obtained by approximating the function by

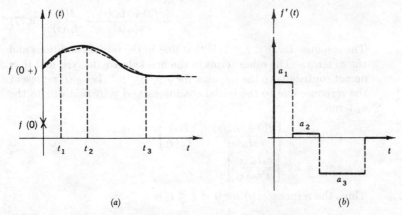

Fig. 3.19. Piecewise approximation. (a) The function; (b) the derivative.

a series of straight lines (shown by dashed lines). This general case shows a step change at the origin. The initial value is designated as $f(0)$, and the value at time $t = 0+$ is indicated by $f(0+)$. For $0 < t \le t_1$ the slope is a_1, for $t_1 < t \le t_2$ the slope is a_2, and for $t_2 < t \le t_3$ the slope is a_3. The derivative of a straight line of slope a is a step of height a; hence, the plot of $f'(t)$ is a series of steps as indicated in Fig. 3.19b. Because $f(t)$ undergoes a step change at the origin, $f'(t)$ has an impulse at the origin. The area of this impulse is equal to the step change $f(0+) - f(0)$. The Laplace transform of $f'(t)$ shown in Fig. 3.19b is

$$\mathcal{L}[f'(t)] = [f(0+) - f(0)] + \frac{a_1}{s} + \frac{a_2 - a_1}{s} e^{-t_1 s}$$
$$+ \frac{a_3 - a_2}{s} e^{-t_2 s} + \frac{(0 - a_3)e^{-t_3 s}}{s}$$

The transform of the function $f(t)$ may be determined from the transform of the derivative by solving Eq. (2.8) for $F(s) = \mathcal{L}[f(t)]$; that is,

$$F(s) = \frac{\mathcal{L}[f'(t)] + f(0)}{s}$$
$$= \frac{f(0+)}{s} + \frac{1}{s^2} [a_1 + (a_2 - a_1)e^{-t_1 s}$$
$$+ (a_3 - a_2)e^{-t_2 s} - a_3 e^{-t_3 s}] \quad (3.76)$$

Substitution of $F(s)$ into Eq. (3.7) gives the general transformed equation

$$Y(s) = \frac{[a_1 + (a_2 - a_1)e^{-t_1 s} + (a_3 - a_2)e^{-t_2 s} - a_3 e^{-t_3 s}]L_m(s)}{s^2 L_n(s)}$$
$$+ \frac{f(0+)L_m(s)}{sL_n(s)} + \frac{I(s)}{L_n(s)} \quad (3.77)$$

The response for $0 < t \le t_1$ is that due to the initial conditions and the a_1 term. The other terms in the brackets are delayed and thus do not contribute to the response for $0 < t \le t_1$. Let $y_i(t)$ represent the response due to the initial conditions and $y_1(t)$ that due to the a_1 term.

$$y_i(t) = \mathcal{L}^{-1}\left[\frac{f(0+)L_m(s)}{sL_n(s)} + \frac{I(s)}{L_n(s)} \right]$$
$$y_1(t) = \mathcal{L}^{-1}\left[\frac{a_1 L_m(s)}{s^2 L_n(s)} \right]$$

Thus, the response $y(t)$ for $0 < t \le t_1$ is

$$y(t) = y_1(t) + y_i(t) \quad (3.78)$$

For the interval $t_1 < t \leq t_2$ the response due to the $(a_2 - a_1)e^{-t_1s}$ term in Eq. (3.77) must be added to the preceding expression. The effect of this term is obtained by multiplying $y_1(t)$ by $(a_2 - a_1)/a_1$ to take care of the coefficients and then substituting $t - t_1$ for t to take care of the time delay. Hence for $t_1 < t \leq t_2$

$$y(t) = y_1(t) + \frac{a_2 - a_1}{a_1} y_1(t - t_1) + y_i(t)$$

Similarly, the contribution of each following delayed term is added to the response as it comes into play.

A more accurate approximation can be made by approximating the original function by parabolas rather than straight lines. Because the derivative of a parabola is a straight line of constant slope, $f'(t)$ will be a series of such straight lines. For this case $f''(t)$ will be a series of steps. The transform $F(s)$ of the original function $f(t)$ is then determined from the transform of $f''(t)$ by use of the relationship for the transform of the second derivative of a function [i.e., Eq. (2.9)].

3.8 THE CONVOLUTION INTEGRAL

The convolution integral has many applications in engineering problems. In the preceding section, an approximate technique was developed for solving differential equations with arbitrary inputs. The convolution integral provides a basis for solving such differential equations exactly. Another use is to determine the solution of a differential equation for which the transform of the input is unknown, as would be the case for an input function which is defined by pointwise data. It may also be used to obtain the inverse transform of the product of two or more transformed functions. Because of its general nature, the convolution integral is employed in investigating the behavior of a system for any general input $f(t)$.

The convolution integral is developed by first considering the plot of the function $f(\lambda)$ versus λ, shown in Fig. 3.20. The function $f(\lambda)$ represents the input, or excitation, to a system. It is necessary to use λ as the general time variable, because, as is shown in Fig. 3.20, the symbol t is employed to designate a specific value of time. The specific time t is the time at which the response $y(t)$ of the system is to be determined.

The input function $f(\lambda)$ may be approximated by a series of pulse functions of width $\Delta\lambda$, as is shown in Fig. 3.20. The height of the pulse which begins at $\lambda = k\Delta\lambda$ is $f(k\Delta\lambda)$. The transform of

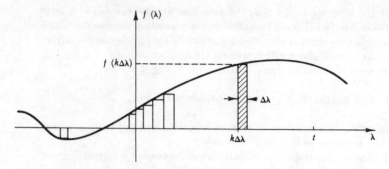

Fig. 3.20.   Approximation of a function by narrow pulses.

this pulse is

$$\frac{f(k\,\Delta\lambda)}{s}\left(e^{-k\Delta\lambda s} - e^{-(k+1)\Delta\lambda s}\right) = \frac{f(k\,\Delta\lambda)}{s}\,e^{-k\Delta\lambda s}(1 - e^{-\Delta\lambda s}) \quad (3.79)$$

Expanding $e^{-\Delta\lambda s}$ in a series gives

$$1 - e^{-\Delta\lambda s} = 1 - \left[1 - \Delta\lambda s + \frac{(\Delta\lambda s)^2}{2!} - \frac{(\Delta\lambda s)^3}{3!} + \cdots\right]$$

For very small values of $\Delta\lambda$

$$1 - e^{-\Delta\lambda s} \approx \Delta\lambda s$$

Substitution of the preceding result into Eq. (3.79) gives, for the transform of the pulse which begins at time $k\,\Delta\lambda$,

$$\Delta\lambda\,f(k\,\Delta\lambda)e^{-k\Delta\lambda s} \quad (3.80)$$

This transform is the same as that of an impulse which occurs at time $k\,\Delta\lambda$ and whose area $\Delta\lambda\,f(k\,\Delta\lambda)$ is the same as that of the pulse. Hence, for very small values of $\Delta\lambda$, a pulse may be approximated by an impulse having the same area.   The fact that a pulse of short duration may be replaced by an impulse whose area is the same as that of the pulse was obtained by independent reasoning at the end of Sec. 3.4.

Let us now investigate the portion of the response due to an impulse of area $\Delta\lambda\,f(k\,\Delta\lambda)$ which occurs at time $k\,\Delta\lambda$.   The response of a system to a unit impulse which occurs at time $\lambda = 0$ is designated as $w(\lambda)$.   A delay of $k\,\Delta\lambda$ is obtained by substituting $\lambda - k\,\Delta\lambda$ for λ.   Hence, the response to a unit impulse which occurs at time $k\,\Delta\lambda$ is designated $w(\lambda - k\,\Delta\lambda)$.   At the specific time $\lambda = t$, the value of the response due to this impulse is $w(t - k\,\Delta\lambda)$.   Thus,

$w(t - k\,\Delta\lambda)$ is the response at time t due to a unit impulse which occurs at time $k\,\Delta\lambda$. Multiplication by the area $\Delta\lambda\,f(k\,\Delta\lambda)$ gives the response at time t due to the impulse of area $\Delta\lambda\,f(k\,\Delta\lambda)$; that is,

$$f(k\,\Delta\lambda)w(t - k\,\Delta\lambda)\,\Delta\lambda$$

The total response at time t is that due to all the impulses which occur prior to time t. Thus

$$y(t) = \sum_{k=-\infty}^{N} f(k\,\Delta\lambda)w(t - k\,\Delta\lambda)\Delta\lambda \tag{3.81}$$

where $N\,\Delta\lambda < t$ indicates the time of the last impulse which occurs before time t. In the limit, $\Delta\lambda = d\lambda$ and $k\,\Delta\lambda = \lambda$, so that the preceding summation becomes the integral

$$y(t) = \int_{-\infty}^{t} f(\lambda)w(t - \lambda)\,d\lambda \tag{3.82}$$

Equation (3.82) is called the *convolution integral*. The physical meaning of the convolution integral is illustrated in Fig. 3.21. The input function $f(\lambda)$ versus λ is shown in Fig. 3.21a. The response $w(\lambda)$ of the system to a unit impulse is shown in Fig. 3.21b. Because the impulse occurs at $\lambda = 0$, then the unit-impulse response $w(\lambda)$ is zero for $\lambda < 0$. In Fig. 3.21c, observe that the plot of $w(-\lambda)$ is obtained by rotating the $w(\lambda)$ plot of Fig. 3.21b about the vertical $\lambda = 0$ axis. Figure 3.21d illustrates that $w(t - \lambda)$ is obtained by shifting the $w(-\lambda)$ plot ahead t units. Multiplication of Fig. 3.21a and d yields the resulting plot of $f(\lambda)w(t - \lambda)$ shown in Fig. 3.21e, whose area is the value of the convolution integral [Eq. (3.82)]. Thus, with the aid of the convolution integral the response $y(t)$ of a system can be determined by knowing the response to a unit impulse $w(\lambda)$ and a plot of the excitation (forcing function) $f(\lambda)$.

Because the lower limit of integration of Eq. (3.82) is $-\infty$, then application of this equation to determine the response requires a knowledge of the past history back to $t = -\infty$. It is next shown that, when the initial conditions are known, it is not necessary to know the past history in order to determine the response. In effect, the behavior of the system for negative values of time is accounted for by the initial conditions.

The general transformed form for a differential equation is given by Eq. (3.7). This may be written as follows,

$$Y(s) = \frac{L_m(s)}{L_n(s)}F(s) + \frac{I(s)}{L_n(s)} = W(s)F(s) + \frac{I(s)}{L_n(s)} \tag{3.83}$$

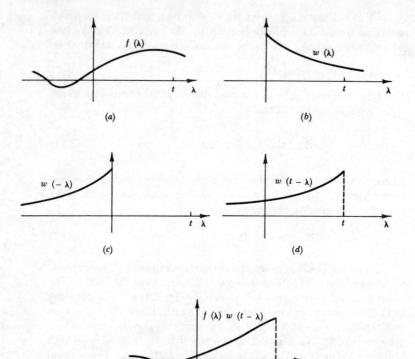

Fig. 3.21. Graphical interpretation of the convolution integral.

where $W(s) = L_m(s)/L_n(s)$. For the case in which $F(s)$ is the transform of a unit impulse $[F(s) = 1]$ and all the initial conditions are zero, then the inverse of Eq. (3.83) yields the unit-impulse response $w(\lambda)$; that is,

$$w(\lambda) = \mathcal{L}^{-1}\left[\frac{L_m(s)}{L_n(s)}\right] = \mathcal{L}^{-1}[W(s)] \qquad (3.84)$$

The general response $y(t)$ for any input is obtained by inverting Eq. (3.83); hence

$$y(t) = \mathcal{L}^{-1}[F(s)W(s)] + \mathcal{L}^{-1}\left[\frac{I(s)}{L_n(s)}\right] \qquad (3.85)$$

Equations (3.82) and (3.85) are two expressions for the response $y(t)$ of a system. Note that Eq. (3.82) may be written in the form

$$y(t) = \int_0^t f(\lambda)w(t-\lambda)\,d\lambda + \int_{-\infty}^0 f(\lambda)w(t-\lambda)\,d\lambda \qquad (3.86)$$

In the following, it is proved that

$$\mathcal{L}^{-1}[F(s)W(s)] = \int_0^t f(\lambda)w(t-\lambda)\,d\lambda \qquad (3.87)$$

Thus, comparison of Eqs. (3.85) and (3.86) shows that

$$\mathcal{L}^{-1}\left[\frac{I(s)}{L_n(s)}\right] = \int_{-\infty}^0 f(\lambda)w(t-\lambda)\,d\lambda \qquad (3.88)$$

Equation (3.88) verifies the fact that the portion of the response due to negative values of time (i.e., past history) is accounted for by the initial conditions.

To prove Eq. (3.87), first take the Laplace transform of the right-hand side of Eq. (3.87).

$$\mathcal{L}\left[\lim_{T\to\infty}\int_0^T f(\lambda)w(t-\lambda)\,d\lambda\right]$$
$$= \int_0^\infty\left[\lim_{T\to\infty}\int_0^T f(\lambda)w(t-\lambda)\,d\lambda\right]e^{-st}\,dt \qquad (3.89)$$

Because $w(t-\lambda)$ is zero for $\lambda > t$ (see Fig. 3.21d), then the upper limit of integration t may be replaced by $\lim T\to\infty$ as shown in Eq. (3.89). Interchanging the order of integration gives

$$\mathcal{L}\left[\int_0^\infty f(\lambda)w(t-\lambda)\,d\lambda\right]$$
$$= \int_0^\infty f(\lambda)\int_0^\infty [w(t-\lambda)e^{-st}\,dt]\,d\lambda \qquad (3.90)$$

The term on the right in brackets is the transform of $w(t-\lambda)$; that is,

$$\int_0^\infty w(t-\lambda)e^{-st}\,dt = e^{-s\lambda}W(s)$$

Substitution of this result into Eq. (3.90) gives

$$\mathcal{L}\left[\int_0^\infty f(\lambda)w(t-\lambda)\,d\lambda\right] = W(s)\int_0^\infty f(\lambda)e^{-s\lambda}\,d\lambda = F(s)W(s)$$

The inverse transform of the preceding equation is

$$\mathcal{L}^{-1}[F(s)W(s)] = \int_0^\infty f(\lambda)w(t-\lambda)\,d\lambda \qquad (3.91)$$

Replacing this upper limit of integration by t rather than ∞ verifies Eq. (3.87).

This result is applicable for any two functions $f(\lambda)$ and $g(\lambda)$ which are Laplace-transformable; i.e.,

$$\mathcal{L}^{-1}[F(s)G(s)] = \int_0^t f(\lambda)g(t-\lambda)\,d\lambda = f(\lambda) * g(\lambda) \tag{3.92}$$

The asterisk is a short-hand notation for the convolution integral. Since the order is inconsequential, Eq. (3.92) may also be expressed in the form

$$\mathcal{L}^{-1}[F(s)G(s)] = \int_0^t f(t-\lambda)g(\lambda)\,d\lambda = g(\lambda) * f(\lambda) \tag{3.93}$$

Because of the nature of convolution, the upper limit of integration in Eqs. (3.92) and (3.93) may be infinity as well as t.

In order to interchange the order of integration (see Sec. 2.3) as indicated by Eqs. (3.89) and (3.90), it is necessary that the following integrals exist:

$$\int_0^\infty f(\lambda)w(t-\lambda)e^{-st}\,d\lambda \tag{3.94}$$

$$\int_0^\infty f(\lambda) \int_0^\infty w(t-\lambda)e^{-st}\,dt\,d\lambda = W(s)\int_0^\infty f(\lambda)e^{-s\lambda}\,d\lambda \tag{3.95}$$

This last integral exists if $f(t)$ and $w(t)$ are both Laplace-transformable. To show that the first integral exists, note that because $f(\lambda)$ and $w(\lambda)$ are of exponential order, then we may write $|f(\lambda)| \leq M_1 e^{\sigma_1 \lambda}$ and $|w(\lambda)| \leq M_2 e^{\sigma_2 \lambda}$. Thus

$$|f(\lambda)w(t-\lambda)e^{-st}| \leq M e^{\sigma_1 \lambda}e^{\sigma_2(t-\lambda)}e^{-\sigma t} \leq M e^{(\sigma_1 - \sigma_2)\lambda}e^{-(\sigma-\sigma_2)t}$$

By selecting $\sigma > \sigma_2$ and $\sigma_1 = \sigma_2$, then this integrand is seen to be of exponential order and thus the integral is uniformly convergent. Hence, the order of integration may be interchanged when $f(\lambda)$ and $w(\lambda)$ are Laplace-transformable.

The convolution integral provides a convenient means for investigating the behavior of a system when the excitation is some general or arbitrary function $f(t)$. For example, in Eq. (3.8) if $f(t)$ is some general function whose transform is $F(s)$ rather than the step whose transform is F_0/s, then Eq. (3.8) becomes

$$Y(s) = \frac{F(s)}{(s+1)(s+2)} = \frac{F(s)}{s+1} - \frac{F(s)}{s+2} \tag{3.96}$$

Inverting the preceding gives

$$\begin{aligned}
y(t) &= \mathcal{L}^{-1}\left[F(s)\,\frac{1}{s+1}\right] - \mathcal{L}^{-1}\left[F(s)\,\frac{1}{s+2}\right] \\
&= \int_0^t f(\lambda)e^{-(t-\lambda)}\,d\lambda - \int_0^t f(\lambda)e^{-2(t-\lambda)}\,d\lambda
\end{aligned} \tag{3.97}$$

This is a general form for the response $y(t)$ which is valid for any excitation $f(\lambda)$. For the case in which $f(\lambda) = F_0$, evaluation of Eq. (3.97) gives

$$y(t) = F_0 e^{-t} \int_0^t e^\lambda \, d\lambda - F_0 e^{-2t} \int_0^t e^{2\lambda} \, d\lambda = \left(\frac{1}{2} - e^{-t} + \frac{e^{-2t}}{2} \right) F_0$$

This verifies the result of Eq. (3.14), in which the same problem was worked by the transformation technique.

By equivalence of Eqs. (3.92) and (3.93), it follows that Eq. (3.97) may also be expressed in the form

$$y(t) = \int_0^t f(t - \lambda) e^{-\lambda} \, d\lambda - \int_0^t f(t - \lambda) e^{-2\lambda} \, d\lambda$$

For $f(\lambda) = F_0$, then $f(t - \lambda) = F_0$. Hence

$$y(t) = F_0 \int_0^t e^{-\lambda} \, d\lambda - F_0 \int_0^t e^{-2\lambda} \, d\lambda = \left(\frac{1}{2} - e^{-t} + \frac{e^{-2t}}{2} \right) F_0$$

As another example of the use of the convolution integral, let it be desired to determine the inverse transform

$$\mathcal{L}^{-1} \left(\frac{1}{s^2} \frac{1}{s + a} \right) \tag{3.98}$$

By letting $F(s) = 1/s^2$ and $G(s) = 1/(s + a)$, then $f(\lambda) = \lambda$ and $g(\lambda) = e^{-a\lambda}$. Thus, the application of Eq. (3.92) gives

$$\mathcal{L}^{-1} \left(\frac{1}{s^2} \frac{1}{s + a} \right) = \int_0^t \lambda e^{-a(t-\lambda)} \, d\lambda = e^{-at} \int_0^t \lambda e^{a\lambda} \, d\lambda$$

$$= e^{-at} \left[\frac{e^{a\lambda}}{a^2} (a\lambda - 1) \right]_0^t = \frac{1}{a^2} (e^{-at} + at - 1) \tag{3.99}$$

As another application, consider the following equation:

$$y(t) = \int_0^t \lambda e^{-(t-\lambda)} \, d\lambda \tag{3.100}$$

Let $f(\lambda) = \lambda$ and $g(t - \lambda) = e^{-(t-\lambda)}$ so that $g(\lambda) = e^{-\lambda}$. Thus, the respective transforms are $F(s) = 1/s^2$ and $G(s) = 1/(s + 1)$. Transforming Eq. (3.100) gives

$$Y(s) = F(s)G(s) = \frac{1}{s^2(s + 1)}$$

Performing a partial fraction expansion and inverting gives the desired result,

$$y(t) = (t - 1) + e^{-t}$$

Integral equations

The convolution integral may also be used to solve integral equations. An integral equation is one in which the unknown function occurs inside an integral. To illustrate the application to integral equations, let it be desired to solve the following equation:

$$f(t) + f'(t) = t + \int_0^t f(\lambda)e^{-(t-\lambda)} \, d\lambda$$

Transforming each term gives

$$F(s) + sF(s) - f(0) = \frac{1}{s^2} + F(s)\frac{1}{s+1}$$

In the transform of the integral, it is to be noted that

$$g(t - \lambda) = e^{-(t-\lambda)}$$

so that $g(\lambda) = e^{-\lambda}$, and thus $G(s) = 1/(s + 1)$. Solving the transformed equation for $F(s)$ gives

$$F(s) = \frac{s+1}{s^3(s+2)} + \frac{(s+1)f(0)}{s(s+2)}$$

The final step is to perform a partial fraction expansion and then invert to obtain the desired result $f(t)$. For the case in which the initial condition is $f(0) = 0$, the result is

$$f(t) = \tfrac{1}{8}[(2t^2 + 2t - 1) + e^{-2t}]$$

As another application of the convolution integral consider the system shown in Fig. 3.22. The differential equation for this system is

$$y(t) = \frac{L_m(D)}{L_n(D)} f(t)$$

Adaptive control systems are systems in which the characteristics of the system to be controlled are continually changing (e.g., the dynamic and aerodynamic characteristics of a missile change considerably with altitude). With adaptive control systems, it is necessary to determine the characteristics of the system to be con-

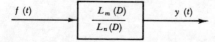

Fig. 3.22. Block diagram of a system.

trolled [that is, $L_m(D)/L_n(D)$] so that the control system may be changed accordingly to yield the best overall operation. By knowing the input $f(t)$ and the output $y(t)$, theoretically the convolution integral [Eq. (3.82)] can be solved for the impulse response $w(\lambda)$. Inverting the impulse response then yields the desired transfer function $L_m(s)/L_n(s)$. Because $w(t - \lambda)$ occurs inside the integral of Eq. (3.82), it is an extremely difficult task to unravel the impulse response. The very interesting topic of adaptive control systems is beyond the scope of this text. The purpose here is to indicate some of the applications of the convolution integral to engineering problems. These applications are many and varied. For example, the convolution integral provides a general formula that simplifies the solution of certain partial differential equations. Still other applications arise in the statistical design of control systems.

PROBLEMS

3.1   Use Laplace transforms to solve each of the following differential equations:

$$(D + 2)(D + 5)y = 10f(t) \qquad\qquad y(0) = y'(0) = 0$$
$$(D + 2)(D + 5)y = (D + 10)f(t) \qquad y(0) = y'(0) = 0$$

The input $f(t)$ is a step of height h such that $f(0) = 0$ and $f(t) = h$ for $t > 0$.

3.2   Use classical methods to verify the solution of Prob. 3.1.

3.3   Same as Prob. 3.1, except that $y(0) = 25$, $y'(0) = 0$, $f(0) = 25$, and $f(t) = 25 + h$ for $t > 0$.

3.4   Use Laplace transforms to solve each of the following differential equations:

$$(D + 2)(D + 5)y = t \qquad\quad y(0) = y'(0) = 0$$
$$(D + 2)(D + 5)y = e^{-t} \qquad y(0) = y'(0) = 0$$
$$(D + 2)(D + 5)y = e^{-2t} \qquad y(0) = y'(0) = 0$$

3.5   Use classical methods to verify the solution of Prob. 3.4.

3.6   Invert each of the following transformed equations:

(a) $Y(s) = \dfrac{3(s + 2)}{(s + 1)(s + 4)}$ (b) $Y(s) = \dfrac{5}{s^2 + 7s + 10}$

(c) $Y(s) = \dfrac{4(s + 2)}{s(s + 3)^2}$ (d) $Y(s) = \dfrac{6}{s^3(s + 2)}$

(e) $Y(s) = \dfrac{s^2 + 3s + 5}{s^2(s + 1)}$

3.7   Invert each of the following transformed equations:

(a) $Y(s) = \dfrac{100}{(s+4)(s^2+6s+25)}$

(b) $Y(s) = \dfrac{s+100}{(s+4)(s^2+6s+25)}$

(c) $Y(s) = \dfrac{4}{s(s^2+4)}$

(d) $Y(s) = \dfrac{s+10}{(s^2+4)(s^2+6s+25)}$

3.8   Equation (3.56) gives the value of the first peak overshoot for a step-function excitation.   Determine the ratio of the peak overshoot at time $t_k = (\pi + k2\pi)/\sqrt{1-\zeta^2}\,\omega_n$ to that at time $t_{k+1} = (3\pi + k2\pi)/\sqrt{1-\zeta^2}\,\omega_n$ (note that this ratio is a function of ζ only).   Determine the quantity known as the *logarithmic decrement* which is the logarithm of this ratio.

3.9   Equation (3.58) gives the response to an impulse of area A.

(a)   Determine an expression for the maximum value of the response.
(b)   Determine the ratio of the peak overshoot at time t_k to that at time t_{k+1}.

3.10   The poles and zeros of $Y(s)$ are plotted in Fig. P3.10.   Thus, the transformed equation is

$$Y(s) = \frac{K(s+3)}{s(s^2+6s+18)}$$

Evaluate the response $y(t)$ graphically from Fig. P3.10.   It is known that $y(t) = 12$ at time $t = \infty$.   Apply the final-value theorem to determine K in the preceding transformed equation.

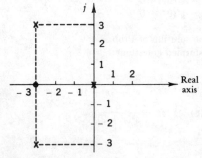

Fig. P3.10

3.11 Verify the solution of Prob. 3.3 by applying the vertical-shifting theorem to Prob. 3.1.

3.12 Solve the differential equation

$$(D^2 + 5D + 6)y = (D + 6)f(t)$$

for each of the following sets of boundary conditions:

(a) $f(0) = y(0) = y'(0) = 0$ and $f(t) = h$ for $t > 0$
(b) $f(0) = y(0) = 0$, $y'(0) = 4$ and $f(t) = h$ for $t > 0$

3.13 If the excitation $f(t)$ in Prob. 3.12 is raised vertically five units and the initial derivative terms are the same, then the boundary conditions become:

(a) $f(0) = 5$, $y(0) = 5$, $y'(0) = 0$ and $f(t) = h + 5$ for $t > 0$
(b) $f(0) = 5$, $y(0) = 5$, $y'(0) = 4$ and $f(t) = h + 5$ for $t > 0$

Use the vertical-shifting theorem to determine the new solution, and then verify these results by working out the Laplace transform solution.

3.14 Solve the differential equation

$$(D^2 + 7D + 12)y = 6f(t)$$

for the case in which the system is initially at a steady-state operating condition and:

(a) $f(0) = 2$ and $f(t) = 2 + h$ for $t > 0$
(b) $f(0) = 2$ and $f(t) = 2 + t$ for $t > 0$

3.15 The system described by each of the following differential equations is initially at a steady-state operating condition. Determine the response $y(t)$. In both cases $f(0) = 2$ and $f(t) = h + 2$ for $t > 0$:

(a) $(D + 1)(D + 2)y = (D + 4)f(t)$
(b) $(D + 1)(D^2 + 3D + 2)y = (D + 4)f(t)$

3.16 Determine the solution for the following differential equation:

$$(D^2 + 6D + 25)y(t) = 50f(t)$$

The initial conditions are $y(0) = 20$ and $y'(0) = 0$. The excitation is $f(0) = 10$ and $f(t) = 10 + h$ for $t > 0$.

3.17 Use Laplace transforms to solve each of the following differential equations:

(a) $(D + 2)y = \sin \omega t$ $\qquad y(0) = 0$
(b) $(D + 2)y = \cos \omega t$ $\qquad y(0) = 0$
(c) $(D + 2)y = e^{j\omega t}$ $\qquad y(0) = 0$

Because $e^{j\omega t} = \cos \omega t + j \sin \omega t$, verify that the real part of the answer to (c) is the answer for (b) and similarly the imaginary part of the answer to (c) is the answer for (a).

3.18 Determine the solution $y(t)$ of the following differential equation. The input is $f(0) = 0$ and $f(t) = 2$ for $t > 0$.

$$(D + 1)y = f(t)$$

Use this general result to determine $y(0)$, and hence the solution for each of the following cases:

(a) At time $t = 1$, the value of $y(t)$ is $y(1) = 2 + e^{-1}$.
(b) At time $t = 1$, the value of $y'(t)$ is $y'(1) = e^{-1}$.

3.19 The transform of a differential equation is

$$Y(s) = \frac{1}{s(s + 1)^2}$$

(a) Use partial fractions to determine the solution $y(t)$.
(b) Verify the answer to part a by using the initial-value theorem to eliminate the differentiation.

3.20 Use the convolution integral to determine the solution of the differential equation

$$(D + r)y(t) = f(t)$$

where r is a constant.

3.21 Determine the solution of each of the following differential equations in terms of the convolution integral. All the initial conditions are zero:

(a) $(D + 5)y = f(t)$
(b) $(D^2 + 5D)y = f(t)$
(c) $(D^2 + 6D + 25)y = f(t)$

3.22 Equation (3.85) may be written in the form

$$y(t) = \mathcal{L}^{-1}[sF(s)G(s)] + \mathcal{L}^{-1}\frac{I(s)}{L_n(s)}$$

where $G(s) = W(s)/s$ would be the transformed equation for the case in which all the initial conditions are zero and the excitation is a unit step function. Because $\mathcal{L}[f'(t)] = sF(s) - f(0)$, then

$$\mathcal{L}^{-1}[sF(s)] = f'(\lambda) + f(0)u_1(\lambda)$$

Regard $sF(s)$ as one quantity, and then apply the convolution integral to show that the response may be expressed in the form

$$y(t) = \int_0^t [f'(\lambda) + f(0)u_1(\lambda)]g(t - \lambda) \, d\lambda + \mathcal{L}^{-1}\left[\frac{I(s)}{L_n(s)}\right]$$

$$= \int_0^t f'(\lambda)g(t - \lambda) \, d\lambda + f(0)g(t) + \mathcal{L}^{-1}\left[\frac{I(s)}{L_n(s)}\right]$$

This result (Duhamel's equation) makes it possible to determine the response to an arbitrary excitation $f(t)$ when the response to a unit step function $g(t)$ is known. Show that Duhamel's equation may be written in the alternative form

$$y(t) = \int_0^t f(\lambda)g'(t - \lambda) \, d\lambda + f(t)g(0) + \mathcal{L}^{-1}\left[\frac{I(s)}{L_n(s)}\right]$$

3.23 Determine the solution of each of the following integral equations:

(a) $f(t) = 1 + \int_0^t f(\lambda) \sin (t - \lambda) \, d\lambda$

(b) $f(t) + 2f'(t) = \int_0^t f(\lambda) \, d\lambda \qquad f(0) = 1$

REFERENCES

1 Kaplan, W.: "Operational Methods for Linear Systems," Addison-Wesley Publishing Company, Inc., Reading, Mass., 1962.

2 Evans, W. R.: Control System Synthesis by Root-Locus Method, *Trans. AIEE*, vol. 69, pp. 66–69, 1950.

3 Evans W. R.: "Control-systems Dynamics," McGraw-Hill Book Company, New York, 1954.

4 Harris, L. D.: "Introduction to Feedback Systems," John Wiley & Sons, Inc., New York, 1961.

5 Savant, Jr., C. J.: "Control System Design," 2d ed., McGraw-Hill Book Company, New York, 1964.

6 Kuo, B. C.: "Automatic Control Systems," Prentice-Hall, Inc., Englewood Cliffs, N.J., 1962.

chapter 4

Engineering applications

A linear system is one whose operation is described by a linear differential equation. Numerous engineering systems are linear. In addition, the behavior of nonlinear systems may often be approximated by linear differential equations. Thus, linear analysis has very extensive applications in engineering problems. As indicated in the preceding chapters, the Laplace transformation lends itself very well to the solution of such problems. In this chapter, more specific considerations which arise in the application of Laplace transforms to the solution of engineering problems are explained.

4.1 MECHANICAL SYSTEMS

The spring, damper, and mass as shown in Fig. 4.1 are the three basic elements for translational motion. The equation of operation for a spring is

$$f_s = Ky \tag{4.1}$$

where f_s is the force acting on the spring, y is the corresponding deflection (i.e., change in length of the spring), and K is the spring constant.

The force f_d acting on a damper is proportional to the velocity. Thus,

$$f_d = B\dot{y} = BDy \tag{4.2}$$

where B is the coefficient of viscous damping.

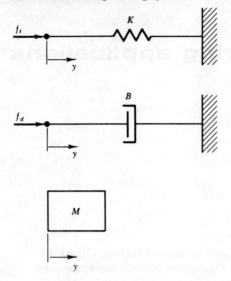

Fig. 4.1. Mechanical components—a spring, damper, and mass.

Because the summation of external forces acting on a mass M is proportional to the acceleration, then

$$\Sigma f_e = M\ddot{y} = MD^2y \tag{4.3}$$

In Fig. 4.2 the mass M represents a large piece of equipment, (e.g., a business machine, industrial equipment, etc.). The ground to which the spring and damper are fixed indicates the support. The force f is the net unbalanced force which occurs during opera-

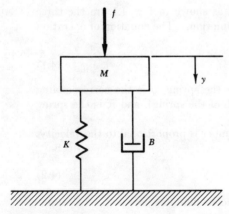

Fig. 4.2. Mass-spring-damper system.

tion of this equipment. If the support is a floor to which the mass is bolted, the force f is transmitted directly to the floor. Thus, sudden shocks and jolts caused by rapid changes in f are transmitted directly to the floor. The purpose of the spring and damper is to smooth out and reduce the shaking, or vibratory, effect of the force f upon the floor.

When the force f is zero, the mass, spring, and damper achieve an equilibrium position in which y is zero. The equation of motion is then obtained by considering all changes which take place from this equilibrium, or reference, operating condition. The change in spring force is Ky and the change in damper force is BDy. Thus, the summation of the change in forces acting on the mass is

$$f - Ky - BDy = MD^2y$$

or

$$f = (MD^2 + BD + K)y \tag{4.4}$$

Because only changes from the reference position are considered, forces which remain constant, such as the gravitational force, are not included. The reason for not including constant forces becomes apparent if Eq. (4.4) is derived by subtracting the force equation at the reference operating condition from the equation for the total forces acting on the system.

The total value of a parameter is obtained by adding the reference value to the change. For example, at the reference operating condition the deflection of the spring is Y_i. Thus, the total spring deflection is the reference value Y_i plus the change y; that is,

$$Y = Y_i + y$$

Similarly, let F_i represent the value of the external force at the reference operating condition. Thus, the total external force is

$$F = F_i + f$$

The equation of operation at the reference operating condition is

$$F_i + Mg = KY_i \tag{4.5}$$

where Mg is the gravitational force acting on the mass. In terms of total values, the equation of operation is

$$F + Mg = (MD^2 + BD + K)Y$$

The substitution of $F = F_i + f$ and $Y = Y_i + y$ gives

$$F_i + f + Mg = (MD^2 + BD + K)(Y_i + y)$$

Because Y_i is a constant, then $MD^2Y_i = BDY_i = 0$. Thus, the preceding equation becomes

$$F_i + f + Mg = KY_i + (MD^2 + BD + K)y \qquad (4.6)$$

Subtracting Eq. (4.5) from Eq. (4.6) verifies Eq. (4.4). Thus, in obtaining the equation of operation for an engineering system, it suffices to consider only the changes with respect to a reference operating condition.

Figure 4.2 shows that the force which is transmitted to the support is that which goes through the spring and damper; hence

$$f_t = (K + BD)y \qquad (4.7)$$

where f_t is the force transmitted to the support. Substituting y from Eq. (4.4) into Eq. (4.7) gives

$$f_t = \frac{(K + BD)f}{MD^2 + BD + K} \qquad (4.8)$$

This is the differential equation which relates the force transmitted to the ground, f_t, to the applied force f. Transforming Eq. (4.8) gives

$$F_t(s) = \frac{(K + Bs)F(s)}{M[s^2 + (B/M)s + (K/M)]} + \frac{I(s)}{M[s^2 + (B/M)s + (K/M)]} \qquad (4.9)$$

where $I(s)$ is the sum of all the initial-condition terms. The basic form of the response is governed by the roots of the characteristic equation

$$s^2 + \frac{B}{M}s + \frac{K}{M} \qquad (4.10)$$

The force f_t which is transmitted to the support is what causes the undesirable shaking, or vibration, of the support. Thus, for any change in the applied, or external, force, it is desired that f_t will attain a constant equilibrium value as soon as possible without excessive overshoots. By referring to Figs. 3.8 and 3.11, it is seen that a damping ratio $\zeta \approx 0.7$ yields this type of response. As shown in Sec. 3.4, the larger ω_n is, the faster the speed of response.

The natural frequency and damping ratio of the characteristic equation for this system are primary factors in governing the response. The equations for ω_n and ζ as given by Eq. (3.52) are

$$\omega_n = \sqrt{\frac{K}{M}}$$

$$\zeta = \frac{B}{2\sqrt{KM}}$$

Ordinarily, the equipment would have a certain mass M. Because $\omega_n = \sqrt{K/M}$, to have a fast speed of response a spring with a large spring constant K should be used. The larger the value of K, the larger B would have to be in order that $\zeta = 0.7$. Realistic values of B and K would be dictated by practical as well as economic considerations. The purpose of this example has been to show that, in the absence of more specific information about the loading, the system parameters are proportioned so that the roots of the characteristic equation are such as to yield good overall response.

4.2 SINUSOIDAL RESPONSE

Engineering equipment is often subjected to a sinusoidal type of loading. For example, suppose that the equipment represented by the mass M in Fig. 4.2 incorporates an electric motor which has an unbalanced mass m. If the center of gravity of this unbalanced mass m is a distance r from the center of rotation, the resulting unbalanced centrifugal force is

$$f_c = mr\omega^2 \tag{4.11}$$

where ω is the speed of rotation. This force may be represented by a rotating vector, as shown in Fig. 4.3. The vertical component is

$$f = mr\omega^2 \sin \omega t \tag{4.12}$$

This vertical component of force tends to set up undesirable vibrations of the floor. Most supports are extremely stiff or rigid with regard to force excitations in the horizontal direction. Consequently, for such engineering applications, it suffices to consider only the vertical component of force transmitted to the support.

The response of a system to a sinusoidally varying excitation is referred to as the *sinusoidal response, frequency response,* or *harmonic response.* Sinusoidal response has extensive applications to engineering situations. It forms the basis for much of the theory of vibration analysis. In the study of automatic control systems, the sinusoidal response provides a very powerful method for determining

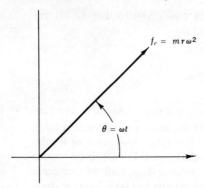

Fig. 4.3. Rotating vector.

the dynamic behavior of the system. Because of the importance of sinusoidal response, a few general equations are now developed.

The general differential equation for a linear system is given by Eq. (3.6). The corresponding transform for this differential equation is given by Eq. (3.7). Inverting Eq. (3.7) to obtain the general response $y(t)$ gives

$$\mathcal{L}^{-1}[Y(s)] = \mathcal{L}^{-1}\left[\frac{L_m(s)}{L_n(s)}F(s)\right] + \mathcal{L}^{-1}\left[\frac{I(s)}{L_n(s)}\right] \tag{4.13}$$

The effect of the initial conditions is investigated by expressing the characteristic function $L_n(s)$ in the factored form,

$$L_n(s) = (s - r_1)(s - r_2) \cdots (s - r_n)$$

where $r_1, r_2, \ldots, r_n$ are the roots of the characteristic equation. Performing a partial fraction expansion on $I(s)/L_n(s)$ gives

$$\frac{I(s)}{L_n(s)} = \frac{I_1}{s - r_1} + \frac{I_2}{s - r_2} + \cdots + \frac{I_n}{s - r_n}$$

where $I_1, I_2, \ldots, I_n$ are the constants which result from the partial fraction expansion. Inverting the preceding expression gives the response due to the initial conditions

$$\mathcal{L}^{-1}\left[\frac{I(s)}{L_n(s)}\right] = I_1 e^{r_1 t} + I_2 e^{r_2 t} + \cdots + I_n e^{r_n t} \tag{4.14}$$

All the roots $r_1, r_2, \ldots, r_n$ of the characteristic equation must have negative real parts (i.e., must lie to the left of the imaginary axis), for otherwise the response would contain an increasing exponential term. An increasing exponential term would cause the response to become infinite regardless of the excitation. Such a system is said to be *unstable*. It should be noted that it makes no difference whether the roots of the characteristic equation are repeated or complex conjugate: the significant thing is that they must all be located to the left of the imaginary axis for the system to be stable. Sinusoidal response is thus limited to stable systems. From Eq. (4.14) it follows that for stable systems the response due to the initial conditions will eventually "die out." In investigating the sinusoidal response of a system, sufficient time is allowed for the effect of the initial conditions to die out.

Because the excitation is a sinusoidal,

$$f(t) = f_0 \sin \omega t \tag{4.15}$$

where f_0 is the amplitude of the driving, or input, sinusoidal. From Euler's equation, it follows that

$$e^{j\omega t} = \cos \omega t + j \sin \omega t \tag{4.16}$$

Thus, the response to $e^{j\omega t}$ yields the response to $\cos \omega t$ and $j \sin \omega t$. The real part of this resulting response function is the response due to $\cos \omega t$, and the imaginary part is the response due to $\sin \omega t$. The transform of $f_0 e^{j\omega t}$ is

$$F(s) = \frac{f_0}{s - j\omega} \tag{4.17}$$

Thus

$$\frac{L_m(s)}{L_n(s)} F(s) = \frac{L_m(s) f_0}{L_n(s)(s - j\omega)} \tag{4.18}$$

Performing a partial fraction expansion gives

$$\frac{L_m(s)}{L_n(s)} F(s) = \frac{K_1}{s - r_1} + \cdots + \frac{K_n}{s - r_n} + \frac{K_{j\omega}}{s - j\omega} \tag{4.19}$$

Inverting the preceding expression yields the corresponding response term in Eq. (4.13); that is,

$$K_1 e^{r_1 t} + K_2 e^{r_2 t} + \cdots + K_n e^{r_n t} + K_{j\omega} e^{j\omega t} \tag{4.20}$$

After sufficient time has elapsed, the only response term that remains in Eq. (4.20) is

$$y(t) = K_{j\omega} e^{j\omega t} \tag{4.21}$$

From Eq. (4.18) the partial-fraction-expansion constant $K_{j\omega}$ is found be

$$K_{j\omega} = \lim_{s \to j\omega} \frac{L_m(s)}{L_n(s)} f_0 = \frac{L_m(j\omega)}{L_n(j\omega)} f_0 \tag{4.22}$$

The term $L_m(j\omega)/L_n(j\omega)$ means that $j\omega$ is substituted for s in the operator $L_m(s)/L_n(s)$. It is equally valid to substitute $j\omega$ for D in the differential operator $L_m(D)/L_n(D)$. The term $L_m(j\omega)/L_n(j\omega)$ is a vector, as shown in Fig. 4.4. Thus, $L_m(j\omega)/L_n(j\omega)$ may be expressed in the form

$$\frac{L_m(j\omega)}{L_n(j\omega)} = \left| \frac{L_m(j\omega)}{L_n(j\omega)} \right| e^{j\phi} \qquad (4.23)$$

where $|L_m(j\omega)/L_n(j\omega)|$ is the length of the vector and ϕ is the angle of the vector. Substituting these results into Eq. (4.21) gives

$$y(t) = f_0 \left| \frac{L_m(j\omega)}{L_n(j\omega)} \right| e^{j(\omega t + \phi)}$$

$$= f_0 \left| \frac{L_m(j\omega)}{L_n(j\omega)} \right| [\cos(\omega t + \phi) + j \sin(\omega t + \phi)] \qquad (4.24)$$

The imaginary part of Eq. (4.24) is the response due to $\sin \omega t$, that is,

$$y(t) = f_0 \left| \frac{L_m(j\omega)}{L_n(j\omega)} \right| \sin(\omega t + \phi)$$

$$= y_0 \sin(\omega t + \phi) \qquad (4.25)$$

where $y_0 = f_0 |L_m(j\omega)/L_n(j\omega)|$ is the amplitude of the resultant sine wave. The ratio of the amplitude of the response y_0 to that of the input f_0 is called the *amplitude ratio*.

$$\frac{y_0}{f_0} = \left| \frac{L_m(j\omega)}{L_n(j\omega)} \right| \qquad (4.26)$$

The phase angle between the response $y_0 \sin(\omega t + \phi)$ and the input $f_0 \sin \omega t$ is ϕ.

$$\phi = \measuredangle \frac{L_m(j\omega)}{L_n(j\omega)} \qquad (4.27)$$

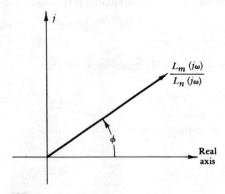

Fig. 4.4. Vector representation of $L_m(j\omega)/L_n(j\omega)$.

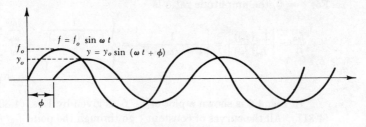

Fig. 4.5. Sinusoidal input f and sinusoidal output y.

Thus, sinusoidal response is determined completely by the vector $L_m(j\omega)/L_n(j\omega)$. The length of this vector is the amplitude ratio, and the angle of this vector is the phase shift. In Fig. 4.5 is shown a sinusoidal input $f_0 \sin \omega t$. After the initial transients have died out, the response $y(t)$ becomes sinusoidal as shown. The resultant sinusoidal is completely determined by the amplitude ratio and the phase shift.

Vibration analysis

For the system shown in Fig. 4.2, the differential equation for the transmitted force is given by Eq. (4.8). For the case in which $f = f_0 \sin \omega t$, then

$$f_t = \frac{(K/M) + (B/M)D}{D^2 + (B/M)D + (K/M)} f_0 \sin \omega t$$

$$= \frac{\omega_n{}^2 + 2\zeta\omega_n D}{D^2 + 2\zeta\omega_n D + \omega_n{}^2} f_0 \sin \omega t \quad (4.28)$$

where $\omega_n{}^2 = K/M$ and $2\zeta\omega_n = B/M$. After sufficient time has elapsed for the initial transients to die out, the response becomes $f_t = f_{t0} \sin (\omega t + \phi)$. The substitution of $j\omega$ for D in the operator gives

$$\frac{L_m(j\omega)}{L_n(j\omega)} = \frac{\omega_n{}^2 + j2\zeta\omega_n\omega}{\omega_n{}^2 - \omega^2 + j2\zeta\omega_n\omega} = \frac{1 + j2\zeta(\omega/\omega_n)}{1 - (\omega/\omega_n)^2 + j2\zeta(\omega/\omega_n)} \quad (4.29)$$

The ratio of the amplitude of the force transmitted to the support f_{t0} to the exciting force is

$$\frac{f_{t0}}{f_0} = \left| \frac{L_m(j\omega)}{L_n(j\omega)} \right| = \sqrt{\frac{1 + (2\zeta\omega/\omega_n)^2}{[1 - (\omega/\omega_n)^2]^2 + (2\zeta\omega/\omega_n)^2}} \qquad \zeta \neq 0 \quad (4.30)$$

123

For $\zeta = 0$, the amplitude ratio is

$$\frac{f_{t0}}{f_0} = \left| \frac{L_m(j\omega)}{L_n(j\omega)} \right| = \left| \frac{1}{1 - (\omega/\omega_n{}^2)} \right| = \begin{cases} \dfrac{1}{1 - (\omega/\omega_n)^2} & \dfrac{\omega}{\omega_n} \leq 1 \\[2mm] \dfrac{1}{(\omega/\omega_n)^2 - 1} & \dfrac{\omega}{\omega_n} \geq 1 \end{cases}$$

$$(4.31)$$

In Fig. 4.6 is shown a plot of f_{t0}/f_0 as given by Eqs. (4.30) and (4.31). All the curves of constant ζ go through the point $f_{t0}/f_0 = 1$ when $\omega/\omega_n = \sqrt{2}$. To have the amplitude of the transmitted force less than that of the exciting force (that is, $f_{t0}/f_0 < 1$), it is necessary that $\omega/\omega_n > \sqrt{2}$. For the region $\omega/\omega_n > \sqrt{2}$, the smaller the value of ζ the smaller the ratio f_{t0}/f_0. A certain amount of damping such as $\zeta = 0.15$ is desirable so that transients will die out in a reasonable length of time.

Figure 4.6 shows that the transmissibility decreases as the ratio ω/ω_n increases. For most applications the unbalanced forces are caused by a motor which rotates at constant ω. Thus, to increase ω/ω_n, it is necessary to decrease $\omega_n = \sqrt{K/M}$. Because M is a constant, the spring rate K must be decreased as much as possible. In practice a ratio ω/ω_n of about 4 yields a substantial reduction in the transmitted force which is attainable with a spring of reasonable proportions.

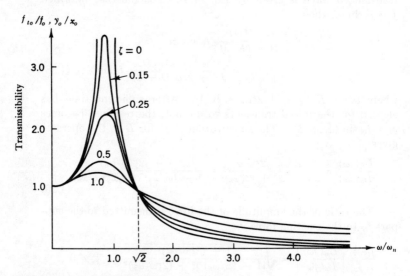

Fig. 4.6. Plot of Eqs. (4.30) and (431).

The phase angle ϕ can be determined from Eq. (4.29), but for vibration analysis it is of little value. This is not the case with feedback control systems, in which ϕ is a very significant factor.

Illustrative example The machine represented by the mass M in Fig. 4.2 has a weight of 386 lb. This machine incorporates a motor which rotates at 1,800 rpm. Determine the spring rate K and coefficient of viscous damping B to yield a damping ratio $\zeta = 0.15$ and a ratio $\omega/\omega_n = 4$. What is the transmissibility ratio for the resulting system?

Solution The natural frequency ω_n is

$$\omega_n = \frac{\omega}{4} = \frac{(1,800)(2\pi)}{(60)(4)} = 15\pi \text{ radians/sec}$$

Because $\omega_n = \sqrt{K/M}$ and $M = W/g = 386/(32.2)(12) = 1$, then the spring constant K is

$$K = M\omega_n{}^2 = (1)(15\pi)^2 = 2,220 \text{ lb/in.}$$

Because $B/M = 2\zeta\omega_n$, the coefficient of viscous damping is

$$B = 2\zeta\omega_n M = (0.3)(15\pi) = 14.15 \text{ lb-sec/in.}$$

The transmissibility ratio may be read directly from Fig. 4.6, or for greater accuracy it may be calculated from Eq. (4.30). For $\zeta = 0.15$ and $\omega/\omega_n = 4$, the transmissibility ratio is 0.105.

For some design applications, it may be more important to minimize the amplitude of motion of the mass in Fig. 4.2 rather than the force transmitted to the support. Such a situation occurs when the machine contains delicate electronic equipment that can be damaged by excessive oscillations. The differential equation relating the motion y and applied force f is given by Eq. (4.4). For the case in which $f = f_0 \sin \omega t$, then

$$y = \frac{f_0 \sin \omega t}{M[D^2 + (B/M)D + K/M]} = \frac{f_0 \sin \omega t}{M(D^2 + 2\zeta\omega_n D + \omega_n{}^2)}$$

The sinusoidal response has the form $y = y_0 \sin(\omega t + \phi)$. Substituting $j\omega$ for D in the differential operator and noting that $\omega_n{}^2 = K/M$ or $1/M = \omega_n{}^2/K$ gives

$$\frac{L_m(j\omega)}{L_n(j\omega)} = \frac{\omega_n{}^2/K}{\omega_n{}^2 - \omega^2 + j2\zeta\omega_n\omega} = \frac{1/K}{1 - (\omega/\omega_n)^2 + j2\zeta(\omega/\omega_n)}$$

$$(4.32)$$

125

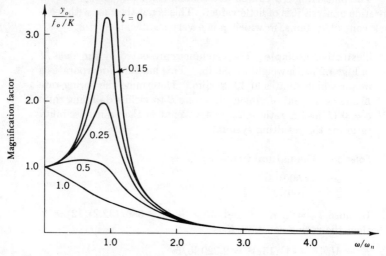

Fig. 4.7. Plot of Eqs. (4.33) and (4.34).

The magnitude of Eq. (4.32) is the ratio of the amplitude of the response y_0 to that of the input f_0. Multiplication of this amplitude ratio y_0/f_0 by the spring constant K yields the nondimensional form

$$\frac{y_0}{f_0/K} = K \left| \frac{L_m(j\omega)}{L_n(j\omega)} \right| = \frac{1}{\sqrt{[1 - (\omega/\omega_n)^2]^2 + (2\zeta\omega/\omega_n)^2}} \qquad \zeta \neq 0$$
$$(4.33)$$

For $\zeta = 0$, then

$$\frac{y_0}{f_0/K} = K \left| \frac{L_m(j\omega)}{L_n(j\omega)} \right| = \left| \frac{1}{1 - (\omega/\omega_n)^2} \right|$$
$$= \begin{cases} \dfrac{1}{1 - (\omega/\omega_n)^2} & \dfrac{\omega}{\omega_n} \leq 1 \\[2mm] \dfrac{1}{(\omega/\omega_n)^2 - 1} & \dfrac{\omega}{\omega_n} \geq 1 \end{cases} \qquad (4.34)$$

The term f_0/K is the deflection that would result if f_0 were a static force. A plot of the nondimensional ratio $y_0/(f_0/K)$ is shown in Fig. 4.7. The requirements for minimizing the amplitude of the motion y_0 are similar to those for minimizing the transmitted force. That is, the ratio ω/ω_n should be about 4 or greater. Figure 4.7 shows that, the larger the damping ratio ζ, the smaller the amplitude y_0. A larger damping ratio also serves the purpose of damping out extraneous disturbances faster. Thus, when motion is the main consideration, a larger damping ratio such as $\zeta = 0.5$ is desirable.

The selection of damping ratio is therefore seen to depend upon whether the motion or the transmitted force is to be minimized.

Illustrative example Determine the amplitude of motion y_0 of the preceding example for the case in which the amplitude of the disturbing force is $f_0 = mr\omega^2 = 222$ lb.

Solution The static deflection due to this force is $f_0/K = 0.1$ in. For $\omega/\omega_n = 4$ and $\zeta = 0.15$, then from Fig. 4.7 or Eq. (4.33) it follows that $y_0/(f_0/K) = 0.079$. Hence the amplitude of motion is $y_0 = 0.0079$.

A different class of vibration problem occurs when the supporting structure is the source of the excitation. A schematic representation of this situation is shown in Fig. 4.8. This situation occurs when it is desired to isolate the motion of delicate electronic equipment in an airplane from the small vibratory motion of the frame. In this case, the position x represents the vibratory motion of the support (i.e., frame) and the position y represents the motion of the delicate equipment, which it is desired to minimize. Another application is that in which it is desired to isolate the motion of the body of a car from irregularities and bumps in the road. In this latter case, the position x represents the vertical motion of the wheels, and y is the corresponding motion of the body of the car.

Summing the forces acting on the mass in Fig. 4.8 gives

$$K(x - y) + BD(x - y) = MD^2y$$

Thus, the position y is

$$y = \frac{(K + BD)x}{MD^2 + BD + K} = \frac{[K/M + (B/M)D]x}{D^2 + (B/M)D + K/M} \tag{4.35}$$

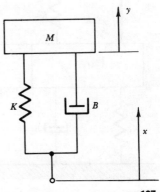

Fig. 4.8. Mass-spring-damper system excited by the support.

127

For a sinusoidal excitation ($x = x_0 \sin \omega t$), the response becomes $y = y_0 \sin (\omega t + \phi)$. The operator for Eq. (4.35) is identical to that for Eq. (4.28). Consequently, the amplitude ratio y_0/x_0 is the same as that for force transmissibility as given by Eqs. (4.30) and (4.31). The plot of Fig. 4.6 is therefore applicable for motion transmissibility y_0/x_0 as well as force transmissibility f_{t0}/f_0.

4.3 SIMULTANEOUS DIFFERENTIAL EQUATIONS

The number of degrees of freedom of a system is equal to the number of coordinates required to specify the position of each element in the system. Because two coordinates y_1 and y_2 are required to locate the position of each element in Fig. 4.9, this system has two degrees of freedom. The equation of motion is obtained by writing the force equation at each coordinate. The summation of forces acting at y_1 is equal to the mass M_1 times the corresponding acceleration D^2y_1, that is,

$$f + K_2(y_2 - y_1) - (K_1 + B_1D)y_1 = M_1D^2y_1$$

Similarly, the force equation at the y_2 coordinate is

$$K_2(y_1 - y_2) = M_2D^2y_2$$

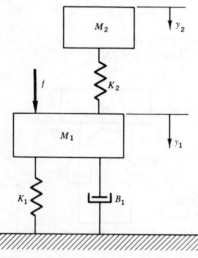

Fig. 4.9. A two-degree-of-freedom system.

The preceding force equations may be written in the form

$$(M_1 D^2 + B_1 D + K_1 + K_2)y_1 - K_2 y_2 = f \tag{4.36}$$
$$-K_2 y_1 + (M_2 D^2 + K_2)y_2 = 0 \tag{4.37}$$

The motion of a two-degree-of-freedom system is governed by two simultaneous differential equations. For n degrees of freedom, there are n simultaneous differential equations.

Using determinants to solve Eqs. (4.36) and (4.37) for y_1 and y_2 gives

$$
y_1 = \frac{\begin{vmatrix} f & -K_2 \\ 0 & M_2 D^2 + K_2 \end{vmatrix}}{\begin{vmatrix} M_1 D^2 + B_1 D + K_1 + K_2 & -K_2 \\ -K_2 & M_2 D^2 + K_2 \end{vmatrix}}
$$

$$
= \frac{(M_2 D^2 + K_2)f}{(M_1 D^2 + B_1 D + K_1 + K_2)(M_2 D^2 + K_2) - K_2{}^2}
$$

$$
= \frac{L_{m_1}(D)f}{L_n(D)} \tag{4.38}
$$

$$
y_2 = \frac{\begin{vmatrix} M_1 D^2 + B_1 D + K_1 + K_2 & f \\ -K_2 & 0 \end{vmatrix}}{\begin{vmatrix} M_1 D^2 + B_1 D + K_1 + K_2 & -K_2 \\ -K_2 & M_2 D^2 + K_2 \end{vmatrix}}
$$

$$
= \frac{L_{m_2}(D)f}{L_n(D)} \tag{4.39}
$$

where

$$
\begin{aligned}
L_{m_1}(D) &= M_2 D^2 + K_2 \\
L_{m_2}(D) &= K_2 \\
L_n(D) &= (M_1 D^2 + B_1 D + K_1 + K_2)(M_2 D^2 + K_2) - K_2{}^2
\end{aligned} \tag{4.40}
$$

Equations (4.38) and (4.39) show that the motions y_1 and y_2 are governed by ordinary differential equations. The characteristic equation $L_n(D)$ is the same for y_1 and y_2. Although the complete motion for a system of n degrees of freedom is specified by n simultaneous differential equations, the motion of each variable may be expressed as an ordinary differential equation. Furthermore, each differential equation for each of the n variables has the same characteristic equation $L_n(D)$. The characteristic equation $L_n(D)$ is the determinant of the coefficients of the simultaneous differential equations.

The transforms $Y_1(s)$ and $Y_2(s)$ may be obtained by transforming Eqs. (4.38) and (4.39). However, considerable computational effort is ordinarily saved by first transforming the original set of simultaneous differential equations and then using determinants to solve for the desired transformed variable. Thus, first transforming Eqs. (4.36) and (4.37) gives

$$(M_1 s^2 + B_1 s + K_1 + K_2)Y_1(s) - K_2 Y_2(s) = F(s) + I_1(s)$$
$$\text{(4.41)}$$

$$-K_2 Y_1(s) + (M_2 s^2 + K_2)Y_2(s) = I_2(s) \qquad \text{(4.42)}$$

where $I_1(s) = M_1[sy_1(0) + y_1'(0)] + B_1 y_1(0)$ is the sum of the initial-condition terms that result from transforming Eq. (4.36) and where $I_2(s) = M_2[sy_2(0) + y_2'(0)]$ is the sum of the initial-condition terms that result from transforming Eq. (4.37).

Because $L_n(s)$ is of the fourth order, these are fourth-order differential equations. Thus, four initial conditions $y_1(0)$, $y_1'(0)$, $y_2(0)$, and $y_2'(0)$ are required. If Eq. (4.38) had been transformed, the four initial conditions that arise would be $y_1(0)$, $y_1'(0)$, $y_1''(0)$, and $y_1'''(0)$. Ordinarily one knows the lower-order initial conditions $y_1(0)$, $y_1'(0)$, $y_2(0)$, and $y_2'(0)$. The value of $y_1''(0)$ may be determined by solving Eq. (4.36) for $y_1''(t)$ and then evaluating at $t = 0$. To determine $y_1'''(0)$, it is necessary to differentiate Eq. (4.36) with respect to time, then solve for $y'''(t)$, and then evaluate at $t = 0$. It should now be apparent that, by first transforming the set of simultaneous differential equations as was done in Eqs. (4.41) and (4.42), the resulting transformed expressions will involve the lowest-order initial conditions, which are ordinarily the initial conditions which are known.

Using determinants to solve Eqs. (4.41) and (4.42) for $Y_1(s)$ and $Y_2(s)$ gives

$$Y_1(s) = \frac{\begin{vmatrix} F(s) + I_1(s) & -K_2 \\ I_2(s) & M_2 s^2 + K_2 \end{vmatrix}}{\begin{vmatrix} M_1 s^2 + B_1 s + K_1 + K_2 & -K_2 \\ -K_2 & M_2 s^2 + K_2 \end{vmatrix}}$$

$$= \frac{L_{m_1}(s)F(s) + I_{y_1}(s)}{L_n(s)} \qquad \text{(4.43)}$$

$$Y_2(s) = \frac{\begin{vmatrix} M_1 s^2 + B_1 s + K_1 + K_2 & F(s) + I_1(s) \\ -K_2 & I_2(s) \end{vmatrix}}{\begin{vmatrix} M_1 s^2 + B_1 s + K_1 + K_2 & -K_2 \\ -K_2 & M_2 s^2 + K_2 \end{vmatrix}}$$

$$= \frac{L_{m_2}F(s) + I_{y_2}(s)}{L_n(s)} \qquad \text{(4.44)}$$

where

$$\begin{aligned}
L_{m_1}(s) &= M_2 s^2 + K_2 \\
I_{y_1}(s) &= (M_2 s^2 + K_2)I_1(s) + K_2 I_2(s) \\
L_{m_2}(s) &= K_2 \\
I_{y_2}(s) &= K_2 I_1(s) + (M_1 s^2 + B_1 s + K_1 + K_2)I_2(s) \\
L_n(s) &= (M_1 s^2 + B_1 s + K_1 + K_2)(M_2 s^2 + K_2) - K_2^2
\end{aligned} \qquad \text{(4.45)}$$

Equations (4.43) and (4.44) are of the same form as Eq. (3.7), which is the transform of an ordinary linear differential equation.

The basic form of the response of Eqs. (4.43) and (4.44) is governed by the location of the roots of the characteristic equation, $L_n(s)$. For more complex and for higher-order systems, it becomes an increasingly more difficult task to determine the roots of the characteristic equation. An important aspect of the theory of vibrations is concerned with this task. Well-known techniques for determining roots include the root-locus method, matrix methods, and numerical methods which lend themselves to the use of digital computers.

The analog computer is very useful for investigating the behavior of engineering systems. The equations which describe the operation of the analog computer are analogous to the equations which describe the operation of the actual system. The variables y_1, Dy_1, D^2y_1, y_2, Dy_2, and D^2y_2 in Eqs. (4.36) and (4.37) are each represented by voltages in the analog computer. The effect of changing a parameter such as the spring constant K_1 may be ascertained by changing the value of the resistor in the analog computer which corresponds to the spring K_1. Thus, the electrical quantities in the analog computer are adjusted by "turning knobs" until satisfactory transient behavior is achieved.

Illustrative example   Let it be desired to determine the response y_1 of the system of Fig. 4.9 for the case in which $f(t)$ is a unit impulse and all the initial conditions are zero. The basic system constants are $K_1 = 100$ lb/in., $B_1 = 5.6$ lb-sec/(in., $W_1 = 77.2$ lb, $M_1 = W_1/g = 77.2/386 = 0.2$ lb-sec^2/in., $K_2 = 32$ lb/in., $W_2 = 0.32W_1$, and thus $M_2 = (0.32)(0.2) = \frac{8}{125}$ lb-sec^2/in.

Solution   From Eqs. (4.43) and (4.44), it follows that the transformed equation for $Y_1(s)$ is

$$Y_1(s) = \frac{(M_2s^2 + K_2)F(s)}{(M_1s^2 + B_1s + K_1 + K_2)(M_2s^2 + K_2) - K_2^2}$$

For a unit impulse $F(s) = 1$. Substitution of numerical values into the preceding expression gives

$$Y_1(s) = \frac{5(s^2 + 500)}{s^4 + 28s^3 + 1,160s^2 + 14,000s + 250,000}$$

Writing the characteristic equation in factored form gives

$$Y_1(s) = \frac{5(s^2 + 500)}{(s^2 + 20s + 500)(s^2 + 8s + 500)}$$

The characteristic equation is thus seen to have two sets of complex conjugate roots, which are $a_1 \pm jb_1 = -10 \pm j20$ and $a_2 \pm jb_2 = -4 \pm j22$. The corresponding response has the form

$$y_1(t) = \frac{1}{b_1} \left| \frac{A(a_1 + jb_1)}{B(a_1 + jb_1)} \right| e^{a_1 t} \sin(b_1 t + \alpha_1)$$

$$+ \frac{1}{b_2} \left| \frac{A(a_2 + jb_2)}{B(a_2 + jb_2)} \right| e^{a_2 t} \sin(b_2 t + \alpha_2)$$

Evaluation gives

$$\frac{A(a_1 + jb_1)}{B(a_1 + jb_1)} = \lim_{s \to -10 + j20} \frac{5(s^2 + 500)}{s^2 + 8s + 500} = \frac{5(200 - j400)}{120 - j240}$$

$$= \frac{(5)(200)}{120} \frac{1 - j2}{1 - j2} = \frac{25}{3} \underline{/0°}$$

and

$$\frac{A(a_2 + jb_2)}{B(a_2 + jb_2)} = \lim_{s \to -4 + j22} \frac{5(s^2 + 500)}{s^2 + 20s + 500} = \frac{5(32 - j176)}{-48 + j264}$$

$$= \frac{(5)(32)(1 - j5.5)}{(48)(-1 + j5.5)} = \frac{10}{3} \underline{/-180°}$$

Thus, the desired result for $y_1(t)$ is

$$y_1(t) = \frac{25}{(3)(20)} e^{-10t} \sin 20t + \frac{10}{(3)(22)} e^{-4t} \sin(22t - 180°)$$

For vibration analysis the frequency response is ordinarily of more interest than the transient behavior. The system of Fig. 4.9 is a vibration absorber in which it is desired to minimize the motion of y_1. For a sinusoidally varying load $f = f_0 \sin \omega t$, the response y_1 will also be sinusoidal. The substitution of $j\omega$ for D in the differential operator of Eq. (4.38) gives

$$\frac{L_{m_1}(j\omega)}{L_n(j\omega)} = \frac{K_2 - M_2\omega^2}{[(K_1 + K_2 - M_1\omega^2) + jB_1\omega](K_2 - M_2\omega^2) - K_2^2}$$

$$(4.46)$$

The magnitude of this expression is the ratio of the amplitude of the response to that of the exciting force. When the mass M_2 and spring K_2 are chosen such that

$$K_2 = M_2\omega^2 \quad \text{or} \quad \omega = \sqrt{\frac{K_2}{M_2}} \tag{4.47}$$

the numerator of Eq. (4.46) vanishes. For this case, the amplitude ratio is zero, and thus the vibration absorber functions perfectly. This absorber is especially suited for synchronous machines in which the angular velocity ω of the exciting force is always constant.

The preceding analysis shows that, although the complete motion of an n-degree-of-freedom system is governed by n differential equations, this set of n equations may be solved simultaneously to obtain a single differential equation for each variable. The investigation of the transient or frequency response of each particular variable is then carried out in the same manner as for a single-degree-of-freedom system.

Matrix methods, numerical methods, and computer techniques for solving simultaneous equations are presented in Chap. 12. Because Chap. 12 is independent of the intervening chapters, it may be taken up after this chapter if so desired.

Series and parallel laws

Considerable time and effort in obtaining the differential equation relating the motion of a coordinate to the applied force is saved by application of series and parallel laws. In Fig. 4.10 it is to be noted that the spring K_1 and the damper B_1 have the same coordinates;

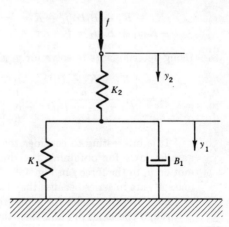

Fig. 4.10. A mechanical network.

i.e., the top is y_1, and the bottom is ground. Mechanical elements with the same coordinates add by the series law of addition; i.e.,

$$Z_1 = K_1 + B_1D \tag{4.48}$$

Elements add by the reciprocal, or parallel, law when the same force is transmitted through each element. The spring K_2 is in parallel with Z_1, and hence the total impedance from ground to y_2 is

$$Z = \frac{1}{1/Z_1 + 1/Z_2} = \frac{1}{1/(K_1 + B_1D) + 1/K_2}$$
$$= \frac{K_2(K_1 + B_1D)}{K_1 + K_2 + B_1D}$$

Because $f = Zy_2$, the differential equation relating the motion y_2 to f is

$$f = \frac{K_2(K_1 + B_1D)y_2}{K_1 + K_2 + B_1D} \tag{4.49}$$

or

$$y_2 = \frac{K_1 + K_2 + B_1D}{K_2(K_1 + B_1D)} f \tag{4.50}$$

This result is verified by writing the force equations at the y_1 and y_2 coordinates of Fig. 4.10 and then solving these equations simultaneously to obtain the differential equation for y_2. That is,

$$(K_1 + B_1D)y_1 = K_2(y_2 - y_1)$$
$$K_2(y_2 - y_1) = f$$

or

$$(K_1 + K_2 + B_1D)y_1 - K_2y_2 = 0$$
$$-K_2y_1 + K_2y_2 = f$$

Using determinants to solve for y_2 gives

$$y_2 = \frac{\begin{vmatrix} K_1 + K_2 + B_1D & 0 \\ -K_2 & f \end{vmatrix}}{\begin{vmatrix} K_1 + K_2 + B_1D & -K_2 \\ -K_2 & K_2 \end{vmatrix}} = \frac{(K_1 + K_2 + B_1D)f}{K_2(K_1 + B_1D)} \tag{4.51}$$

It is interesting to consider the application of these series and parallel laws for obtaining the differential equation relating the motion y_1 to the force f in Fig. 4.9. The spring K_1, damper B_1, and mass M_1 are in series because they have the same coordinates y_1 and ground. Another route for going from ground to the y_1 coordinate is through the mass M_2 and then through the spring K_2. That is, we go from ground to y_2 through the mass M_2 and from y_2 to y_1 through the spring K_2. By letting Z_2 represent the impedance of this last route from ground to y_1, then

$$f = [(M_1D^2 + B_1D + K_1) + Z_2]y_1 \tag{4.52}$$

The mass M_2 and spring K_2 add by the parallel law because all the force in the spring K_2 is transmitted to the mass M_2. Thus

$$Z_2 = \frac{1}{1/M_2D^2 + 1/K_2} = \frac{K_2M_2D^2}{M_2D^2 + K_2}$$

Substituting Z_2 into Eq. (4.52) and solving for y_1 gives

$$y_1 = \frac{(M_2D^2 + K_2)f}{(M_1D^2 + B_1D + K_1)(M_2D^2 + K_2) + K_2M_2D^2} \tag{4.53}$$

Equation (4.53) is the same result as that obtained in Eq. (4.38) [note that $K_2{}^2$ may be canceled out of $L_n(D)$ in Eq. (4.38)]. In applying series and parallel laws, it is necessary that the assumed positive sense of each coordinate be the same as that of the applied force f acting on the system. Note that, in Figs. 4.9 and 4.10, the positive sense of y_1 and y_2 is the same as that for f.

Any linear mechanical system may be represented as a circuit composed of springs, dampers, and masses. For example, by obtaining the linearized equations of operation for hydraulic or pneumatic devices such as motors and valves, it is possible to investigate their behavior by studying the equivalent spring-mass-damper circuit.

In a similar manner, any linear electrical system may be represented by a circuit composed of resistors, inductors, and capacitors. This topic is considered in the next section.

4.4 ELECTRICAL SYSTEMS

The resistor, inductor, and capacitor are the three basic elements of electrical systems. The representation for each of these elements is shown in Fig. 4.11. By designating e_R as the voltage drop across a resistor, e_L the voltage drop across an inductor, and e_C the voltage drop across a capacitor, then the current-voltage relationships for a resistor, inductor, and a capacitor are

$$e_R = Ri \tag{4.54}$$

$$e_L = L\frac{di}{dt} = LDi \tag{4.55}$$

$$e_C = \int \frac{i}{C}\, dt \tag{4.56}$$

135

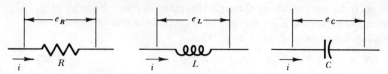

Fig. 4.11. Electrical components—resistor, inductor, and capacitor.

Differentiation of both sides of Eq. (4.56) with respect to time gives

$$De_C = \frac{1}{C}i$$

or

$$e_C = \frac{1}{D}\frac{i}{C} \tag{4.57}$$

Comparison of Eqs. (4.56) and (4.57) shows that $1/D$ (the reciprocal of the differential operator) means integration with respect to time.

The transform for the voltage drop $e_C = \int(i/C)\,dt$ across a capacitor is obtained by application of Eq. (2.23), which is the equation for obtaining the transform of an indefinite integral. The function to be integrated, i/C, corresponds to $f(t)$, so that $F(s) = I(s)/C$. Thus, application of Eq. (2.23) gives

$$\mathcal{L}(e_C) = \mathcal{L}\left(\int \frac{i}{C}\,dt\right) = \frac{1}{s}\left[\frac{I(s)}{C} + e_C(0)\right] \tag{4.58}$$

where $e_C(0)$ is the initial value of the voltage across the capacitor.

An alternative technique for obtaining the transform of an indefinite integral is first to differentiate to eliminate the integration and then obtain the transform of the resulting function. Thus, differentiating Eq. (4.57) gives

$$De_C = \frac{i}{C}$$

The transform of this function is

$$sE_C(s) - e_C(0) = \frac{I(s)}{C}$$

Solving the preceding expression for $E_C(s)$ verifies the result of Eq. (4.58).

Loop analysis

The basis for a loop analysis is Kirchhoff's first law, which states that, for any loop, the summation of the voltage drops across each

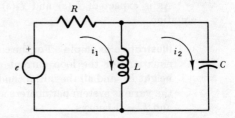

Fig. 4.12. An electrical network—loop analysis.

element is equal to the summation of the applied voltages. In Fig. 4.12 is shown an electrical circuit which has two loops. Application of Kirchhoff's law to loop 1 gives

$$e = Ri_1 + LD(i_1 - i_2) \tag{4.59}$$

Similarly, for loop 2

$$0 = LD(i_2 - i_1) + \frac{1}{D}\frac{i_2}{C} \tag{4.60}$$

The operation of this circuit is described by two simultaneous differential equations. The number of equations is equal to the number of loops. Transforming Eq. (4.59) gives

$$E(s) = RI_1(s) + Ls[I_1(s) - I_2(s)] - L[i_1(0) - i_2(0)]$$

or

$$(R + Ls)I_1(s) - LsI_2(s) = E(s) + L[i_1(0) - i_2(0)] \tag{4.61}$$

where $i_1(0) - i_2(0)$ is the initial value of the current flowing through the inductor.

Similarly, transforming Eq. (4.60) yields

$$0 = sL[I_2(s) - I_1(s)] - L[i_2(0) - i_1(0)] + \frac{1}{s}\left[\frac{I_2(s)}{C} + e_C(0)\right]$$

or

$$sLI_1(s) - \left(\frac{1}{Cs} + sL\right)I_2(s) = L[i_1(0) - i_2(0)] + \frac{e_C(0)}{s} \tag{4.62}$$

where $e_C(0)$ is the initial voltage across the capacitor. Using determinants to solve Eqs. (4.61) and (4.62) for $I_1(s)$ and $I_2(s)$ gives

$$I_1(s) = \frac{(s^2 + 1/Lc)E(s) + (1/C)[i_1(0) - i_2(0)] - se_C(0)}{R[s^2 + (1/RC)s + 1/LC]}$$

$$\tag{4.63}$$

$$I_2(s) = \frac{s^2E(s) - sR[i_1(0) - i_2(0)] - (s + R/L)e_C(0)}{R[s^2 + (1/RC)s + 1/LC]} \tag{4.64}$$

As is expected, $I_1(s)$ and $I_2(s)$ have the same characteristic equation.

Illustrative example   For the circuit of Fig. 4.12, determine the equation for the loop current i_1. The source e is a step of height E_0, and all the initial conditions are zero.   The values of the various system parameters are $R = 5$ ohms, $C = 0.1$ farad, and $L = 2$ henrys.

Solution   Substitution of these values into Eq. (4.63) gives

$$I_1(s) = \frac{(s^2 + 5)E_0}{5s(s^2 + 2s + 5)}$$

The response has the form

$$i_1(t) = K_1 + \frac{1}{b} \left| \frac{A(a + jb)}{B(a + jb)} \right| e^{at} \sin(bt + \alpha)$$

The value of K_1 is

$$K_1 = \lim_{s \to 0} \frac{(s^2 + 5)E_0}{5(s^2 + 2s + 5)} = \frac{E_0}{5}$$

The roots of the quadratic are $a \pm jb = -1 \pm j2$.   Thus,

$$\frac{A(a + jb)}{B(a + jb)} = \lim_{s \to -1+j2} \frac{(s^2 + 5)E_0}{5s} = \frac{(-3 - j4 + 5)E_0}{5(-1 + j2)}$$

$$= \frac{(2 - j4)E_0}{5(-1 + j2)} = \frac{2E_0/180°}{5}$$

Substitution of these results into the response expression gives

$$i_1(t) = \frac{E_0}{5}[1 + e^{-t} \sin(2t + 180°)]$$

Node analysis

A node analysis is an application of Kirchhoff's second law that the summation of the currents entering and leaving a node (i.e., junction) is zero.   Figure 4.13 is the same circuit as Fig. 4.12.   The voltage at the nodal point of Fig. 4.13 is designated as v.   Application of Kirchhoff's second law to this nodal point gives

$$i_a = i_b + i_c \tag{4.65}$$

Each branch current in Eq. (4.65) may be expressed in terms of voltages in the circuit ($e - v = Ri_a$, $v = LDi_b$, and $v = i_c/DC$).

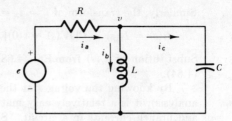

Fig. 4.13. An electrical network—nodal analysis.

Thus, Eq. (4.65) becomes

$$\frac{e - v}{R} = \frac{v}{DL} + CDv \tag{4.66}$$

The transform of the term $i_b = v/DL$ is obtained by first differentiating to eliminate the integration; thus

$$Di_b = \frac{v}{L}$$

The transform is

$$sI_b(s) - i_b(0) = \frac{V(s)}{L}$$

Solving for $I_b(s) = \mathcal{L}(i_b) = \mathcal{L}(v/DL)$ gives

$$\mathcal{L}\left(\frac{v}{DL}\right) = \frac{1}{s}\left[\frac{V(s)}{L} + i_b(0)\right] \tag{4.67}$$

The transform of Eq. (4.66) is

$$\frac{E(s)}{R} - \frac{V(s)}{R} = \frac{1}{s}\left[\frac{V(s)}{L} + i_b(0)\right] + sCV(s) - Cv(0)$$

Solving the preceding expression for $V(s)$ gives

$$V(s) = \frac{sE(s)/RC + sv(0) - i_b(0)/C}{s^2 + (1/RC)s + 1/LC} \tag{4.68}$$

By comparison of Figs. 4.12 and 4.13, it is seen that $v(0) = e_c(0)$, $i_a = i_1$, $i_b = i_1 - i_2$, and $i_c = i_2$. The transformed expression for the current $i_a = i_1 = (e - v)/R$ is obtained as follows:

$$I_a(s) = I_1(s) = \frac{E(s) - V(s)}{R} \tag{4.69}$$

Substituting $V(s)$ from Eq. (4.68) into Eq. (4.69) yields the same result for $I_1(s)$ as that obtained by the loop analysis, i.e., Eq. (4.63).

Similarly, the transform of $i_c = i_2 = CDv$ is

$$I_c(s) = I_2(s) = C[sV(s) - v(0)] \tag{4.70}$$

Substitution of $V(s)$ from Eq. (4.68) into Eq. (4.70) verifies Eq. (4.64).

By knowing the voltages at the nodes (as obtained in a node analysis), it is a relatively easy matter to obtain the various loop and branch currents in a circuit. Similarly, by knowing the loop currents (as obtained in a loop analysis), the various voltages in a circuit may be determined. A given network may be analyzed by either the loop method or the nodal method. The choice depends upon whether there are fewer loop equations or fewer nodal equations and also upon the particular variables to be evaluated.

The series RLC circuit shown in Fig. 4.14 has but one loop equation and two nodal equations. The loop equation is

$$e = \left(LD + R + \frac{1}{DC}\right)i \tag{4.71}$$

The voltage across the capacitor is $v_2 = i/DC$. Thus, the transform of this term is $[I(s)/C + v_2(0)]/s$. Proceeding to transform Eq. (4.71) and solving for $I(s)$ gives

$$I(s) = \frac{sE(s) + sLi(0) - v_2(0)}{Ls^2 + Rs + 1/C} \tag{4.72}$$

As is shown in Fig. 4.14, the voltage at the first node is v_1, and that at the second node is v_2. The two nodal equations follow directly from the current equations, $i_a = i_b$ and $i_b = i_c$; that is,

$$\frac{e - v_1}{R} = \frac{v_1 - v_2}{LD}$$

$$\frac{v_1 - v_2}{LD} = CDv_2 \tag{4.73}$$

These two node equations may be transformed and then solved

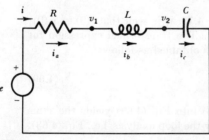

Fig. 4.14. A series RLC circuit.

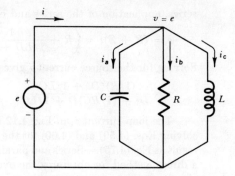

Fig. 4.15. A parallel RLC circuit.

simultaneously to obtain $V_1(s)$ and $V_2(s)$. The loop method is obviously better suited to the solution of this particular problem.

A parallel RLC circuit is shown in Fig. 4.15. The solution is most readily obtained by a node analysis, for there is but one node, whereas there are three loops. The node voltage is $v = e$. The current equation is $i = i_a + i_b + i_c$. Expressing these currents in terms of voltages yields the differential equation of operation for this circuit.

$$i = \left(CD + \frac{1}{R} + \frac{1}{LD} \right) e \tag{4.74}$$

Series and parallel laws

The overall equation relating the source voltage e and current i may be obtained directly by application of series and parallel laws. Electrical elements add by the series law when the same current flows through them. Each of the elements in Fig. 4.14 is in series; thus application of the series law of addition yields directly Eq. (4.71). Electrical elements add by the parallel law when the same voltage drop exists across each element. This is the case for each of the elements in Fig. 4.15. Thus, application of the parallel, or reciprocal, law of addition gives

$$e = \frac{i}{CD + 1/R + 1/LD}$$

Solving for i verifies the result of Eq. (4.74).

For the circuit of Fig. 4.12, it is to be noted that the inductor and capacitor are in parallel. The impedance of this parallel combination is designated as Z. Because Z and the resistance R are in series, application of the series and parallel laws gives

$$e = (R + Z)i = \left(R + \frac{1}{1/LD + CD} \right) i$$

Solving for the source current i gives

$$i = \frac{(1/R)(D^2 + 1/LC)}{D^2 + (1/RC)D + 1/LC} e \qquad (4.75)$$

The loop current i_1 in Fig. 4.12 is the source current i. Thus, solving Eqs. (4.59) and (4.60) for the loop current i_1 yields the same result as Eq. (4.75). Series and parallel laws are thus seen to provide a direct method for obtaining the overall differential equation for a circuit.

4.5 DUALS AND ANALOGS

For a given circuit it is possible to construct a different circuit which has the same differential equation of operation. The new circuit is referred to as the *dual* of the original circuit.

Analogs are two different types of systems, such as an electrical and a mechanical system, which have the same differential equation of operation. For a given circuit, it is possible to construct two analogs.

Duals

Comparison of Eqs. (4.71) and (4.74) shows that the differential equations for the circuits of Figs. 4.14 and 4.15 have the same form. Two different electrical circuits whose differential equations have the same form are called duals. The quantities which occupy corresponding positions in Eqs. (4.71) and (4.74) are *dual elements*. These dual elements are tabulated in Table 4.1. The subscript d

**Table 4.1
Dual elements
for electrical
circuits**

$e \sim i_d$
$i \sim e_d$
$L \sim C_d$
$R \sim 1/R_d$
$C \sim L_d$

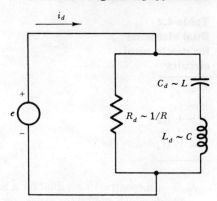

Fig. 4.16. Dual of electrical network shown in Fig. 4.13.

for each of the elements on the right-hand side of the table refers to the parameter in the dual circuit. Thus, current i_d in the dual circuit corresponds to voltage e in the original circuit, etc.

It is to be noted that the elements of Fig. 4.14 are connected in series and that the elements of Fig. 4.15 are connected in parallel. To construct the dual of an electrical circuit, first change the connections so that elements that are in series in the original circuit are in parallel in the dual. Similarly, elements that are in parallel in the original circuit are connected in series in the dual. Then, in accordance with Table 4.1, inductors are replaced by capacitors, resistors are replaced by resistors whose value is $1/R$, and capacitors are replaced by inductors.

Illustrative example   Construct the dual for the electrical network shown in Fig. 4.13.

Solution   The inductor L and capacitor C, which are in parallel in Fig. 4.13, are replaced by a capacitor C_d and an inductor L_d in series, as shown in Fig. 4.16. The resistor R, which is in series with this unit in Fig. 4.13, is replaced by a resistor R_d, which is in parallel with the capacitor and inductor of Fig. 4.16. Application of the series and parallel laws to obtain the equation of operation for the circuit of Fig. 4.16 gives

$$e_d = \frac{1}{\dfrac{1}{R_d} + \dfrac{1}{L_d D + (1/C_d D)}} i_d = \frac{R_d(D^2 + 1/C_d L_d) i_d}{D^2 + \dfrac{R_d}{L_d} D + \dfrac{1}{C_d L_d}} \qquad (4.76)$$

Comparison of Eq. (4.76) for the dual and Eq. (4.75) for the original circuit verifies the dual relationships given in Table 4.1.

**Table 4.2
Dual elements
for mechanical
circuits**

$f \sim \dot{y}_d$
$\dot{y} \sim f_d$
$M \sim 1/K_d$
$B \sim 1/B_d$
$K \sim 1/M_d$

In constructing a dual for a mechanical circuit, mechanical elements which add by the series law are replaced by a parallel arrangement, and vice versa, in accordance with Table 4.2. This technique for constructing the dual of a mechanical system is illustrated in the following example.

Illustrative example Let it be desired to construct the dual for the mechanical system shown in Fig. 4.10.

Solution The spring K_1 and damper B_1, which add by the series law in Fig. 4.10, are replaced by a mass M_{1d} and damper B_{1d}, which are in parallel in Fig. 4.17. The spring K_2 of Fig. 4.10 is in parallel with K_1 and B_1; hence this spring is replaced by a mass M_{2d}, which is in series with M_{1d} and B_{1d} of Fig. 4.17. Application of the series and parallel laws to obtain the equation of operation for the circuit of Fig. 4.17 gives

$$f_d = \left(M_{d2}D^2 + \frac{1}{\dfrac{1}{B_{d1}D} + \dfrac{1}{M_{d1}D^2}} \right) y_d$$

$$= \frac{\left(\dfrac{1}{M_{d1}} + \dfrac{1}{M_{d2}} + \dfrac{1}{B_{d1}} D \right) D\ddot{y}_d}{\dfrac{1}{M_{d2}} \left(\dfrac{1}{M_{d1}} + \dfrac{1}{B_{d1}} D \right)} \tag{4.77}$$

Differentiation of both sides of Eq. (4.51) gives

$$\dot{y}_2 = \frac{(K_1 + K_2 + B_1D)Df}{K_2(K_1 + B_1D)} \tag{4.78}$$

Equation (4.77), which describes the operation of the dual, has exactly the same form as Eq. (4.78), which is the equation for the

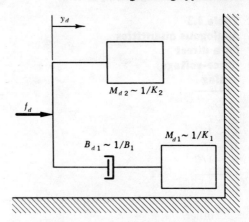

$M_{d2} \sim 1/K_2$

f_d

$B_{d1} \sim 1/B_1$

$M_{d1} \sim 1/K_1$

y_d

Fig. 4.17. Dual of mechanical network shown in Fig. 4.10.

original circuit. Comparison of these equations shows the validity of the dual relationships given in Table 4.2.

It should be pointed out that it is not always possible to construct the dual of a mechanical circuit. For example, the dual of two springs in series would be two masses in parallel. Because it is physically impossible to have two masses in parallel, this dual is meaningless. Because electrical elements can be connected in any parallel or series arrangement, the dual for an electrical circuit can always be constructed.

Analogs

It is possible to construct two electrical circuits whose differential equations have the same form as that for a given mechanical system. Each electrical circuit is called an *electrical analog* for the mechanical system. Similarly, two mechanical analogs may be constructed for an electrical circuit.

For a given circuit, one analog is obtained by replacing series elements by analogous elements in series, and, similarly, the parallel elements are replaced by analogous elements also in parallel. This type of analog is called a *direct analog*. The second analog is obtained by replacing the series elements in the given circuit by analogous elements in parallel, and the parallel elements are replaced by analogous elements in series. Because the connections are inverted, this type of analog is called an *inverse analog*. As would

145

**Table 4.3
Analogous quantities
for a direct
(force-voltage)
analog**

$f \sim e$
$\dot{y} \sim i$
$M \sim L$
$B \sim R$
$K \sim 1/C$

be expected, these two analogs for the given circuit are duals of each other.

Equation (4.4), which describes the operation of the series spring-mass-damper system of Fig. 4.2, may be written in the form

$$f = \left(MD + B + \frac{K}{D} \right) \dot{y} \tag{4.79}$$

Equation (4.71), which describes the operation of the series RLC circuit of Fig. 4.14, has the same form as Eq. (4.79). Termwise comparison of the equations for these two series circuits [i.e., Eqs. (4.71) and (4.79)] reveals the analogous quantities given in Table 4.3.

To construct a direct analog between electrical and mechanical systems, series elements are replaced by series elements in accordance with the analogous relationships of Table 4.3. Similarly, parallel elements are replaced by parallel elements in accordance with the table. Because force and voltage are analogous quantities, this type of analog is referred to as a *force-voltage analog*.

Illustrative example Determine the direct (force-voltage) analog for the mechanical circuit of Fig. 4.10.

Solution The spring K_1 and damper B_1, which are in series, are replaced by a series capacitor C_1 and resistor R_1 as shown in Fig. 4.18. Next, the parallel spring K_2 is replaced by a parallel capacitor C_2. The equation of operation for this electrical analog is

$$e = \frac{i}{\dfrac{1}{R_1 + 1/C_1 D} + C_2 D} = \frac{(1/C_2)(1/C_1 + R_1 D)}{\left(\dfrac{1}{C_1} + \dfrac{1}{C_2} + R_1 D \right)} \frac{i}{D} \tag{4.80}$$

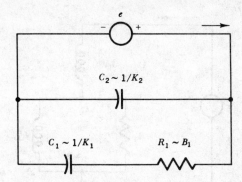

Fig. 4.18. Direct (force-voltage) analog of Fig. 4.10.

Termwise comparison of Eq. (4.51), which describes the operation of the mechanical circuit, and Eq. (4.80) substantiates the analogous quantities given in Table 4.3. Because velocity is analogous to current, then position $y = \dot{y}/D$ is analogous to charge $Q = i/D$.

The other analog which exists between mechanical and electrical circuits is the inverse analog. To construct an inverse analog, series elements are replaced by parallel elements, and vice versa. Equation (4.79), which describes the operation of the series mechanical circuit of Fig. 4.2, and Eq. (4.74), which describes the parallel electrical circuit of Fig. 4.15, have the same form. Comparison of corresponding terms in the equation for the series circuit [Eq. (4.79)] and the equation for the parallel circuit [Eq. (4.74)] indicates the analogous relationships given in Table 4.4. Because force and current are analogous quantities, this inverse analog is referred to as a *force-current analog.*

**Table 4.4
Analogous quantities
for an inverse
(force-current)
analog**

$$f \sim i$$
$$\dot{y} \sim e$$
$$M \sim C$$
$$B \sim 1/R$$
$$K \sim 1/L$$

147

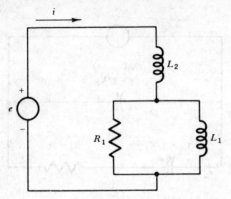

Fig. 4.19. Inverse (force-current) analog of Fig. 4.10.

Illustrative example Determine the inverse (force-current) analog for the mechanical circuit of Fig. 4.10.

Solution First replace the series spring K_1 and damper B_1 by an inductor and resistor in parallel, as shown in Fig. 4.19. Next the parallel spring K_2 is replaced by a series inductor L_2. The equation of operation for the circuit of Fig. 4.19 is

$$e = \left(L_2 D + \frac{1}{1/R_1 + 1/L_1 D} \right) i$$

Solving for i gives

$$i = \frac{(1/L_2)[1/L_1 + (1/R_1)D]}{1/L_1 + 1/L_2 + (1/R_1)D} \frac{e}{D} \tag{4.81}$$

Termwise comparison of Eqs. (4.49) and (4.81) substantiates the analogous relationships in Table 4.4. Because velocity is analogous to voltage, then position $y = \dot{y}/D$ is analogous to e/D. Figures 4.18 and 4.19, which are the two analogs for Fig. 4.10, are dual circuits.

The preceding techniques may also be employed for developing duals and analogs (direct and inverse) for other types of systems such as torsional or rotating systems, thermal systems, fluid flow, etc.

4.6 STATE-SPACE METHOD

The state-space technique is an extremely powerful method for investigating engineering systems. It lends itself very well to the analysis of more complex systems, and may also be adapted to the analysis of nonlinear systems.

The basis for the state-space method is that an nth-order system may be represented by n first-order equations. For example, a third-order system is described as follows:

$$\begin{aligned}
\dot{x}_1 &= a_{11}x_1 + a_{12}x_2 + a_{13}x_3 + f_1(t) \\
\dot{x}_2 &= a_{21}x_1 + a_{22}x_2 + a_{23}x_3 + f_2(t) \\
\dot{x}_3 &= a_{31}x_1 + a_{32}x_2 + a_{33}x_3 + f_3(t)
\end{aligned} \tag{4.82}$$

where $\dot{x}_1 = dx_1/dt$, $\dot{x}_2 = dx_2/dt$, and $\dot{x}_3 = dx_3/dt$. The terms x_1, x_2, and x_3 are called *state variables*. The terms $f_1(t)$, $f_2(t)$, and $f_3(t)$ are forcing functions. The a coefficients are constants to be evaluated.

The equation for an inductor and a capacitor may be expressed in the form

$$\dot{i} = \frac{di}{dt} = \frac{1}{L}v_L$$

$$\dot{v} = \frac{dv}{dt} = \frac{1}{C}i_c$$

where v_L is the voltage drop across the inductor and i_c is the current flowing through the capacitor.

For electrical circuits the state variables are taken as all the inductor currents i and all the capacitor voltages v. To illustrate this method, let it be desired to determine the equation of operation for the circuit of Fig. 4.13. Because there is but one inductor and one capacitor, this circuit is described by two equations. One state variable is i, the current through the inductor, and the other state variable is v, the voltage drop across the capacitor. Thus

$$\dot{i} = \frac{1}{L}v_L = a_{11}i + a_{12}v + f_1(t) \tag{4.83}$$

$$\dot{v} = \frac{1}{C}i_c = a_{21}i + a_{22}v + f_2(t) \tag{4.84}$$

The constant a_{11} in Eq. (4.83) is obtained by letting $i = 1$, $v = 0$, and $f_1(t) = 0$, in which case $a_{11} = v_L/L$. For $i = 1$, the inductor is replaced by a 1-amp current source. For $v = 0$, the capacitor is short-circuited. For $f_1(t) = 0$, the forcing function (i.e., supply voltage e) is short-circuited. The corresponding circuit diagram is shown in Fig. 4.20a. Because the capacitor is short-circuited in Fig. 4.20a, then $v_L = 0$, in which case $a_{11} = v_L/L = 0$.

The constant a_{21} is evaluated by letting $i = 1$, $v = 0$, and $f_2(t) = 0$, in which case Eq. (4.84) becomes $a_{21} = i_c/C$. Figure

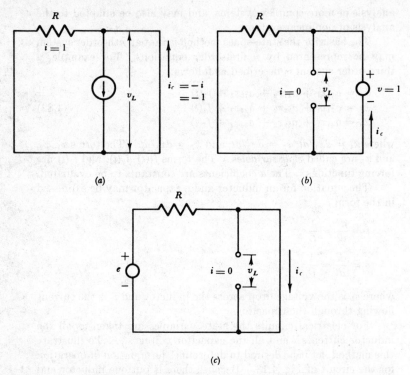

Fig. 4.20. State-space analysis of Fig. 4.13.

4.20a serves as the equivalent circuit for evaluating a_{21} as well as a_{11}. Because $i_c = -1$, then $a_{21} = -1/C$. The preceding manner of evaluating these constants is summarized in the following short-hand notation:

$$a_{11} = \frac{v_L}{L} = 0 \qquad \left\{ \begin{array}{l} i = 1 \\ v = 0 \\ f_1(t) = 0 \end{array} \right.$$

$$a_{21} = \frac{i_c}{C} = -\frac{1}{C} \qquad \left\{ \begin{array}{l} i = 1 \\ v = 0 \\ f_2(t) = 0 \end{array} \right.$$

The boundary conditions are specified in the brackets.

The constant a_{12} is evaluated by letting $i = 0$, $v = 1$, and $f_1(t) = 0$ in Eq. (4.83); thus $a_{12} = v_L/L$. For $i = 0$, the inductor is open-circuited. For $v = 1$, the capacitor is replaced by a unity voltage source. For $f_1(t) = 0$, the source is short-circuited. The corresponding circuit diagram is shown in Fig. 4.20b. Because $v = v_L = 1$, then $a_{12} = v_L/L = 1/L$.

The constant a_{22} is evaluated by letting $i = 0$, $v = 1$, and $f_2(t) = 0$, in which case $a_{22} = i_c/C$. The equivalent circuit is Fig. 4.20b. Because $i_c = -v/R = -1/R$, then $a_{22} = -1/RC$. The shorthand notation for a_{12} and a_{22} is

$$a_{12} = \frac{v_L}{L} = \frac{1}{L} \qquad \begin{cases} i = 0 \\ v = 1 \\ f_1(t) = 0 \end{cases}$$

$$a_{22} = \frac{i_c}{C} = -\frac{1}{RC} \qquad \begin{cases} i = 0 \\ v = 1 \\ f_2(t) = 0 \end{cases}$$

The term $f_1(t)$ is determined by letting $i = 0$ and $v = 0$ in Eq. (4.83), in which case $f_1(t) = v_L/L$. The corresponding circuit is shown in Fig. 4.20c. Because v_L is zero, then $f_1(t)$ is zero also. In a similar manner, $f_2(t)$ is evaluated by letting $i = 0$ and $v = 0$ in Eq. (4.84), in which case $f_2(t) = i_c/C$. From the equivalent circuit of Fig. 4.20c, it follows that $i_c = e/R$. Thus, $f_2(t) = i_c/C = e/RC$. The notational form for $f_1(t)$ and $f_2(t)$ is

$$f_1(t) = \frac{v_L}{L} = 0 \qquad \begin{cases} i = 0 \\ v = 0 \end{cases}$$

$$f_2(t) = \frac{i_c}{C} = \frac{e}{RC} \qquad \begin{cases} i = 0 \\ v = 0 \end{cases}$$

Substitution of the values for all the constants into Eqs. (4.83) and (4.84) gives

$$i = \frac{1}{L} v \tag{4.85}$$

$$\dot{v} = -\frac{1}{C} i - \frac{1}{RC} v + \frac{1}{RC} e \tag{4.86}$$

Equations (4.85) and (4.86) are the two first-order equations which describe the operation of this circuit. Transforming these equations gives

$$sI(s) - \frac{1}{L} V(s) = i(0)$$

$$\frac{1}{C} I(s) + \left(s + \frac{1}{RC} \right) V(s) = \frac{E(s)}{RC} + v(0) \tag{4.87}$$

These transformed equations may now be solved by the use of determinants to determine the transform $I(s)$ of the inductor current and the transform $V(s)$ of the capacitor voltage. Because the capacitor voltage v is also the node voltage v (see Fig. 4.13), solving Eqs. (4.87) for $V(s)$ yields the same result as Eq. (4.68). By knowing the state variables i and v, any other unknown currents or voltages in the circuit can be determined.

With the state-space method initial conditions are the initial current through each inductor and the initial voltage across each capacitor. An important feature of this method is that the a coefficients are determined from resistive networks only. The reason for this is that all inductors are open-circuited and capacitors are short-circuited. In practice, a very significant advantage of the state-space method is the ease with which analog computers can be set up to solve the state-space equations.

The application of the state-space method to mechanical systems follows the same pattern as that for electrical systems. The equation for a mass and for a spring may be expressed as follows:

$$\dot{v} = \frac{dv}{dt} = \frac{1}{M} f_M$$

$$\dot{f} = \frac{df}{dt} = K \frac{dy}{dt} = K v_K$$

where f_M is the summation of the external forces acting on the mass and $dy/dt = v_K$ is the rate of change of length of the spring (i.e., the velocity of one end relative to the other). As the force on the spring increases, the spring becomes shorter. Thus, v_K is positive when the spring is becoming shorter and negative when the spring is becoming longer. The state variables are the velocity v of each mass and the force f acting on each spring.

This method is now illustrated for the mass-spring-damper system of Fig. 4.21a. The applied force is designated f_a. Because

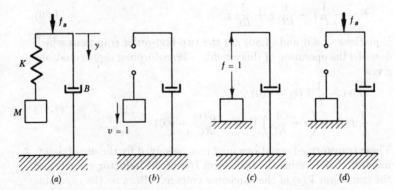

Fig. 4.21. State-space analysis.

this system has one spring and one mass, it is described by two equations of the form

$$\dot{v} = \frac{1}{M} f_M = a_{11}v + a_{12}f + f_1(t) \tag{4.88}$$

$$\dot{f} = Kv_K = a_{21}v + a_{22}f + f_2(t) \tag{4.89}$$

The equations for a_{11} and a_{21} are

$$a_{11} = \frac{1}{M} f_M = 0 \qquad \left\{ \begin{array}{l} v = 1 \\ f = 0 \\ f_1(t) = 0 \end{array} \right.$$

$$a_{21} = Kv_K = -K \qquad \left\{ \begin{array}{l} v = 1 \\ f = 0 \\ f_2(t) = 0 \end{array} \right.$$

The velocity of the mass is $v = 1$, and the spring force is $f = 0$. Because there is no force transmitted through the spring, it is simply removed. For $f_1(t) = 0$ and $f_2(t) = 0$, there is no excitation or applied force. The equivalent system is shown in Fig. 4.21b. There is no external force acting on the mass; so $f_M = 0$, and thus $a_{11} = f_M/M = 0$. The velocity of one end of the spring relative to the other is $v_K = -v = -1$. The minus occurs because the spring is elongating. Thus, $a_{21} = Kv_K = -K$.

The equations for a_{12} and a_{22} are

$$a_{12} = \frac{1}{M} f_M = \frac{1}{M} \qquad \left\{ \begin{array}{l} v = 0 \\ f = 1 \\ f_1(t) = 0 \end{array} \right.$$

$$a_{22} = Kv_K = -\frac{K}{B} \qquad \left\{ \begin{array}{l} v = 0 \\ f = 1 \\ f_2(t) = 0 \end{array} \right.$$

The equivalent circuit for $v = 0, f = 1$, and $f_1(t) = f_2(t) = 0$ is shown in Fig. 4.21c. Because the velocity v of the mass is zero, the mass is grounded. The spring is replaced by the force $f = 1$. The force on the mass is $f_M = f = 1$, and thus $a_{12} = (1/M)f_M = 1/M$. The velocity v_K of the spring is the velocity of the damper (the mass end of the spring is grounded). For the damper $f = B\, dy/dt = Bv_K$. Figure 4.21a shows that the positive sense for f and y and thus all forces and coordinates are downward. Because the force f acting on the damper in Fig. 4.21c is up, $f = -1 = Bv_K$, or $v_K = -1/B$. This negative sign for v_K checks the fact that the spring is elongating.

Finally the equations for $f_1(t)$ and $f_2(t)$ are

$$f_1(t) = \frac{1}{M} f_M = 0 \qquad \begin{cases} v = 0 \\ f = 0 \end{cases}$$

$$f_2(t) = Kv_K = \frac{K}{B} f_a \qquad \begin{cases} v = 0 \\ f = 0 \end{cases}$$

The equivalent circuit for $v = f = 0$ is shown in Fig. 4.21d. The value of $f_1(t)$ is zero because there is no force f_M acting on the mass. The velocity v_K of the spring is the velocity of the damper, which is $f_a = Bv_K$ or $v_K = f_a/B$. Substitution of the preceding results into Eqs. (4.88) and (4.89) yields the state-space equations which describe the operation of this system.

$$\dot{v} = \frac{1}{M} f$$

$$\dot{f} = -Kv - \frac{K}{B} f + \frac{K}{B} f_a \tag{4.90}$$

These equations may now be transformed or solved on an analog computer. In applying the state-space method to mechanical systems, masses are grounded and springs removed so that only dampers remain in the equivalent circuits.

4.7 STRUCTURES

The Laplace transform method is well suited to the solution of structural problems in strength of materials. The boundary conditions for the structure appear as initial conditions in the Laplace transform. In the classical method, separate equations must be written for each interval between loads, whereas with Laplace transforms one equation suffices.

The general beam equations are

$$EI \frac{d^4y}{dx^4} = f(x) \tag{4.91}$$

$$EI \frac{d^3y}{dx^3} = V(x) \tag{4.92}$$

$$EI \frac{d^2y}{dx^2} = M(x) \tag{4.93}$$

where E is the modulus of elasticity, I is the moment of inertia, x is the horizontal distance along the length of the beam, y is the beam

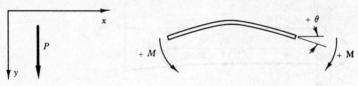

Fig. 4.22.   Sign conventions.

deflection, $f(x)$ is the loading per unit length, $V(x)$ is the shear, and $M(x)$ is the moment. The sign convention is illustrated in Fig. 4.22. The deflection y and the loading $f(x)$ are positive in the downward direction. The positive moment M and beam slope θ are also indicated in Fig. 4.22. For structural problems the space coordinate x replaces time t in the Laplace transformation.

A uniform load P/δ, which acts over a small width δ, is shown in Fig. 4.23a. As δ approaches zero, this loading becomes a concentrated load P. Hence, as indicated in Fig. 4.23a, a concentrated load P may be represented by an impulse whose area is P. Another common type of loading is a couple, or moment, M. In Fig. 4.23b are shown two equal and opposite concentrated loads M/δ. Because these loads are a distance δ apart, they are equivalent to a moment M. The transform of this moment is the transform of the upward load (M/δ) of Fig. 4.23b, minus that of the downward load, which is offset a distance δ. By designating this force couple, which represents the moment, as f_M, then

$$\mathcal{L}(f_M) = \frac{M}{\delta} - \frac{M}{\delta}\, e^{-\delta s} = \frac{M}{\delta}\, (1 - e^{-\delta s}) \tag{4.94}$$

The term $e^{-\delta s}$ indicates a delay of a distance δ along the beam (when time is the variable, a time delay a is indicated as e^{-as}). By writing

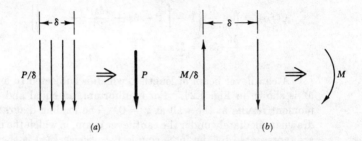

Fig. 4.23.   Loading representations.   (a) Concentrated load; (b) moment.

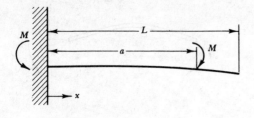

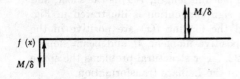

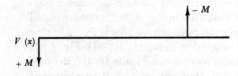

Fig. 4.24. Cantilever beam with a moment.

$e^{-\delta s}$ in its series expansion form and taking the limit as δ approaches zero, the preceding expression for the transform of the force couple becomes

$$\mathcal{L}(f_M) = \lim_{\delta \to 0} \frac{M}{\delta} \left\{ 1 - \left[1 - \delta s + \frac{(\delta s)^2}{2!} - \cdots \right] \right\}$$

$$\approx \frac{M}{\delta} (\delta s) = Ms \tag{4.95}$$

A cantilever beam of length L which is subjected to a moment M is shown in Fig. 4.24. For equilibrium, an equal and opposite moment results at the wall at $x = 0$. The loading diagram $f(x)$ is drawn immediately under the cantilever beam, in which the moments are represented by the force couple f_M. Shear $V(x)$ is the integral of the load along the length of the beam; thus, from Eqs. (4.91) and

(4.92), it follows that

$$\frac{d}{dx} V(x) = DV(x) = f(x)$$

or

$$V(x) = \frac{1}{D} f(x) = \int_0^x f(x)\, dx + V(0) \tag{4.96}$$

Integration of the first (downward) impulse at the wall yields a constant shear M/δ. The upward part of this couple reduces the shear to zero at $x = \delta$. Because the height is M/δ and the width is δ, the shear at $x = 0$ may be represented by an impulse of area M as shown in the shear diagram at $x = 0$.

From Eqs. (4.92) and (4.93), it follows that $M(x)$ is the integral of the shear diagram.

$$M(x) = \int_0^x V(x)\, dx + M(0) \tag{4.97}$$

The area of the impulse at the origin of the shear diagram is M. Because there is no shear between $x = 0$ and $x = a$, the value of the moment $M(x) = M$ remains constant throughout this region. The moment is reduced to zero at $x = a$.

To obtain the equation for the deflection y, first transform Eq. (4.91). Thus

$$s^4 Y(s) - s^3 y(0) - s^2 y'(0) - s y''(0) - y'''(0) = \frac{F(s)}{EI} \tag{4.98}$$

At the wall, the deflection $y(0)$ and slope $y'(0)$ are both zero. The initial conditions are evaluated at $x = 0-$ just as, when time is the coordinate, initial conditions are evaluated at $t = 0-$. For simplicity, the initial conditions at $x = 0-$ are designated as $y(0)$, $y'(0)$, $y''(0)$, $y'''(0)$, $M(0)$, $V(0)$, etc. The moment and shear diagrams show that $M(0) = V(0) = 0$. Thus, from Eqs. (4.92) and (4.93), it follows that $y''(0) = y'''(0) = 0$.

The transform of the force couple f_M for a moment is Ms. Thus, the transform $F(s)$ of the loading $f(x)$ is

$$\mathcal{L}[f(x)] = F(s) = Ms - Mse^{-as} \tag{4.99}$$

Substitution of the preceding results into Eq. (4.98) gives

$$Y(s) = \frac{M}{EIs^3} (1 - e^{-as}) \tag{4.100}$$

The delayed term e^{-as} does not come into play for $0 < x \le a$. Thus

$$y = \frac{Mx^2}{2EI} \qquad 0 < x \le a \tag{4.101}$$

For $x > a$, the effect of the delayed term must be added to the preceding result.

$$y = \frac{Mx^2}{2EI} - \frac{M(x-a)^2}{2EI} = \frac{aM}{2EI}(2x - a) \qquad x > a \qquad (4.102)$$

These results could also have been obtained from either the shear or the moment diagram. For example, when the moment is used, Eq. (4.93) is transformed.

$$s^2Y(s) - sy(0) - y'(0) = \frac{M(s)}{EI} \qquad (4.103)$$

From the moment diagram, the transform $M(s)$ is

$$M(s) = \frac{M}{s} - \frac{M}{s}e^{-as}$$

Because $y(0)$ and $y'(0)$ are zero, substituting these results into Eq. (4.103) yields the same result for $Y(s)$ as Eq. (4.100). As illustrated by this example, it is not necessary to construct the shear and moment diagrams, for the deflection y may be obtained directly from the loading diagram.

A cantilever beam which is subjected to a concentrated load is shown in Fig. 4.25. The equal and opposite load reaction $-P$ is provided by the wall. These loads constitute a couple so that for equilibrium a moment $M = bP$ exists at the wall. The transformed equation is the same as that given by Eq. (4.98). From Fig. 4.25, it follows that the transform $F(s)$ of the load is

$$F(s) = Pbs - P(1 - e^{-bs})$$

Because $y(0) = y'(0) = y''(0) = y'''(0) = 0$, the transformed equation for $Y(s)$ is

$$Y(s) = \frac{1}{EI}\left[\frac{Pb}{s^3} - \frac{P}{s^4}(1 - e^{-bs})\right] \qquad (4.104)$$

The delayed term e^{-bs} does not come into play for $0 < x \leq b$; thus

$$y(x) = \frac{1}{EI}\left(\frac{Pbx^2}{2} - \frac{Px^3}{3!}\right) = \frac{Px^2}{6EI}(3b - x) \qquad (4.105)$$

For $x > b$, adding the contribution of the delayed term gives

$$y(x) = \frac{1}{EI}\left[\frac{Pbx^2}{2} - \frac{Px^3}{6} + \frac{P(x-b)^3}{6}\right] = \frac{Pb^2}{6EI}(3x - b) \qquad (4.106)$$

Adding the deflection equation for the moment and that for the concentrated load (with the appropriate equation used for the posi-

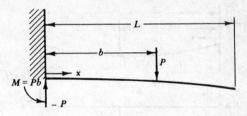

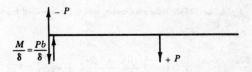

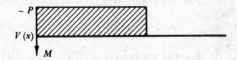

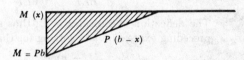

Fig. 4.25. Cantilever beam with a concentrated load.

tion along the beam) gives the equation for a beam subjected to both
a moment (Fig. 4.24) and a concentrated load (Fig. 4.25).

A simply supported beam which has a concentrated load P is
shown in Fig. 4.26. The fact that some of the initial conditions are
known at the support $x = L$ rather than at $x = 0$ presents no particular difficulties. The transform $F(s)$ for the loading is

$$F(s) = -R_1 + Pe^{-as}$$

Because there is no deflection at the left support, then $y(0) = 0$. The
initial conditions $y''(0)$ and $y'''(0)$ are also zero, for there is no
moment or shear at $x = 0$. Substitution of these results into Eq.
(4.98) gives

$$s^4 Y(s) - s^2 y'(0) = -\frac{R_1}{EI} + \frac{Pe^{-as}}{EI}$$

159

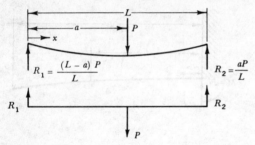

Fig. 4.26. Simply supported beam with a concentrated load.

or

$$Y(s) = \frac{y'(0)}{s^2} - \frac{R_1}{s^4 EI} + \frac{P}{s^4 EI} e^{-as} \qquad (4.107)$$

For the region $0 < x \leq a$, the deflection is

$$y(x) = xy'(0) - \frac{x^3 R_1}{6EI} \qquad 0 < x \leq a \qquad (4.108)$$

For the region $a < x \leq L$, the deflection is

$$y(x) = xy'(0) - \frac{x^3 R_1}{6EI} + \frac{(x-a)^3 P}{6EI} \qquad a < x \leq L \qquad (4.109)$$

The deflection is zero at $x = L$; thus substituting $x = L$ into the preceding expression and solving for the slope $y'(0)$ gives

$$y'(0) = \frac{L^2 R_1}{6EI} - \frac{(L-a)^3 P}{6LEI} \qquad (4.110)$$

The substitution of this value of $y'(0)$ and $R_1 = P(L - a)/L$ into Eq. (4.108) gives

$$y(x) = \frac{x(L-a)(2aL - a^2 - x^2)P}{6LEI} \qquad 0 < x \leq a \qquad (4.111)$$

Similarly, substitution of the values of $y'(0)$ and R_1 into Eq. (4.109) gives

$$y(x) = \frac{[x(L-a)(2aL - a^2 - x^2) + L(x-a)^3]P}{6LEI}$$
$$a < x \leq L \qquad (4.112)$$

A major feature of the transform method of solution is that the boundary conditions for the engineering situation appear directly as initial conditions in the transformed equation. Also, the same general procedure [i.e., (1) transforming, (2) inserting boundary conditions, and (3) inverting to obtain the response] is followed for the solution of all problems, including partial differential equations and differential equations with variable coefficients.

PROBLEMS

4.1   For the mass-spring-damper system shown in Fig. P4.1, $M = 0.5$ lb-sec^2/in., $B = 6$ lb-sec/in., and $K = 50$ lb/in.   A force F_0 is suddenly applied at time $t = 0$.

(a) Determine ζ and ω_n.
(b) Solve the differential equation for the motion y (all the initial conditions are zero).

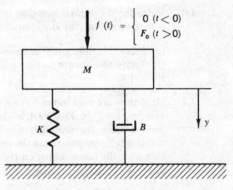

$$f(t) = \begin{cases} 0 & (t < 0) \\ F_0 & (t > 0) \end{cases}$$

Fig. P4.1

4.2   A weight W is dropped from a distance h upon a platform supported by a spring and damper, as shown in Fig. P4.2.   The initial velocity when the mass contacts the platform is $\dot{y}(0) = \sqrt{2gh}$.   The position y is measured from the extended position of the spring before contact. Summing the forces on the mass gives

$$Mg - Ky - BDy = MD^2y$$

The gravitational force Mg must be included unless the position y is measured from the equilibrium position in which $W = Ky_{ST}$, where y_{ST} is the initial static deflection of the spring.   For the case in which $M = 0.5$ lb-sec^2/in., $B = 6$ lb-sec/in., and $K = 50$ lb/in.:

(a) Determine ζ and ω_n.
(b) Solve the differential equation for the motion y.

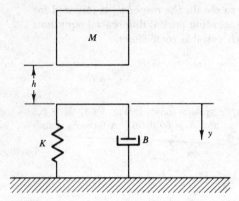

Fig. P4.2

4.3 Write the differential equation for Prob. 4.2 for the case in which y is measured from the static equilibrium position.

(a) Solve this differential equation for the motion y.
(b) Verify this answer to part a by substituting $y + Mg/K$ for y in Prob. 4.2.

4.4 Drop tests are used in the testing of landing gear for airplanes, space vehicles, etc. In Fig. P4.4 is shown a schematic diagram at the instant when the unit first touches ground. At this instant the spring is fully extended, and the velocity of the mass is $\dot{y}(0) = \sqrt{2gh}$. Summing the forces acting on the mass gives

$$Mg - Ky - BDy = MD^2y$$

Solve this differential equation for the case in which $M = 0.5$ lb-sec²/in., $B = 6$ lb-sec/in., and $K = 50$ lb/in.

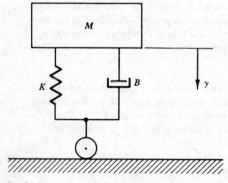

Fig. P4.4

4.5 In Fig. P4.5 is shown a slight modification of the drop-testing equipment discussed in Prob. 4.4. The frame which is fastened to the mass compresses the spring an amount y_P from its free length. The purpose of this frame is to preload the spring to prevent minor oscillations of the wheel (undercarriage) during descent. At contact the frame separates from the bottom side of the spring and damper. Hence the summation of forces acting on the mass is $Mg - K(y + y_P) - BD(y + y_P)$, where y_P, the initial preload, is a constant. The initial velocity of the mass is $\dot{y}(0) = \sqrt{2gh}$. Write the differential equation for this system, and solve it for the case in which $M = 0.5$ lb-sec²/in., $B = 6$ lb-sec/in., and $K = 50$ lb/in.

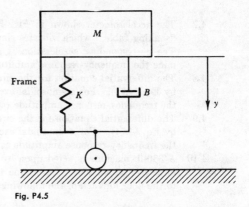

Fig. P4.5

4.6 In Fig. P4.6 is shown an accelerometer, which is used to measure the acceleration of the frame (i.e., box) $\ddot{x}$. The position of the mass relative to the frame is designated by y. This position y is measured

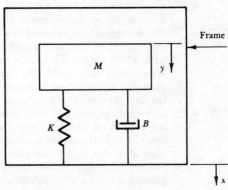

Fig. P4.6

from the static-equilibrium position of the mass-spring-damper system (i.e., when y is zero, the spring is deflected $\delta_{ST} = W/K$ in. from its free length). Summing the forces acting on the mass gives $-(K + BD)y = M(\ddot{x} + \ddot{y})$ or $(MD^2 + BD + K)y = -M\ddot{x}$, where $\ddot{x} + \ddot{y}$ is the acceleration of the mass. For $M = 0.01$ lb-sec^2/in.:

(a) Determine B and K such that $\zeta = 0.60$ and $\omega_n = 100$ rad/sec (a large ω_n yields a fast response).

(b) Solve the differential equation for the case in which the accelerometer is initially at rest and then is allowed to free-fall so that $\ddot{x} = g$ (note that the spring will be extended to its free length after the initial transients of the free fall die out).

4.7   The accelerometer shown in Fig. P4.6 is fastened to a vibrating "shaking table," which vibrates sinusoidally so that $x = x_0 \sin \omega t$. The corresponding acceleration is $a = \ddot{x} = -x_0\omega^2 \sin \omega t$. Determine the frequency-response amplitude ratio y_0/x_0.

4.8   The differential equation for the circuit shown in Fig. 4.14 is given by Eq. (4.71). For a sinusoidal excitation $e = e_0 \sin \omega t$, determine the frequency-response amplitude ratio i_0/e_0.

4.9   The differential equation for the circuit shown in Fig. 4.15 is given by Eq. (4.74). For a sinusoidal excitation $e = e_0 \sin \omega t$, determine the frequency-response amplitude ratio i_0/e_0.

4.10   A 386-lb machine is acted upon by a sinusoidally varying shaking force $f = 25 \sin 200t$. Determine the vibratory force transmitted to the floor for each of the following cases:

(a) The machine is rigidly fastened to the floor.
(b) The machine is mounted on springs such that $K = 1,600$ lb/in. and ζ is negligible.

Determine the amplitude of motion for case b.

4.11   A new air-conditioning unit for an office incorporates a blower which is driven by an electric motor at a speed of 1,800 rpm. Although the blower has been balanced to a high degree, the office workers say that the vibrations are annoying. Investigation shows that the unit, which weighs 386 lb, has been rigidly mounted on the floor. What spring rate would you recommend for a suspension system (damping negligible) such that the unbalanced force transmitted to the floor is reduced by a factor of 10?

4.12   A new lightweight (4-lb) transistorized "electronic box" has replaced the old (16-lb) "tubes'n junk box" in the 0-2-B airplane. It is believed that the excessive vibration of the new unit may be the cause of trouble, rather than poor electronic design. Further investigation reveals that the steel spring isolators ($K = 600$ lb/in.), which did such a fine job of vibration isolation in the old unit, are also used for the new unit. This equipment is located in the airplane where the amplitude of vibration is 0.10 in., the frequency 3,000 cpm, and damping is negligible.

(a) Determine the amplitude of vibration for the old unit, and also the amplitude for the new unit.

(b) What spring rate do you recommend such that the new unit will have the same transmissibility as the old unit?

4.13   In Fig. P4.13 is shown a schematic diagram of an automobile traveling over a "washboard" type of road such that the motion x is roughly sinusoidal. The vertical distance from peak to valley is 1 ft, and the horizontal distance from valley to valley is 22 ft. For $M = 10$ lb-sec^2/in., $K = 640$ lb/in., and $\zeta = 0.4$, use charts to determine the amplitude of motion y_0 for the case in which the forward velocity V is:

(a) 60 mph.
(b) 20 mph.

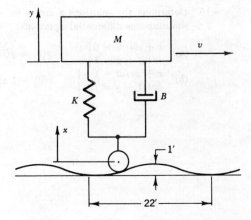

Fig. P4.13

4.14   In the illustrative example problem of Sec. 4.3, all of the initial conditions are zero, and $f(t)$ is a unit impulse. Solve this example for the case in which $y_1(0) = 1, y_1'(0) = 0, y_2(0) = 0, y_2'(0) = 0$, and $f(t) = 0$.

4.15   In Fig. P4.15 is shown a two-degree-of-freedom system. Determine two differential equations which describe the motion of this system. Transform these equations, and then solve for the transform $Y(s)$. What is the characteristic equation for this system?

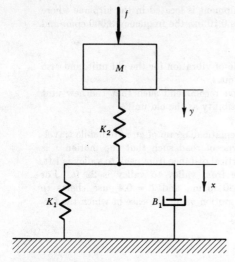

Fig. P4.15

4.16 Determine the solutions x and y for each of the following sets of simultaneous differential equations:

(a) $\left.\begin{array}{l} x + dy/dt = 0 \\ dx/dt + y = 1 \end{array}\right\}$ $x(0) = y(0) = 0$

(b) $\left.\begin{array}{l} x + dy/dt = 1 \\ dx/dt + y = f(t) \end{array}\right\}$ $x(0) = 1$ and $y(0) = 0$

4.17 For each of the electrical circuits shown in Fig. P4.17a to c, the switch is open for $t < 0$ and then closed for $t > 0$. Determine the differential equation relating e and i for each of these circuits, transform to obtain $I(s)$, and then invert to determine the response $i(t)$. (For circuits which have switches, write the differential equation for the circuit with the switch in its position for $t > 0$. Initial conditions are determined with the switch in its position for $t < 0$.)

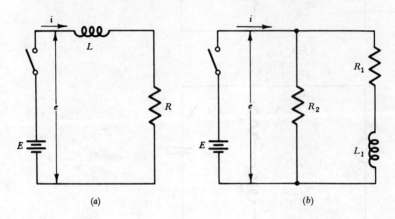

(a)   (b)

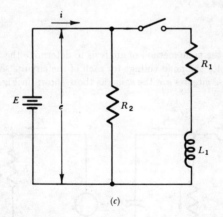

(c)

Fig. P4.17

167

4.18 Use the loop method of analysis to determine the transformed equation for the loop currents for each of the circuits shown in Fig. P4.18.

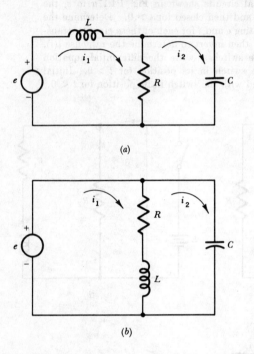

(a)

(b)

Fig. P4.18

4.19 Use the node method of analysis to determine the transformed equation for the node voltage for each of the circuits shown in Fig. P4.19 (these circuits are the same as those shown in Fig. P4.18).

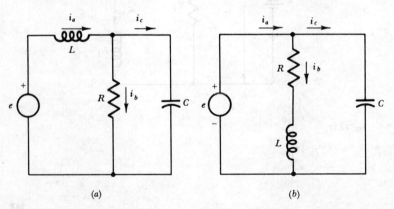

(a)   (b)

Fig. P4.19

4.20   Use parallel and series laws to determine the differential equation relating f and y for each of the mass-spring-damper systems shown in Fig. P4.20.   Then for each system:

(a) Construct the mechanical dual.
(b) Construct the direct (force-voltage) electrical analog.
(c) Construct the inverse (force-current) electrical analog.

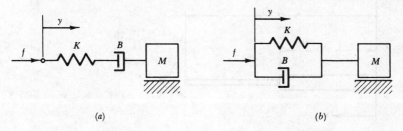

(a)   (b)

Fig. P4.20

4.21   Use parallel and series laws to determine the differential equation relating supply voltage e to supply current i for each of the electrical circuits shown in Fig. P4.19.   Then for each circuit:

(a) Construct the dual circuit.
(b) Construct the direct (force-voltage) mechanical analog.
(c) Construct the inverse (force-current) mechanical analog.

4.22   Determine the state-space equations for:

(a) Each of the electrical circuits shown in Fig. P4.19.
(b) Each of the mechanical circuits shown in Fig. P4.20.

4.23   A simply supported beam with an overhung load is shown in Fig. P4.23.   Construct the loading diagram for this beam, and then determine the equation for the deflection y for the regions $0 < x \leq a$, $a < x \leq a + b$, and $a + b < x \leq L$.

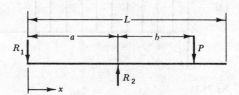

Fig. P4.23

169

4.24 In Fig. P4.24 is shown a cantilever beam which is loaded by a moment M at the end L. For $0 < x < a$ the moment of inertia of the beam is I_1, and for $a < x < L$ the moment of inertia is I_2. Construct the moment diagram for this beam, and then divide by the appropriate EI in each region to obtain the M/EI diagram. Use this M/EI diagram to determine the equation for the deflection for the regions $0 < x < a$ and $a < x < L$.

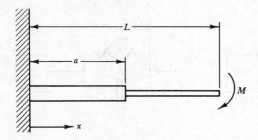

Fig. P4.24

4.25 A cantilever beam has a uniform load of w lb/in. which extends a distance a along the beam, as illustrated in Fig. P4.25. The reactions are $P = wa$ and $M = Pa/2 = wa^2/2$. Draw the loading diagram, and then determine the equation for the beam deflection in the interval:

(a) $0 < x \leq a$
(b) $a < x \leq L$

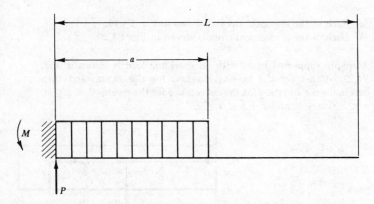

Fig. P4.25

4.26 In Fig. P4.26 is shown a column of length L which is subjected to an axial load P. The moment at the end $x = L$ is $M(L)$. The differential equation is

$$\frac{d^2y}{dx^2} = \frac{M}{EI} = \frac{M(L) + P[y(L) - y]}{EI}$$

where, from Fig. P4.26, the moment at the station x is $M = M(L) + P[y(L) - y]$. Because there is no deflection at $x = 0$ [that is,

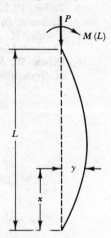

Fig. P4.26

$y(0) = 0$], transforming and solving for $Y(s)$ gives

$$Y(s) = \frac{y'(0)}{s^2 + \alpha^2} + \frac{M(L) + Py(L)}{s(s^2 + \alpha^2)EI}$$

where $\alpha^2 = P/EI$. Invert the preceding expression to obtain $y(x)$, and then determine $y(L)$. Use this result for $y(L)$ to determine the critical, or buckling, load P_c for each of the following cases:

(a) The $x = 0$ end is clamped so that $y'(0) = y(0) = 0$, and the $x = L$ end is free. (Note what happens when α is such that $\cos \alpha L = 0$.)
(b) Both ends are pinned so that $y(0) = y(L) = 0$.
(c) Both ends are clamped so that $y(0) = y'(0) = 0$ and $y(L) = y'(L) = 0$.

REFERENCES

1 Raven, F. H.: "Automatic Control Engineering," McGraw-Hill Book Company, New York, 1961.

2 Aseltine, J. A.: "Transform Method in Linear System Analysis," McGraw-Hill Book Company, New York, 1958.
3 Thomson, W. T.: "Laplace Transformation," 2d ed., Prentice-Hall, Inc., Englewood Cliffs, N.J., 1960.
4 Goldman, S.: "Transformation Calculus and Electrical Transients," Prentice-Hall, Inc., Englewood Cliffs, N.J., 1949.
5 Carson, J. R.: "Electric Circuit Theory and Operational Calculus," McGraw-Hill Book Company, New York, 1926.
6 Balise, P. L.: Demonstrating Analogous Mechanical and Electrical Response, *ISA J.*, February-March, 1960.
7 Koenig, H. E., and W. A. Blackwell: "Electromechanical System Theory," McGraw-Hill Book Company, New York, 1961.
8 Lynch, W. A., and J. G. Truxal: "Introductory System Analysis," McGraw-Hill Book Company, New York, 1961.

chapter 5

Complex-variable theory

The theory of complex variables provides the basis for the development of additional methods of operational mathematics. These additional methods furnish powerful tools for the solution of partial differential equations, differential equations with variable coefficients, and numerous other equations which describe the operation of more complex engineering situations.

5.1 COMPLEX NUMBERS

To solve quantitative problems, several types of numbering systems have been devised. The simplest is the *natural number system*, which consists of the integers 0, 1, 2, ... A more complete numbering system is obtained by including negative numbers and points between adjacent natural numbers. Such additional points are defined as numbers of the form P/Q, where P and Q are natural numbers (for example, $P/Q = \frac{1}{3}$ is a fraction). This more complete system is called the *rational numbering system*. This system does not include quantities such as $\sqrt{2}$ or π (that is, irrational numbers). The system which includes all possible points (natural, rational, and irrational numbers) is referred to as the *real number system*. The continuous chain of real number points is referred to as the *real number line*.

The real number system is unable to describe certain quantities such as the square root of negative real numbers (for example, $\sqrt{-1}$). Such quantities may be specified by the *complex number system*. While the real number system may be regarded as a continuous line

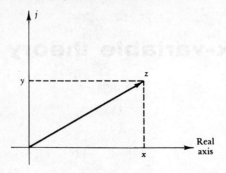

Fig. 5.1. A vector in the complex plane.

of points, the complex system may be considered as a continuous plane of points which includes the real number line. This real line is the horizontal axis. The perpendicular line through the real 0 is the vertical, or imaginary, axis. The point located one unit up the imaginary axis is defined as $\sqrt{-1} = j$.

Thus, the point located two units up the imaginary axis is $2j = 2\sqrt{-1} = \sqrt{-4}$, etc. As shown in Fig. 5.1, a complex number z has a real component x and an imaginary component jy. Thus

$$z = x + jy \tag{5.1}$$

Of particular significance to engineering applications is the fact that a point $z = x + jy$ in the complex plane (see Fig. 5.1) may be regarded as a vector. For the vector z, the magnitude of the horizontal component is x, and the magnitude of the vertical component is y. If x is three units and y is two units, then $z = 3 + j2$. The symbol j distinguishes the vertical from the horizontal component. The horizontal, or x, axis is referred to as the *real axis*, and the vertical, or y, axis is called the *imaginary axis*.

As is illustrated in Fig. 5.2a, the addition of two complex numbers or two vectors $z_1 = x_1 + jy_1$ and $z_2 = x_2 + jy_2$ follows the polygon rule for vector addition. Thus, the sum $z_1 + z_2$ is

$$
\begin{aligned}
z_1 &= x_1 + jy_1 \\
z_2 &= x_2 + jy_2 \\
\hline
z_1 + z_2 &= (x_1 + x_2) + j(y_1 + y_2)
\end{aligned}
\tag{5.2}
$$

The polygon rule also applies for vector subtraction. The vector difference $z_2 - z_1$ is

$$
\begin{aligned}
z_2 &= x_2 + jy_2 \\
-z_1 &= -(x_1 + jy_1) \\
\hline
z_2 - z_1 &= (x_2 - x_1) + j(y_2 - y_1)
\end{aligned}
\tag{5.3}
$$

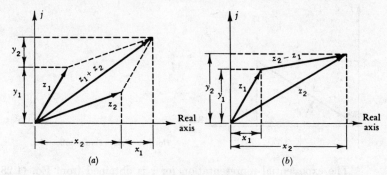

Fig. 5.2. (a) Addition of vectors. (b) Subtraction of vectors.

As is illustrated in Fig. 5.2b, the vector difference $z_2 - z_1$ proceeds from the tip of z_1 to the tip of z_2. From Fig. 5.3, it is apparent that, if $z = z_1 + z_2$, then

$$|z| = |z_1 + z_2| \leq |z_1| + |z_2| \tag{5.4}$$

Similarly, it follows that, if $z = z_1 + z_2 + \cdots + z_n$,

$$|z| \leq \sum_{k=1}^{n} |z_k| \tag{5.5}$$

As illustrated in Fig. 5.4, the polar form for z is

$$z = x + jy = r(\cos \theta + j \sin \theta) \tag{5.6}$$

where

$r = \sqrt{x^2 + y^2}$ = length of vector
$x = r \cos \theta$ = horizontal component
$y = r \sin \theta$ = vertical component

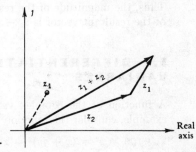

Fig. 5.3. Illustration of $|z_1 + z_2| \leq |z_1| + |z_2|$.

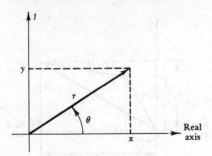

Fig. 5.4. Polar representation of z.

The exponential representation for z is obtained from Eq. (1.28), that is,

$$\cos \theta + j \sin \theta = e^{j\theta} \tag{5.7}$$

Substituting this result into Eq. (5.6) yields for the exponential form

$$z = re^{j\theta} \tag{5.8}$$

As discussed in Sec. 1.2, $e^{j\theta}$ is a unit vector which is rotated counterclockwise an angle θ from the real axis. Thus, $re^{j\theta}$ is a vector of length r which is rotated counterclockwise through the angle θ. The multiplication of two complex numbers $z_1 = r_1 e^{j\theta_1}$ and $z_2 = r_2 e^{j\theta_2}$ is defined as follows:

$$z_1 z_2 = (r_1 e^{j\theta_1})(r_2 e^{j\theta_2}) = r_1 r_2 e^{j(\theta_1 + \theta_2)} \tag{5.9}$$

In multiplication, the length of the resultant vector is the product of the magnitudes $r_1 r_2$ of the vectors being multiplied. The angle of the resultant vector is the sum of the angles $(\theta_1 + \theta_2)$ of the vectors being multiplied.

Similarly, division is defined as follows:

$$\frac{z_2}{z_1} = \frac{r_2 e^{j\theta_2}}{r_1 e^{j\theta_1}} = \frac{r_2}{r_1} e^{j(\theta_2 - \theta_1)} \tag{5.10}$$

Thus, the magnitude of the resultant vector is r_2/r_1, and the angle of the resultant vector is $\theta_2 - \theta_1$.

5.2 DIFFERENTIATION OF COMPLEX VARIABLES

A function of the complex variable z is designated as $f(z)$. For example, consider the function

$$f(z) = z^2 = (x + jy)^2 = x^2 - y^2 + j2xy \tag{5.11}$$

The function $f(z)$ has a real part $u(x,y) = x^2 - y^2$ and an imaginary part $v(x,y) = 2xy$. In general, a function $f(z)$ may be expressed in terms of its real part $u(x,y)$ and its imaginary part $v(x,y)$; that is,

$$f(z) = u(x,y) + jv(x,y) \tag{5.12}$$

The derivative of a function of a complex variable is

$$f'(z) = \frac{df(z)}{dz} = \lim_{\Delta z \to 0} \frac{\Delta f(z)}{\Delta z} \tag{5.13}$$

where $\Delta z = \Delta x + j\,\Delta y$.

To have a unique derivative, it is necessary that the value of $f'(z)$ be independent of the manner in which Δz approaches zero. Because $f(z)$ is a function of x and y, then

$$\Delta f(z) = \frac{\partial f(z)}{\partial x}\,\Delta x + \frac{\partial f(z)}{\partial y}\,\Delta y$$

Substitution of the preceding expression for $\Delta f(z)$ into Eq. (5.13) gives

$$f'(z) = \lim_{\substack{\Delta x \to 0 \\ \Delta y \to 0}} \frac{\dfrac{\partial f(z)}{\partial x}\,\Delta x + \dfrac{\partial f(z)}{\partial y}\,\Delta y}{\Delta x + j\,\Delta y} = \frac{\dfrac{\partial f(z)}{\partial x} + \dfrac{\partial f(z)}{\partial y}\,m}{1 + jm} \tag{5.14}$$

where

$$m = \lim_{\substack{\Delta x \to 0 \\ \Delta y \to 0}} \frac{\Delta y}{\Delta x}$$

Because Δx and Δy may approach zero independently, then

$$\Delta z = \Delta x + j\,\Delta y$$

may approach zero along any path, as is indicated in Fig. 5.5. In order that the derivative $f'(z)$ be independent of the manner in which Δz approaches zero (i.e., independent of the path along which Δz approaches zero), then Eq. (5.14) must be independent of the value of m. Thus, there can be no change of $f'(z)$ with respect to m [that is, $\partial f'(z)/\partial m = 0$]. Differentiating Eq. (5.14) with respect to m and setting the result equal to zero gives

$$\frac{\partial f'(z)}{\partial m} = \frac{(1 + jm)[\partial f(z)/\partial y] - \{\partial f(z)/\partial x + [\partial f(z)/\partial y]m\}j}{(1 + jm)^2}$$

$$= \frac{\partial f(z)/\partial y - j[\partial f(z)/\partial x]}{(1 + jm)^2} = 0$$

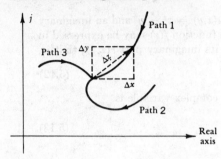

Fig. 5.5. Different paths along which Δz approaches zero.

The preceding equation is satisfied when

$$j \frac{\partial f(z)}{\partial x} = \frac{\partial f(z)}{\partial y} \tag{5.15}$$

The partial derivative of Eq. (5.12) with respect to x is

$$\frac{\partial f(z)}{\partial x} = \frac{\partial u}{\partial x} + j \frac{\partial v}{\partial x}$$

Similarly the partial derivative of Eq. (5.12) with respect to y is

$$\frac{\partial f(z)}{\partial y} = \frac{\partial u}{\partial y} + j \frac{\partial v}{\partial y}$$

Substitution of the preceding expressions for $\partial f/\partial x$ and $\partial f/\partial y$ into Eq. (5.15) gives

$$j \frac{\partial u}{\partial x} - \frac{\partial v}{\partial x} = \frac{\partial u}{\partial y} + j \frac{\partial v}{\partial y}$$

The real and imaginary parts of the preceding equation must be equal; hence

$$\frac{\partial u}{\partial x} = \frac{\partial v}{\partial y} \tag{5.16}$$

$$\frac{\partial v}{\partial x} = - \frac{\partial u}{\partial y} \tag{5.17}$$

Equations (5.16) and (5.17) are known as the *Cauchy-Riemann equations*. In order that the derivative $f'(z)$ have a unique value which is independent of the manner in which Δz approaches zero, it is necessary that the real and imaginary parts of $f(z)$ satisfy the Cauchy-Riemann equations.

For the function $f(z) = z^2$,

$$u = x^2 - y^2 \qquad v = 2xy$$
$$\frac{\partial u}{\partial x} = 2x = \frac{\partial v}{\partial y} \qquad \frac{\partial v}{\partial x} = 2y = -\frac{\partial u}{\partial y}$$

Hence, $f(z) = z^2$ satisfies the Cauchy-Riemann equations. Because these equations are satisfied for all values of x and y, the derivative exists for all values of z. The value of the derivative is obtained by the usual methods of differentiation.

$$f'(z) = \lim_{\Delta z \to 0} \frac{\Delta f(z)}{\Delta z} = \frac{d}{dz} z^2 = 2z$$

The rules for differentiating functions of a real variable apply equally well for obtaining the derivative of a function of a complex variable. Some well-known derivatives are

$$\frac{d(z^n)}{dz} = nz^{n-1} \qquad \frac{d(\sin z)}{dz} = \cos z$$
$$\frac{d(e^{az})}{dz} = ae^{az} \qquad \frac{d(\cos z)}{dz} = -\sin z \qquad (5.18)$$
$$\frac{d(\ln z)}{dz} = \frac{1}{z} \qquad \frac{d(\sinh z)}{dz} = \cosh z$$

Taking the partial derivative of Eq. (5.16) with respect to x and taking the partial derivative of Eq. (5.17) with respect to y, and then adding, gives

$$\frac{\partial^2 u}{\partial x^2} + \frac{\partial^2 u}{\partial y^2} = 0$$

In a similar manner, taking the partial derivative of Eq. (5.16) with respect to y and the partial derivative of Eq. (5.17) with respect to x, and then subtracting, shows that

$$\frac{\partial^2 v}{\partial x^2} + \frac{\partial^2 v}{\partial y^2} = 0$$

The preceding partial differential equations are known as *Laplace's differential equation*. Both the real part u and the imaginary part v of an analytic function satisfy Laplace's differential equation. This fact forms the basis for the solution of Laplace's equation by potential theory, as is explained in Sec. 9.3.

In working with functions that are expressed in polar coordinates, it is more convenient to use the polar form of the Cauchy-Riemann equations. The polar form of $f(z)$ is

$$f(z) = f(re^{j\theta}) = f(r,\theta) = u(r,\theta) + jv(r,\theta) \qquad (5.19)$$

where $u(r,\theta)$ is the real part and $v(r,\theta)$ is the imaginary part. In polar form Δz and $\Delta f(z)$ are

$$\Delta z = e^{j\theta} \Delta r + jre^{j\theta} \Delta\theta \tag{5.20}$$

$$\Delta f(z) = \frac{\partial f}{\partial r} \Delta r + \frac{\partial f}{\partial \theta} \Delta\theta$$

$$= \left(\frac{\partial u}{\partial r} + j\frac{\partial v}{\partial r}\right) \Delta r + \left(\frac{\partial u}{\partial \theta} + j\frac{\partial v}{\partial \theta}\right) \Delta\theta \tag{5.21}$$

Substitution of these results into Eq. (5.13) gives

$$f'(z) = \lim_{\Delta z \to 0} \frac{\Delta f(z)}{\Delta z} \approx \frac{\left(\dfrac{\partial u}{\partial r} + j\dfrac{\partial v}{\partial r}\right) \Delta r + \left(\dfrac{\partial u}{\partial \theta} + j\dfrac{\partial v}{\partial \theta}\right) \Delta\theta}{e^{j\theta} \Delta r + jre^{j\theta} \Delta\theta} \tag{5.22}$$

For the derivative to exist, it must have the same value for a path along which Δr is zero as it does for a path along which $\Delta\theta$ is zero. For the case in which Δr is zero, Eq. (5.22) becomes

$$f'(z) = \frac{\partial u/\partial\theta + j(\partial v/\partial\theta)}{jre^{j\theta}} = \frac{(1/r)(\partial v/\partial\theta) - j(1/r)(\partial u/\partial\theta)}{e^{j\theta}}$$

For a path along which $\Delta\theta$ is zero,

$$f'(z) = \frac{\partial u/\partial r + j(\partial v/\partial r)}{e^{j\theta}}$$

In order that the two preceding equations be equal, it is necessary that the real parts be equal to one another and also that the imaginary parts be equal to one another. Equating the real parts gives

$$\frac{\partial u}{\partial r} = \frac{1}{r}\frac{\partial v}{\partial\theta} \tag{5.23}$$

Similarly, equating the imaginary parts gives

$$\frac{\partial v}{\partial r} = -\frac{1}{r}\frac{\partial u}{\partial\theta} \tag{5.24}$$

Equations (5.23) and (5.24) are the polar form of the Cauchy-Riemann equations. It is interesting to consider the application of these equations to the logarithmic function:

$$\ln z = \ln re^{j\theta} = \ln r + \ln e^{j\theta} = \ln r + j\theta$$

Because $u = \ln r$ and $v = \theta$, it follows that

$$\frac{\partial u}{\partial r} = \frac{1}{r} \qquad \frac{1}{r}\frac{\partial v}{\partial \theta} = \frac{1}{r}$$

$$\frac{\partial v}{\partial r} = 0 \qquad -\frac{1}{r}\frac{\partial u}{\partial \theta} = 0$$

Hence, the Cauchy-Riemann equations are satisfied.

Analytic functions

As is illustrated in Fig. 5.6, the neighborhood of a point z_0 consists of all the points located within a very small circle having z_0 as its center (the neighborhood includes the center z_0 but not the boundary). If δ is the radius of this small circle, then for any point z in the neighborhood of z_0

$$|z - z_0| < \delta \tag{5.25}$$

The derivative of a function exists at a point z if the Cauchy-Riemann equations are satisfied at the point z. A function $f(z)$ is said to be analytic at a point z_0 if the derivative exists at the point z_0 and at every point z in the neighborhood of z_0. That is, a function is analytic at the point z_0 if the Cauchy-Riemann equations are satisfied at every point in the neighborhood of z_0.

A function $f(z)$ is said to be analytic in some region R if the function is analytic at every point in R. As is illustrated by the shaded areas of Fig. 5.7a and b, a region R consists of all the points located within certain prescribed boundaries. The region of Fig. 5.7a is said to be *simply connected*, and that of Fig. 5.7b is said to be *multiply connected*. A simply connected region has the property

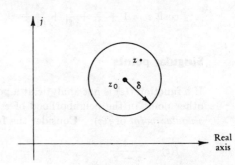

Fig. 5.6. The neighborhood of a point z_0.

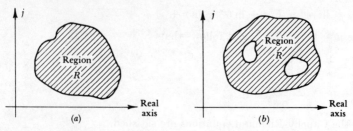

Fig. 5.7.   Regions. (a) A simply connected region; (b) a multiply connected region.

that any simple closed curve drawn within R encloses points belonging to R only.   In a multiply connected region, it is possible to draw at least one closed curve in R that encloses points not belonging to R.

The function $f(z) = z^2$ is analytic for all finite values of z and is thus referred to as an *entire function* (i.e., analytic in the entire z plane).   The sum or product of two analytic functions is also an analytic function.   A function which can be represented by an infinite series such as

$$f(z) = a_0 + a_1 z + a_2 z^2 + a_3 z^3 + \cdots \tag{5.26}$$

is also an entire function if the series converges for all finite values of z.   Thus, the following are entire functions:

$$e^z = 1 + z + \frac{z^2}{2!} + \frac{z^3}{3!} + \cdots$$

$$\sin z = z - \frac{z^3}{3!} + \frac{z^5}{5!} - \frac{z^7}{7!} + \cdots$$

$$\cos z = 1 - \frac{z^2}{2!} + \frac{z^4}{4!} - \frac{z^6}{6!} + \cdots \tag{5.27}$$

$$\sinh z = z + \frac{z^3}{3!} + \frac{z^5}{5!} + \frac{z^7}{7!} + \cdots$$

$$\cosh z = 1 + \frac{z^2}{2!} + \frac{z^4}{4!} + \frac{z^6}{6!} + \cdots$$

Singular points

If a function $f(z)$ is not analytic at a point z_0 but is analytic at every other point in the neighborhood of z_0, then the point z_0 is called a *singular point* of $f(z)$.   Consider the function

$$f(z) = \frac{1}{z - z_0} \tag{5.28}$$

The derivative does not exist at the point $z = z_0$; thus this function has a singular point at z_0. Next consider the function

$$f(z) = \frac{z + 1}{(z + 2)(z - 4)(z + 5)^3} \tag{5.29}$$

The singular points at -2, 4, and -5 are called *poles*. The poles at -2 and 4 are first-order poles. Because of the repeated pole at $z = -5$, this pole is called a *pole of order* 3 or a *third-order pole*. Each pole of a function is also a singular point of the function. The value of z which makes $f(z)$ vanish [that is, $f(z) = 0$] is called a *zero* of $f(z)$. Thus, in Eq. (5.29), the point $z = -1$ is the zero of $f(z)$.

5.3 INTEGRATION IN THE COMPLEX PLANE

In Fig. 5.8 is shown the continuous path C which joins the points z_0 and z_n. The path C is subdivided into n subintervals Δz_k, where $k = 1, 2, \ldots, n$. In the subinterval Δz_k, the point ζ_k lies on the path between z_{k-1} and z_k. Consider the summation

$$\sum_{k=1}^{n} f(\zeta_k)\, \Delta z_k \tag{5.30}$$

The limit of this sum as Δz_k becomes very small such that the path of integration approaches the curve C is called the *line integral* of $f(z)$ along C; that is,

$$\int_C f(z)\, dz = \lim_{\substack{\Delta z_k \to 0 \\ n \to \infty}} \sum_{k=1}^{n} f(\zeta_k)\, \Delta z_k \tag{5.31}$$

Illustrative example Evaluate the line integral for the case in which $f(z) = 1$ and C is any curve connecting z_0 and z_n.

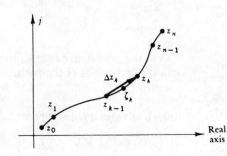

Fig. 5.8. A continuous path subdivided into intervals.

183

Solution Direct application of Eq. (5.31), in which $f(z) = 1$, and thus $f(\zeta_k) = 1$, gives

$$\int_C dz = \lim_{\substack{\Delta z_k \to 0 \\ n \to \infty}} \sum_{k=1}^{n} \Delta z_k = \sum_{k=1}^{n} (z_k - z_{k-1})$$

$$= [(z_1 - z_0) + (z_2 - z_1) + \cdots$$
$$+ (z_{n-1} - z_{n-2}) + (z_n - z_{n-1})]$$

$$= z_n - z_0$$

For the case in which $f(z) = K$ a constant, it follows directly that

$$K \int_C dz = K(z_n - z_0) \tag{5.32}$$

Thus, the integral of a constant along a path does not depend upon the shape of the path C; rather, its value is determined by the end points z_0 and z_n. The integral of a constant around a closed path (that is, $z_n = z_0$) is zero.

Illustrative example Evaluate the line integral for the case in which $f(z) = z$ and C is any curve.

Solution The mean, or average, value of ζ_k in Fig. 5.8 is

$$\zeta_k = \frac{z_k + z_{k-1}}{2}$$

Because $f(z) = z$, then $f(\zeta_k) = \zeta_k$. Thus

$$\int_C z\, dz = \sum_{k=1}^{n} \zeta_k \, \Delta z_k = \sum_{k=1}^{n} \frac{(z_k + z_{k-1})(z_k - z_{k-1})}{2}$$

$$= \frac{z_1^2 - z_0^2}{2} + \frac{z_2^2 - z_1^2}{2} + \cdots + \frac{z_{n-1}^2 - z_{n-2}^2}{2}$$

$$+ \frac{z_n^2 - z_{n-1}^2}{2}$$

$$= \frac{z_n^2 - z_0^2}{2} \tag{5.33}$$

As in the preceding example, the value of this integral depends on the end points of the path. For a closed path, the integral is zero.

Illustrative example Show that

$$\left| \int_C f(z)\, dz \right| \leq ML$$

where L is the length of the path of integration and M is a constant such that $|f(z)| \le M$ at any place on C.

Solution

$$\left| \int_C f(z)\, dz \right| \le \int_C |f(z)|\, |dz| \le M \int_C |dz| \le ML \tag{5.34}$$

Because $|dz|$ is the length of an infinitesimal chord, the last integral approaches the length L of the path.

When the path of integration C is known, the line integral may be computed directly by integration.

Illustrative example   Evaluate the following line integral,

$$\int_C \frac{dz}{z - z_0}$$

where C is a circle of radius ρ about the point z_0. The circle is to be traversed in a counterclockwise direction as θ goes from 0 to 2π in Fig. 5.9.

Solution   For this case $z - z_0 = \rho e^{j\theta}$. Because z_0 and ρ are constant, then $dz = j\rho e^{j\theta}\, d\theta$. Because $dz/(z - z_0) = j\, d\theta$, then

$$\int_C \frac{dz}{z - z_0} = j \int_0^{2\pi} d\theta = 2\pi j \tag{5.35}$$

Illustrative example   Evaluate the following line integral:

$$\int_C (z - z_0)^m\, dz \qquad m \ne -1$$

The path of integration is to be the same as in the preceding example.

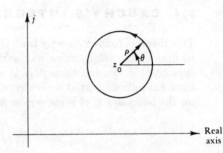

Fig. 5.9.   Circular path of integration.

Solution For this case $f(z) = (z - z_0)^m$, $z - z_0 = \rho e^{j\theta}$, and $dz = j\rho e^{j\theta}\, d\theta$. Thus

$$\int_C (z - z_0)^m\, dz = \int_0^{2\pi} (\rho e^{j\theta})^m (j\rho e^{j\theta})\, d\theta$$

$$= j\rho^{m+1} \int_0^{2\pi} e^{j(m+1)\theta}\, d\theta = 0 \tag{5.36}$$

Basic properties of line integrals

The following basic properties of line integrals may be ascertained directly from Eq. (5.31), which expresses the line integral as a summation.

Reversing the direction of integration changes the sign of the integral.

$$\int_{z_0}^{z_n} f(z)\, dz = - \int_{z_n}^{z_0} f(z)\, dz \tag{5.37}$$

If P is a point on the curve between z_0 and z_n, the integral may be represented as the sum from z_0 to P and from P to z_n. That is,

$$\int_{z_0}^{z_n} f(z)\, dz = \int_{z_0}^{P} f(z)\, dz + \int_{P}^{z_n} f(z)\, dz \tag{5.38}$$

Constants may be taken out from under the integral sign,

$$\int_C K f(z)\, dz = K \int_C f(z)\, dz \tag{5.39}$$

The integral of the sum of two functions $f_1(z)$ and $f_2(z)$ is equal to the sum of the integral of each function.

$$\int_{z_0}^{z_n} [f_1(z) + f_2(z)]\, dz = \int_{z_0}^{z_n} f_1(z)\, dz + \int_{z_0}^{z_n} f_2(z)\, dz \tag{5.40}$$

These properties are the same as those for ordinary integrals.

5.4 CAUCHY'S INTEGRAL THEOREM

This theorem follows directly from Green's lemma, which is a fundamental lemma relating line and surface integrals. Green's lemma says that, if two functions $M(x,y)$ and $N(x,y)$ are continuous and have continuous partial derivatives $\partial M/\partial y$ and $\partial N/\partial x$ within and on the boundary C of some region R, then

$$\int_C M(x,y)\, dx + N(x,y)\, dy = \iint_R \left(\frac{\partial N}{\partial x} - \frac{\partial M}{\partial y} \right) dx\, dy \tag{5.41}$$

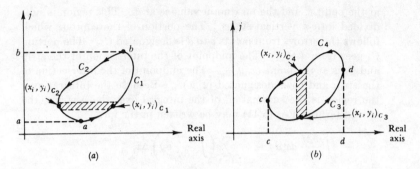

Fig. 5.10. A region R subdivided by: (a) horizontal strips; (b) vertical strips.

where C is the boundary of the region R. In Fig. 5.10a is shown a region R. The point a is the minimum value of y, and b is the maximum. The boundary C_1 follows the arrows from point a to b as indicated. The region R may be subdivided into n horizontal strips, as shown in Fig. 5.10a. The midpoint of the intersection of this strip and C_1 has the coordinates $(x_i, y_i)_{C_1}$. The midpoint of the intersection of the strip and C_2 is designated $(x_i, y_i)_{C_2}$. The integral over the entire region is the summation of the integrals over the n strips. Thus, the first term on the right-hand side of Eq. (5.41) may be written in the form

$$\iint_R \frac{\partial N}{\partial x} \, dx \, dy = \sum_{i=1}^{n} \left(\int \frac{\partial N}{\partial x} \, dx \right) \Delta y_i = \sum_{i=1}^{n} \left[N(x,y) \right]_{(x_i,y_i)_{C_2}}^{(x_i,y_i)_{C_1}} \Delta y_i$$

$$= \sum_{i=1}^{n} N(x_i, y_i)_{C_1} \, \Delta y_i - \sum_{i=1}^{n} N(x_i, y_i)_{C_2} \, \Delta y_i \qquad (5.42)$$

In the first summation the function N takes on values of x and y along the contour C_1. This is the line integral of $N \, dy$ along C_1. As i goes from 1 to n in the last summation, C_2 is traversed in a direction opposite to the arrows. Thus, the second summation is the negative of the line integral along C_2. However, the minus sign in front of this last summation yields a plus sign for the integration along C_2; that is,

$$\iint_R \frac{\partial N}{\partial x} \, dx \, dy = \int_{C_1} N(x,y) \, dy + \int_{C_2} N(x,y) \, dy = \int_C N(x,y) \, dy$$

$$(5.43)$$

In Fig. 5.10b is shown the same region R bounded by the same contour C that is shown in Fig. 5.10a. The minimum value of x is at the point c, and the maximum value is at d. This region is sub-divided into n vertical strips. The portion of the contour which follows the arrows from points c to d is designated C_3. The portion C_4 goes from d to c. The midpoint of the intersection of the strip and C_3 is at the point $(x_i,y_i)_{c_3}$. The midpoint of the intersection of the strip and C_4 is designated $(x_i,y_i)_{c_4}$. Because the integral over the region is the summation of the integral over the n strips, the second term in Eq. (5.41) may be written in the form

$$
\begin{aligned}
- \iint_R \frac{\partial M}{\partial y}\, dx\, dy &= - \sum_{i=1}^{n} \left(\int \frac{\partial M}{\partial y}\, dy \right) \Delta x_i \\
&= - \sum_{i=1}^{n} \left[M(x,y) \right]_{(x_i,y_i)_{c_3}}^{(x_i,y_i)_{c_4}} \Delta x_i \\
&= \sum_{i=1}^{n} M(x_i,y_i)_{c_3} \Delta x_i - \sum_{i=1}^{n} M(x_i,y_i)_{c_4} \Delta x_i \\
&= \int_{C_3} M(x,y)\, dx + \int_{C_4} M(x,y)\, dx \\
&= \int_C M(x,y)\, dx
\end{aligned}
\tag{5.44}
$$

Adding Eqs. (5.43) and (5.44) verifies Green's lemma as given by Eq. (5.41).

In Fig. 5.11a is shown a region which is intersected in two different places by the horizontal strip. This region is subdivided into two parts by the dotted line from e to f. Green's lemma can now be applied to each subregion. When the results are added, the integration over the partition (e to f) cancels out because it has been traversed in opposite directions, as indicated by the arrows in Fig. 5.11a. A similar analysis shows that Green's lemma is also applicable to a region which has one or more islands, as indicated in Fig.

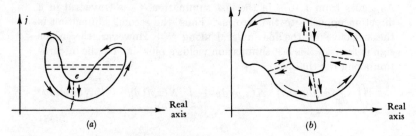

Real axis

Real axis

(a)   (b)

Fig. 5.11. Extension of Green's lemma to more general regions.

5.11*b*.　It may be shown that Green's lemma is applicable for any contour C which does not intersect itself as does a figure 8. A contour which does not intersect, or cross, itself is called a *simple closed curve*, or a *Jordan curve*.

Cauchy's integral theorem may be stated as follows: If $f(z)$ is analytic within and on the boundary of a region R, then

$$\int_C f(z)\, dz = 0 \tag{5.45}$$

To prove Cauchy's integral theorem, substitute into Eq. (5.45) $f(z) = u(x,y) + jv(x,y)$ and $dz = dx + jdy$; that is,

$$\int_C f(z)\, dz = \int_C [u(x,y) + jv(x,y)](dx + j\, dy)$$

$$= \int_C [u(x,y)\, dx - v(x,y)\, dy]$$

$$+ j \int_C [v(x,y)\, dx + u(x,y)\, dy] \quad (5.46)$$

This expresses the integral of a complex variable as the sum of two real integrals.　The first term on the right is the real part, and the last term is the imaginary part.

By letting $u = M$ and $v = -N$ in Eq. (5.41), Green's lemma becomes

$$\int_C u(x,y)\, dx - v(x,y)\, dy = - \iint_R \left(\frac{\partial u}{\partial y} + \frac{\partial v}{\partial x} \right) dx\, dy \tag{5.47}$$

The Cauchy-Riemann equation given by Eq. (5.17) may be written in the form

$$\frac{\partial u}{\partial y} + \frac{\partial v}{\partial x} = 0 \tag{5.48}$$

Thus, the first integral on the right-hand side of Eq. (5.46) is seen to vanish for regions in which $f(z)$ satisfies the Cauchy-Riemann equations.

In a similar manner, letting $v = M$ and $u = N$ in Green's lemma

$$\int_C v\, dx + u\, dy = \iint_R \left(\frac{\partial u}{\partial x} - \frac{\partial v}{\partial y} \right) dx\, dy \tag{5.49}$$

The Cauchy-Riemann equation given by Eq. (5.16) may be written in the form

$$\frac{\partial u}{\partial x} - \frac{\partial v}{\partial y} = 0 \tag{5.50}$$

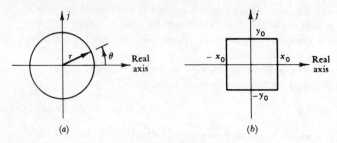

Fig. 5.12. Paths of integration. (a) Circular; (b) rectangular.

The preceding shows that the last integral in Eq. (5.46) also vanishes for regions in which $f(z)$ is analytic. Hence, Cauchy's integral theorem given by Eq. (5.45) has been verified.

An alternative way of stating Cauchy's integral theorem is as follows: If $f(z)$ is analytic in a region R, then

$$\int_C f(z)\, dz = 0 \tag{5.51}$$

where C is any simple closed curve in R.

Illustrative example Evaluate the integral of the function $f(z) = z^2$ around the circular path $z = re^{j\theta}$ shown in Fig. 5.12a.

Solution The derivative is $dz = jre^{j\theta}\, d\theta$, and the value of the function along the path is $z^2 = r^2 e^{2j\theta}$. Thus

$$\int_C z^2\, dz = jr^3 \int_0^{2\pi} e^{3j\theta}\, d\theta = \frac{r^3}{3}\, [e^{3j\theta}]_0^{2\pi}$$

$$= \frac{r^3}{3}\, (1 - 1) = 0$$

Illustrative example Evaluate the integral of the function $f(z) = z^2$ around the rectangular path shown in Fig. 5.12b.

Solution Along the right side $z = x_0 + jy$, $z^2 = x_0^2 - y^2 + j2x_0 y$, and $dz = j\, dy$. Note that the value of x remains constant along the right side. Similarly, determining z, z^2, and dz for the top, left, and bottom sides gives

$$\int_C z^2\, dz = j \int_{-y_0}^{y_0} (x_0^2 - y^2 + j2x_0 y)\, dy + \int_{x_0}^{-x_0} (x^2 - y_0^2$$

$$+ j2xy_0)\, dx + j \int_{y_0}^{-y_0} (x_0^2 - y^2 - j2x_0 y)\, dy$$

$$+ \int_{-x_0}^{x_0} (x^2 - y_0^2 - j2xy_0)\, dx$$

$$= j^2 \int_{-y_0}^{y_0} (4x_0 y)\, dy - j \int_{-x_0}^{x_0} 4xy_0\, dx = 0$$

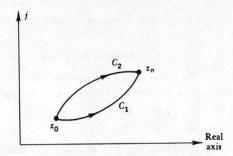

Fig. 5.13. Different paths connecting z_0 and z_n.

Because z^2 is an entire function, by Cauchy's integral theorem it follows that the integral of this function around any closed path must be zero. Thus, application of Cauchy's integral theorem would eliminate the computational effort in the preceding examples.

Suppose that $f(z)$ is analytic within and on the boundary of the curves defined by C_1 and C_2 in Fig. 5.13. Application of Cauchy's integral theorem gives

$$\int_{C_1} f(z)\, dz - \int_{C_2} f(z)\, dz = 0$$

or

$$\int_{C_1} f(z)\, dz = \int_{C_2} f(z)\, dz \tag{5.52}$$

Thus, for any region R in which $f(z)$ is analytic, the value of the integral $\int f(z)\, dz$ is independent of the path in R. The integral does depend on the end points z_0 and z_n.

Exact differential equations

Consider the function

$$\phi(x,y) = C \tag{5.53}$$

where C is a constant and ϕ is some function of the two variables x and y. The derivative of ϕ is

$$d\phi = \frac{\partial \phi}{\partial x} dx + \frac{\partial \phi}{\partial y} dy = 0 \tag{5.54}$$

Equation (5.54) is an exact differential equation whose solution is $\phi(x,y) = C$. An exact differential equation is characterized by the fact that the partial derivative with respect to y of the dx coefficient is equal to the partial derivative with respect to x of the dy coefficient.

$$\frac{\partial}{\partial y}\left(\frac{\partial \phi}{\partial x}\right) = \frac{\partial}{\partial x}\left(\frac{\partial \phi}{\partial y}\right) = \frac{\partial^2 \phi}{\partial x \partial y} \tag{5.55}$$

By letting $M = \partial\phi/\partial x$ and $N = \partial\phi/\partial y$, then Eq. (5.54) takes the form

$$d\phi = M \, dx + N \, dy = 0 \qquad (5.56)$$

In order that an equation which has the form of Eq. (5.56) be an exact differential equation, it is necessary that Eq. (5.55) be satisfied.

$$\frac{\partial M}{\partial y} = \frac{\partial N}{\partial x} \qquad (5.57)$$

The solution ϕ of an exact differential equation is obtained by integration of *either* of the following relationships:

$$M = \frac{\partial\phi}{\partial x} \qquad N = \frac{\partial\phi}{\partial y} \qquad (5.58)$$

Illustrative example   Let it be desired to solve the following differential equation.

$$y(2x - y + 3) \, dx + (x^2 - 2xy + 3x + 2) \, dy = 0$$

This is proved to be an exact differential equation, because Eq. (5.57) is satisfied.

$$\frac{\partial M}{\partial y} = 2x - 2y + 3 = \frac{\partial N}{\partial x}$$

Solution   Integration of the first of Eqs. (5.58) gives

$$\phi(x,y) = \int M \, dx = \int y(2x - y + 3) \, dx = x^2 y - xy^2 + 3xy + g(y)$$

In the preceding integration with respect to x, the resulting constant of integration is some function of y [that is, $g(y)$].   To determine $g(y)$, differentiate the preceding expression with respect to y.   Thus

$$\frac{\partial\phi(x,y)}{\partial y} = x^2 - 2xy + 3x + g'(y)$$

Because

$$N = \frac{\partial\phi(x,y)}{\partial y} = x^2 - 2xy + 3x + 2$$

then equating these two expressions for $\partial\phi(x,y)/\partial y$ gives

$$g'(y) = 2$$

Integration yields the desired function $g(y)$.

$$g(y) = 2y + C_1$$

There is no need to evaluate C_1, because the solution is expressed in terms of the arbitrary constant C; that is,

$$\phi(x,y) = x^2y - xy^2 + 3xy + 2y = C$$

The alternative method of obtaining this result is to integrate the second of Eqs. (5.58); that is,

$$\phi(x,y) = \int N \, dy = \int (x^2 - 2xy + 3x + 2) \, dy$$
$$= x^2y - xy^2 + 3xy + 2y + h(x)$$

The derivative with respect to x is

$$\frac{\partial \phi(x,y)}{\partial x} = 2xy - y^2 + 3y + h'(x)$$

Because

$$M = \frac{\partial \phi(x,y)}{\partial x} = 2xy - y^2 + 3y$$

then equating these two expressions for $\partial \phi(x,y)/\partial x$ gives $h'(x) = 0$ and $h(x) = C_2$. This then yields the same result.

Exact differential equations occur frequently in engineering problems. For example, Eq. (5.46) shows that the integral of a function of a complex variable may be expressed as the sum of two real integrals, and, when $f(z)$ is analytic, then each of the integrands is exact. Green's lemma [Eq. (5.41)] shows that the integral of an exact differential around a closed path C is zero if $\partial M/\partial y = \partial N/\partial x$ everywhere within and on C. Hence, Eq. (5.46) vanishes if $f(z)$ is analytic within and on C (that is, Cauchy's integral theorem).

If a closed path C is subdivided into two parts C_1 and C_2 such that

$$\int_C M \, dx + N \, dy = \int_{C_1} M \, dx + N \, dy - \int_{C_2} M \, dx + N \, dy = 0$$

then

$$\int_{C_1} M \, dx + N \, dy = \int_{C_2} M \, dx + N \, dy \tag{5.59}$$

Thus, the integral from one point to another is independent of the particular path if $\partial M/\partial y = \partial N/\partial x$ everywhere within and on the closed path defined by C_1 and C_2. The value of this line integral is determined entirely by the location of the starting point and the end point.

5.5 CAUCHY'S INTEGRAL FORMULA

Cauchy's integral formula and Cauchy's integral theorem are two of the most fundamental results in the theory of complex variables. Cauchy's integral formula states that, if $f(z)$ is analytic in a region R and if z_0 is any point in R, then

$$\int_C \frac{f(z)}{z - z_0} \, dz = 2\pi j f(z_0) \tag{5.60}$$

where C is any simple closed curve in R that encloses z_0. The integration is to be taken in the positive sense around C.

This theorem is proved in the following manner: Suppose that $f(z)$ is analytic within and on the boundary C of Fig. 5.14. Thus, the function $f(z)/(z - z_0)$ is analytic within and on C except at the point $z = z_0$. A small circle of radius δ is drawn about the point z_0. The function $f(z)/(z - z_0)$ is analytic in the region between C and C_0 and also on the boundaries C and C_0. By Cauchy's integral theorem, it follows that

$$\int_C \frac{f(z)}{z - z_0} \, dz + \int_a^b \frac{f(z)}{z - z_0} \, dz + \int_{C_0} \frac{f(z)}{z - z_0} \, dz + \int_b^a \frac{f(z)}{z - z_0} \, dz = 0 \tag{5.61}$$

The integral along the slit from a to b is the negative of that from b to a; hence these two terms cancel out in Eq. (5.61).

By writing $f(z)$ in the form $f(z) = f(z_0) + [f(z) - f(z_0)]$, then the integral around the C_0 boundary may be expressed as

$$\int_{C_0} \frac{f(z)}{z - z_0} \, dz = \int_{C_0} \frac{f(z_0)}{z - z_0} \, dz + \int_{C_0} \frac{f(z) - f(z_0)}{z - z_0} \, dz \tag{5.62}$$

For values of z on C_0, then $z - z_0 = \delta e^{j\theta}$, $dz = j\delta e^{j\theta} \, d\theta$, and $dz/(z - z_0) = j \, d\theta$. Thus, the first term on the right-hand side of

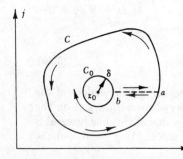

Fig. 5.14. A contour C which encloses one singular point, z_0.

Eq. (5.62) may be written in the form

$$\int_{C_0} \frac{f(z_0)}{z - z_0} \, dz = f(z_0) \int_0^{-2\pi} j \, d\theta = -2\pi j f(z_0) \tag{5.63}$$

Because $f(z)$ is analytic, then for any ϵ no matter how small ($\epsilon > 0$) it is possible to determine a δ such that $|f(z) - f(z_0)| < \epsilon$. Thus, the last term of Eq. (5.62) is shown to vanish as follows:

$$\left| \int_{C_0} \frac{f(z) - f(z_0)}{z - z_0} \, dz \right| \leq \epsilon \left| \int_{C_0} \frac{dz}{z - z_0} \right|$$

$$\leq \epsilon \left| \int_0^{-2\pi} j \, d\theta \right| = \epsilon \, |-2\pi j| \approx 0 \tag{5.64}$$

Substitution of the preceding results into Eq. (5.61) gives

$$\int_C \frac{f(z)}{z - z_0} \, dz = - \int_{C_0} \frac{f(z)}{z - z_0} \, dz = 2\pi j f(z_0) \tag{5.65}$$

Hence, Cauchy's integral formula as given by Eq. (5.60) has been verified.

It is to be noted in Fig. 5.14 that, as one traverses the boundaries C and C_0 following the directions of the arrows, the region to be enclosed is seen to the left. A boundary is traversed in the positive sense when the region to be enclosed is to the left as one walks along the boundary. Reversing the direction of traversing the path changes the sign of the integral. Thus, in applying Cauchy's integral formula care must be taken to traverse the path in the positive sense. In applying Cauchy's integral theorem, it makes no difference in which direction the path is traversed because the value of the integral is zero. However, it is customary always to traverse paths in the positive sense.

Cauchy's integral formula reveals the interesting property that, if a function $f(z)$ is analytic everywhere within and on a contour C, then the value of the function $f(z_0)$ at any point z_0 within C is completely determined by the values of $f(z)$ on the contour C.

Illustrative example Use Cauchy's integral formula to evaluate the integral $\int_C (1) \, dz/(z - z_0)$, where C is any simple closed path about z_0.

Solution Because $f(z) = 1$, then $f(z_0) = 1$. Application of Cauchy's integral formula gives

$$\int_C \frac{(1) \, dz}{z - z_0} = 2\pi j f(z_0) = 2\pi j$$

This result substantiates the solution indicated by Eq. (5.35). Because $f(z)$ is an entire function, this result is valid for any simple closed curve C in the z plane which encloses z_0, and not just a circle of radius ρ such as was employed in obtaining Eq. (5.35).

Extension of Cauchy's integral formula

Successive differentiation with respect to z_0 of Cauchy's integral formula [Eq. (5.60)] gives

$$\int_C \frac{f(z)}{(z-z_0)^2}\, dz = 2\pi j f'(z_0) \tag{5.66}$$

$$\int_C \frac{f(z)}{(z-z_0)^3}\, dz = \frac{2\pi j}{2!} f''(z_0) \tag{5.67}$$

$$\int_C \frac{f(z)}{(z-z_0)^k}\, dz = \frac{2\pi j}{(k-1)!} f^{(k-1)}(z_0) \tag{5.68}$$

Equations (5.65) to (5.68) may all be written in the same form,

$$\int_C \frac{f(z)}{(z-z_0)^k}\, dz = 2\pi j R \qquad k = 1, 2, 3, \ldots \tag{5.69}$$

where

$$R = \frac{1}{(k-1)!} \lim_{z \to z_0} \frac{d^{k-1}}{dz^{k-1}} f(z) \tag{5.70}$$

The function $f(z)$ is analytic within and on the boundary of the region enclosed by C. The term R is called the residue. The residue for a first-order pole is

$$R = f(z_0) \tag{5.71}$$

Illustrative example Determine

$$\int_C \frac{z\, dz}{z-z_0}$$

where C is any simple closed path which encloses z_0.

Solution Because $f(z) = z$, then $R = f(z_0) = z_0$. Thus

$$\int_C \frac{z}{z-z_0}\, dz = 2\pi j R = 2\pi j z_0$$

Illustrative example   Determine

$$\int_C \frac{e^z \, dz}{(z - z_0)^2}$$

where C is any simple closed path which encloses z_0.

Solution   For a second-order pole, $k = 2$ in Eq. (5.70).   The residue for $f(z) = e^z$ and $k = 2$ is

$$R = \frac{1}{(2 - 1)!} \lim_{z \to z_0} \left(\frac{d}{dz} e^z \right) = e^{z_0}$$

Thus

$$\int_C \frac{e^z}{(z - z_0)^2} \, dz = 2\pi j R = 2\pi j e^{z_0}$$

5.6   THE RESIDUE THEOREM

In Fig. 5.15 is shown a contour C which encloses two singular points z_1 and z_2.   Let it be desired to evaluate the following integral around C.

$$\int_C \frac{f(z) \, dz}{(z - z_1)(z - z_2)^3} = \int_C F(z) \, dz \tag{5.72}$$

The function $f(z)$ is analytic within and on C.   The function $F(z)$ has a first-order pole at z_1 and a pole of order 3 at z_2.   The boundary which encloses z_1 is C_1, and that which encloses z_2 is C_2.   In integrating around C_1 and C_2, the inner partition (shown dotted) is traversed in opposite directions.   This cancels the effect of integrating over

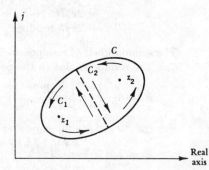

Fig. 5.15.   A contour C which encloses two singular points z_1 and z_2.

Real axis

the inner partition; thus,

$$\int_C F(z)\,dz = \int_{C_1} F(z)\,dz + \int_{C_2} F(z)\,dz \tag{5.73}$$

The only singular point of $F(z)$ within C_1 is that due to the first-order pole at z_1; hence

$$\int_{C_1} F(z)\,dz = 2\pi j R_1 \tag{5.74}$$

where

$$R_1 = \lim_{z \to z_1} [(z - z_1)F(z)]$$

Similarly, the only singularity within C_2 is that due to the pole of order 3 at z_2; hence

$$\int_{C_2} F(z)\,dz = 2\pi j R_2 \tag{5.75}$$

where

$$R_2 = \frac{1}{(3-1)!} \lim_{z \to z_2} \frac{d^{3-1}}{dz^{3-1}} [(z - z_2)^3 F(z)]$$

Substitution of Eqs. (5.74) and (5.75) into Eq. (5.73) shows that, for a contour which encloses two poles of the function $F(z)$, then

$$\int_C F(z)\,dz = 2\pi j (R_1 + R_2)$$

where R_1 and R_2 are the residues due to each of the poles of $F(z)$ within the contour C. In general, for a contour C which encloses n singular points of $F(z)$ as shown in Fig. 5.16

$$\begin{aligned}
\int_C F(z)\,dz &= \int_{C_1} F(z)\,dz + \int_{C_2} F(z)\,dz + \cdots + \int_{C_n} F(z)\,dz \\
&= 2\pi j (R_1 + R_2 + \cdots + R_n) \\
&= 2\pi j \Sigma \text{ residues} \tag{5.76}
\end{aligned}$$

The summation is of residues due to poles of $F(z)$ within the contour C. The residue R_i due to the pole at z_i which is of order k is

$$R_i = \frac{1}{(k-1)!} \lim_{z \to z_i} \frac{d^{k-1}}{dz^{k-1}} [(z - z_i)^k F(z)] \tag{5.77}$$

For the case of a simple, or first-order, pole $(k = 1)$, the residue is

$$R_i = \lim_{z \to z_i} (z - z_i)F(z) \tag{5.78}$$

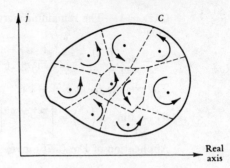

Fig. 5.16. A contour C which encloses many singular points.

An alternative technique for evaluating a first-order residue R_i is obtained by writing $F(z)$ in the form

$$F(z) = \frac{P(z)}{Q(z)} \tag{5.79}$$

where $Q(z)$ is the term in the denominator of $F(z)$ that contains the singular point and $P(z)$ is the remaining portion of $F(z)$. Because $Q(z_i) = 0$, then Eq. (5.78) may be written as follows,

$$R_i = \lim_{z \to z_i} \frac{P(z)}{Q(z)/(z - z_i)} = \lim_{z \to z_i} \frac{P(z)}{[Q(z) - Q(z_i)]/(z - z_i)} = \frac{P(z_i)}{Q'(z_i)} \tag{5.80}$$

where

$$\lim_{z \to z_i} \frac{Q(z) - Q(z_i)}{z - z_i} = \lim_{z \to z_i} \frac{\Delta Q(z)}{\Delta z} = Q'(z)$$

Equation (5.80) is particularly useful when the singularity is not in factored form and hence $z - z_i$ cannot be canceled out.

Illustrative example Evaluate $\int_C e^z \, dz / z(z + 1)^2$, where C is a contour which encloses the singular points at $z = 0$ and -1.

Solution Application of Eq. (5.78) to determine the residue at $z = 0$ gives

$$R_1 = \lim_{z \to 0} z F(z) = \frac{e^z}{(z + 1)^2} \Big|_{z=0} = 1$$

This residue can also be obtained by application of Eq. (5.80). The term which contains the first-order pole is $Q(z) = z$, so that

199

$Q'(z) = 1$. The remaining part of $F(z)$ is $P(z) = e^z/(z + 1)^2$; thus

$$R_1 = \lim_{z \to 0} \frac{P(z)}{Q'(z)} = \frac{e^z}{(z + 1)^2}\Big|_{z=0} = 1$$

The residue R_2 at $z = -1$ is

$$R_2 = \lim_{z \to -1} \frac{d}{dz}\left(\frac{e^z}{z}\right) = \left[\frac{ze^z - e^z}{z^2}\right]_{z=-1} = -2e^{-1}$$

Application of Eq. (5.76) gives

$$\int_C \frac{e^z}{z(z + 1)^2}\, dz = 2\pi j(1 - 2e^{-1})$$

Illustrative example Evaluate $\int_C dz/(1 - e^z)$, where C is a contour which encloses the singular point at $z = 0$.

Solution The residue at $z = 0$ is obtained by application of Eq. (5.80), in which $P(z) = 1$, $Q(z) = 1 - e^z$, and $Q'(z) = -e^z$. Thus,

$$R = \lim_{z \to 0} \frac{P(z)}{Q'(z)} = -\frac{1}{e^z}\Big|_{z=0} = -1$$

Application of Eq. (5.76) gives

$$\int_C \frac{dz}{1 - e^z} = -2\pi j$$

Illustrative example Evaluate $\int_C dz/[(z + \pi/4) \sin z]$, where C is a contour which encloses the singular points at $z = -\pi/4$ and $z = 0$ (for $z = 0$, then $\sin z = 0$).

Solution Application of Eq. (5.78) to evaluate the residue R_1 due to the first-order pole at $z = -\pi/4$ gives

$$R_1 = \lim_{z \to -\pi/4} \frac{1}{\sin z} = -\frac{1}{\sqrt{2}/2}$$

Note that z is in radian units when trigonometric functions are being evaluated.

Because $\sin z$ is not in factored form, Eq. (5.80) must be used to evaluate the residue R_2 at $z = 0$. The term which contains the first-order pole is $Q(z) = \sin z$ so that $Q'(z) = \cos z$.

The remaining portion of $F(z)$ is $P(z) = 1/(z + \pi/4)$; thus

$$R_2 = \lim_{z \to 0} \frac{P(z)}{Q'(z)} = \left[\frac{1}{(z + \pi/4) \cos z} \right]_{z=0} = \frac{4}{\pi}$$

Application of Eq. (5.76) gives

$$\int_C \frac{dz}{(z + \pi/4) \sin z} = 2\pi j \left(\frac{4}{\pi} - \frac{2}{\sqrt{2}} \right)$$

5.7 TAYLOR SERIES

Very interesting and useful properties of analytic functions are obtained from the Taylor series expansion for such functions.

In Fig. 5.17 is shown a circular boundary C with center at z_0. The point $z_0 + \Delta z_0$ represents any point within C. If $f(z)$ is analytic within and on C, application of Cauchy's integral formula gives

$$f(z_0 + \Delta z_0) = \frac{1}{2\pi j} \int_C \frac{f(z)}{z - (z_0 + \Delta z_0)} \, dz \tag{5.81}$$

The denominator of the integrand may be written in the form

$$\frac{1}{z - (z_0 + \Delta z_0)} = \frac{1}{(z - z_0)[1 - \Delta z/(z \quad z_0)]} \tag{5.82}$$

The sum S of a geometric progression in r is

$$S = 1 + r + r^2 + \cdots + r^n \tag{5.83}$$

Multiplication of the preceding expression by r gives

$$rS = r + r^2 + \cdots + r^n + r^{n+1} \tag{5.84}$$

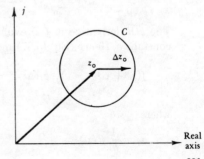

Fig. 5.17.   Circular contour C with center at z_0.

201

Subtracting Eq. (5.84) from Eq. (5.83) gives

$$(1 - r)S = 1 - r^{n+1}$$

For the case in which $|r| < 1$, then $r^{n+1} \approx 0$ as n becomes infinite. Therefore

$$S = \frac{1}{1 - r} = 1 + r + r^2 + \cdots \tag{5.85}$$

In Eq. (5.81), the integration indicates that z takes on values along C. Hence $z - z_0$ is the radius of circle C. Because Δz_0 is less than the radius,

$$\left| \frac{\Delta z_0}{z - z_0} \right| < 1$$

Hence

$$\frac{1}{1 - \Delta z_0 / (z - z_0)} = 1 + \frac{\Delta z_0}{z - z_0} + \frac{\Delta z_0^2}{(z - z_0)^2} + \cdots$$

Equation (5.82) may be written in the form

$$\frac{1}{z - (z_0 + \Delta z_0)} = \frac{1}{z - z_0} \left[1 + \frac{\Delta z_0}{z - z_0} + \frac{\Delta z_0^2}{(z - z_0)^2} + \cdots \right]$$

Substitution of this result into Eq. (5.81) gives

$$f(z_0 + \Delta z_0) = \frac{1}{2\pi j} \int_C \frac{f(z)}{z - z_0} \, dz + \frac{\Delta z_0}{2\pi j} \int_C \frac{f(z)}{(z - z_0)^2} \, dz$$
$$+ \frac{\Delta z_0^2}{2\pi j} \int_C \frac{f(z)}{(z - z_0)^3} \, dz + \cdots \tag{5.86}$$

Substitution of Eqs. (5.66) to (5.68) into the preceding expression gives

$$f(z_0 + \Delta z_0) = f(z_0) + f'(z_0) \, \Delta z_0 + \frac{1}{2!} f''(z_0) \, \Delta z_0^2 + \cdots$$
$$= \sum_{k=0}^{\infty} \frac{1}{k!} f^k(z_0) \, \Delta z_0^k \tag{5.87}$$

The kth derivative of $f(z)$ evaluated at $z = z_0$ [that is, $f^k(z_0)$] is a constant. Hence, Eq. (5.87) may be written in the form

$$f(z_0 + \Delta z_0) = \sum_{k=0}^{n} a_k (\Delta z_0)^k = a_0 + a_1 \, \Delta z_0 + a_2 \, \Delta z_0^2 + \cdots \tag{5.88}$$

where

$$a_k = \frac{1}{k!} f^k(z_0)$$

For the case in which z_0 is zero, the preceding Taylor series reduces to the Maclaurin expansion. Equation (5.87) reveals the interesting characteristic that an analytic function can always be represented by a power series. In addition, it follows that an analytic function has derivatives of all orders.

The circle of convergence is the largest circle about z_0 for which $f(z)$ is analytic. The radius of this circle is called the *radius of convergence*. The Taylor series diverges when $z_0 + \Delta z_0$ lies on or outside the circle of convergence.

The integrals in Eq. (5.86) are the coefficients of the Taylor series expansion

$$a_k = \frac{1}{2\pi j} \int_C \frac{f(z)}{(z - z_0)^{k+1}} \, dz \tag{5.89}$$

Ascertaining the coefficients of the Taylor expansion by independent means yields a unique technique for evaluating these contour integrals. This feature is illustrated in the following example.

Illustrative example· Determine the Taylor series expansion about the point $z_0 = 3$ for the function

$$f(z) = \frac{z - 4}{z(z - 1)}$$

Solution The partial fraction expansion for $f(z)$ is

$$f(z) = \frac{4}{z} - \frac{3}{z - 1}$$

Each term of a partial fraction expansion may be expressed as a geometric series. To do this, first write the denominator of each partial fraction in terms of Δz_0. For an expansion about $z_0 = 3$, then $\Delta z_0 = z - z_0 = z - 3$. Thus, the preceding equation is written in the following form:

$$f(z_0 + \Delta z_0) = \frac{4}{3 + (z - 3)} - \frac{3}{2 + (z - 3)}$$

Each partial fraction may now be rearranged in the form given by Eq. (5.85). That is,

$$f(z_0 + \Delta z_0) = \frac{4}{3} \frac{1}{1 + (z - 3)/3} - \frac{3}{2} \frac{1}{1 + (z - 3)/2}$$

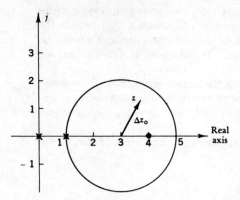

Fig. 5.18. Taylor series expansion about the point $z_0 = 3$.

Expansion into a geometric series gives

$$f(z_0 + \Delta z_0) = \frac{4}{3}\left[1 - \frac{z-3}{3} + \left(\frac{z-3}{3}\right)^2 - \cdots\right]$$
$$-\frac{3}{2}\left[1 - \frac{z-3}{2} + \left(\frac{z-3}{2}\right)^2 - \cdots\right]$$
$$= -\tfrac{1}{6} + \tfrac{11}{36}(z-3) - \tfrac{49}{216}(z-3)^2 + \cdots \quad (5.90)$$

The first geometric series converges for $|(z-3)/3| < 1$ or $|\Delta z_0| = |z - 3| < 3$. The second series converges for $|(z-3)/2| < 1$ or $|\Delta z_0| = |z - 3| < 2$. The smaller value $|\Delta z_0| < 2$ is the limiting value. The circle of convergence is illustrated in Fig. 5.18. The poles at $z = 0$ and $z = 1$ are indicated by ×'s, and the zero at $z = 4$ is indicated by a small circle. The circle of convergence with center at $z_0 = 3$ must have a radius less than 2, for otherwise the singular point at $z = 1$ is enclosed (it makes no difference whether or not the zero at $z = 4$ is enclosed). Thus, the Taylor series expansion given by Eq. (5.90) is valid for values of z within the circle shown in Fig. 5.18 (that is, $z = 3 + \Delta z_0$, where $|\Delta z_0| < 2$).

Replacing the coefficients in the Taylor series of Eq. (5.90) by their integral representation gives

$$\frac{1}{2\pi j}\int_C \frac{f(z)}{z-3}\,dz = a_0 = -\frac{1}{6}$$
$$\frac{1}{2\pi j}\int_C \frac{f(z)}{(z-3)^2}\,dz = a_1 = \frac{11}{36}$$
$$\frac{1}{2\pi j}\int_C \frac{f(z)}{(z-3)^3}\,dz = a_2 = -\frac{49}{216}$$

where C is any circle about $z_0 = 3$ that lies within the circle of convergence. Each of the preceding integrals can be verified by the theory of residues. For example, evaluation of the first integral by residues gives

$$\frac{1}{2\pi j} \int_C \frac{f(z)}{z - 3} \, dz = \lim_{z \to 3} \frac{z - 4}{z(z - 1)} = -\frac{1}{6}$$

Thus, determining the coefficients of the Taylor series by independent means provides an alternative method for evaluating residues, and vice versa.

5.8 LAURENT SERIES

The Laurent series is a power-series expansion of a function around a point at which the function is not analytic (i.e., a singular point). It should be recalled that the Taylor series is an expansion about a point at which $f(z)$ is analytic.

The point z_0 in Fig. 5.19a is the center of the two concentric circles C_1 and C_2. The function $F(z)$ is analytic within and on C_1 except for a singular point at z_0. The point $z_0 + \Delta z_0$ lies within the ring bounded by C_1 and C_2. Because $F(z)$ is analytic in the region between C_1 and C_2, application of Cauchy's integral formula along the boundaries indicated by the arrows in Fig. 5.19a gives

$$F(z_0 + \Delta z_0) = \frac{1}{2\pi j} \left[\int_{C_1} \frac{F(z)}{z - (z_0 + \Delta z_0)} \, dz \right. \\ \left. + \int_{C_2} \frac{F(z)}{z - (z_0 + \Delta z_0)} \, dz \right] \quad (5.91)$$

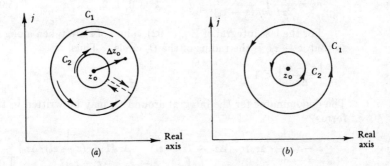

Fig. 5.19. A circle C_1 which encloses a singular point at z_0.

Reversing the sense of integration along C_2 as indicated by the arrows in Fig. 5.19b gives

$$F(z_0 + \Delta z_0) = \frac{1}{2\pi j}\left[\int_{C_1} \frac{F(z)}{z - (z_0 + \Delta z_0)}\,dz \right.$$
$$\left. - \int_{C_2} \frac{F(z)}{z - (z_0 + \Delta z_0)}\,dz\right] \quad (5.92)$$

For the first integral, the values of z are taken along C_1, so that $z - z_0$ is the radius of the C_1 circle. Thus,

$$\left|\frac{\Delta z_0}{z - z_0}\right| < 1$$

The denominator for the integral around C_1 may be written in the form

$$\frac{1}{z - (z_0 + \Delta z_0)} = \frac{1}{(z - z_0)\left(1 - \dfrac{\Delta z_0}{z - z_0}\right)}$$
$$= \frac{1}{z - z_0}\left[1 + \frac{\Delta z_0}{z - z_0} + \frac{\Delta z_0^2}{(z - z_0)^2} + \cdots\right]$$

Thus,

$$\frac{1}{2\pi j}\int_{C_1}\frac{F(z)}{z - (z_0 + \Delta z_0)}\,dz = \frac{1}{2\pi j}\left[\int_{C_1}\frac{F(z)}{z - z_0}\,dz \right.$$
$$\left. + \Delta z_0 \int_{C_1}\frac{F(z)}{(z - z_0)^2}\,dz + \Delta z_0^2 \int_{C_1}\frac{F(z)}{(z - z_0)^3}\,dz + \cdots\right]$$

Hence, the integral around C_1 may be expressed in the form

$$\frac{1}{2\pi j}\int_{C_1}\frac{F(z)}{z - (z_0 + \Delta z_0)}\,dz = \sum_{k=0}^{\infty} a_k \Delta z_0^k \quad (5.93)$$

where

$$a_k = \frac{1}{2\pi j}\int_{C_1}\frac{F(z)}{(z - z_0)^{k+1}}\,dz \qquad k = 0, 1, 2, \ldots \quad (5.94)$$

For the last integral of Eq. (5.92) values of z are taken along C_2 so that $|z - z_0|$ is the radius of the C_2 circle. Thus,

$$\left|\frac{z - z_0}{\Delta z_0}\right| < 1$$

The denominator for the integral around C_2 may be written in the form

$$\frac{1}{z - (z_0 + \Delta z_0)} = \frac{-1}{\Delta z_0 - (z - z_0)} = \frac{-1}{\Delta z_0[1 - (z - z_0)/\Delta z_0]}$$
$$= -\frac{1}{\Delta z_0}\left[1 + \frac{z - z_0}{\Delta z_0} + \frac{(z - z_0)^2}{\Delta z_0^2} + \cdots\right]$$

The preceding series converges for $|(z - z_0)/\Delta z_0| < 1$, which is the case for values of z on C_2. Thus

$$-\frac{1}{2\pi j} \int_{C_2} \frac{F(z)}{z - (z_0 + \Delta z_0)} \, dz = \frac{1}{2\pi j} \left[\frac{1}{\Delta z_0} \int_{C_2} F(z) \, dz \right.$$

$$\left. + \frac{1}{\Delta z_0{}^2} \int_{C_2} \frac{F(z)}{(z - z_0)^{-1}} \, dz + \frac{1}{\Delta z_0{}^3} \int_{C_2} \frac{F(z)}{(z - z_0)^{-2}} \, dz + \cdots \right]$$

$$= \sum_{k=-1}^{-\infty} a_k \, \Delta z_0{}^k \quad (5.95)$$

where

$$a_k = \frac{1}{2\pi j} \int_{C_2} \frac{F(z)}{(z - z_0)^{k+1}} \, dz \qquad k = -1, -2, -3, \ldots \quad (5.96)$$

Substituting the results of Eqs. (5.93) and (5.95) into Eq. (5.92) gives

$$F(z_0 + \Delta z_0) = \sum_{k=0}^{\infty} a_k \, \Delta z_0{}^k + \sum_{k=-1}^{-\infty} a_k \, \Delta z_0{}^k = \sum_{k=-\infty}^{\infty} a_k \Delta z_0{}^k \quad (5.97)$$

In Fig. 5.20 is shown a contour C which is drawn between C_1 and C_2. The function to be integrated in Eq. (5.94), $F(z)/(z - z_0)^{k+1}$, is analytic between C and C_1; thus the integration may be taken around C rather than C_1. Similarly, the integration in Eq. (5.96) may also be taken around C rather than C_2. Hence, Eqs. (5.94) and (5.96) reduce to the common form

$$a_k = \frac{1}{2\pi j} \int_C \frac{F(z)}{(z - z_0)^{k+1}} \, dz \qquad (5.98)$$

As was the case for the Taylor series, evaluation of the coefficients of the Laurent series by independent means yields the value of the corresponding integrals.

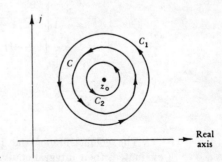

Fig. 5.20. A contour C between C₁ and C₂.

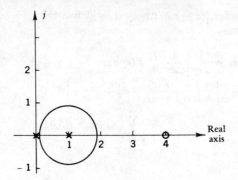

Fig. 5.21. Laurent expansion about $z_0 = 1$.

Illustrative example Effect a Laurent expansion about the point $z_0 = 1$ for the function

$$F(z) = \frac{z - 4}{z(z - 1)}$$

As shown in Fig. 5.21, the function $F(z)$ is analytic in the region $0 < |\Delta z_0| < 1$. The pole at the origin is outside this ring.

Solution Because $\Delta z_0 = z - 1$, the expansion is to be in terms of $z - 1$ factors. The Laurent expansion is obtained by first factoring the $z - 1$ term from $F(z)$ and then expanding the remaining part of $F(z)$ in the same manner as for a Taylor series expansion (i.e., expand each term of the partial fraction expansion in a geometric series). Thus,

$$\begin{aligned}
F(z_0 + \Delta z_0) &= \frac{1}{z - 1} \frac{z - 4}{z} = \frac{1}{z - 1}\left(1 - \frac{4}{z}\right) \\
&= \frac{1}{z - 1}\left[1 - \frac{4}{1 + (z - 1)}\right] \\
&= \frac{1}{z - 1}\{1 - 4[1 - (z - 1) \\
&\qquad\qquad + (z - 1)^2 - (z - 1)^3 + \cdots]\} \\
&= \frac{-3}{z - 1} + 4 - 4(z - 1) + 4(z - 1)^2 - \cdots
\end{aligned}$$

$$(5.99)$$

The coefficients of the Laurent expansion having been evaluated, their integral equivalents are also known.

$$\frac{1}{2\pi j} \int_C F(z)\, dz = a_{-1} = -3$$

$$\frac{1}{2\pi j} \int_C \frac{F(z)}{z-1}\, dz = a_0 = +4$$

$$\frac{1}{2\pi j} \int_C \frac{F(z)}{(z-1)^2}\, dz = a_1 = -4$$

where the contour C encloses the pole at $z_0 = 1$.

Figure 5.21 also shows that, with $z_0 = 1$ as the center, then $F(z)$ is also analytic in the outer ring, $|\Delta z_0| > 1$. To construct the Laurent expansion for this region, in which $|\Delta z_0| = |z - 1| > 1$, write Eq. (5.99) in the form

$$F(z_0 + \Delta z_0) = \frac{1}{z-1}\left[1 - \frac{4}{1 + (z-1)}\right]$$

$$= \frac{1}{z-1}\left[1 - \frac{4}{z-1}\frac{1}{1 + 1/(z-1)}\right]$$

The last term above may be written as a geometric series which converges for $|1/\Delta z_0| = |1/(z-1)| < 1$. Hence

$$F(z_0 + \Delta z_0) = \frac{1}{z-1}\left\{1 - \frac{4}{z-1}\left[1 - \frac{1}{z-1}\right.\right.$$

$$\left.\left. + \frac{1}{(z-1)^2} - \cdots\right]\right\}$$

$$= \frac{1}{z-1} - \frac{4}{(z-1)^2} + \frac{4}{(z-1)^3} - \frac{4}{(z-1)^4} + \cdots$$

The integral equivalents are

$$\frac{1}{2\pi j} \int_C F(z)\, dz = a_{-1} = 1$$

$$\frac{1}{2\pi j} \int_C \frac{F(z)}{(z-1)^{-1}}\, dz = a_{-2} = -4$$

$$\frac{1}{2\pi j} \int_C \frac{F(z)}{(z-1)^{-2}}\, dz = a_{-3} = +4$$

where the contour C now encloses the pole at the origin as well as at $z_0 = 1$.

The integral expression for the a_{-1} coefficient of a Laurent expansion reveals the very interesting property that a_{-1} is the summation of the residues due to poles located inside C. That is, for $k = -1$, Eq. (5.96) becomes

$$a_{-1} = \frac{1}{2\pi j} \int_C F(z) \, dz = \sum \text{residues of } F(z) \qquad (5.100)$$

where the summation is of residues due to poles of $F(z)$ within the contour C.

For the Laurent expansion in which $0 < |\Delta z_0| < 1$, the residue due to the pole at $z = 1$ is

$$a_{-1} = \lim_{z \to 1} (z - 1) \frac{z - 4}{z(z - 1)} = -3$$

For the Laurent expansion in which $1 < |\Delta z_0| < \infty$, the summation of the residues due to the poles at $z = 0$ and $z = 1$ is

$$a_{-1} = \lim_{z \to 1} \frac{z - 4}{z} + \lim_{z \to 0} \frac{z - 4}{z - 1} = -3 + 4 = 1$$

For the case in which the Laurent expansion is such that the contour C encloses but one singular point which is a simple pole, then $F(z)$ may be written in the form

$$F(z) = \frac{f(z)}{z - z_0}$$

where $f(z)$ is analytic within and on C. The a_{-2} coefficient of the Laurent expansion is

$$a_{-2} = \frac{1}{2\pi j} \int_C \frac{f(z) \, dz}{(z - z_0)(z - z_0)^{-1}} = \frac{1}{2\pi j} \int_C f(z) \, dz = 0$$

Similarly, it may be shown that the $a_{-3}, a_{-4}, \ldots$ coefficients also vanish. Thus, the series has the form

$$F(z_0 + \Delta z_0) = \frac{a_{-1}}{z - z_0} + a_0 + a_1(z - z_0) + a_2(z - z_0)^2 + \cdots$$

For a second-order pole at z_0, the series has the form

$$F(z_0 + \Delta z_0) = \frac{a_{-2}}{(z - z_0)^2} + \frac{a_{-1}}{z - z_0} + a_0 + a_1(z - z_0) + \cdots$$

In general, for a pole of order k, the series has the form

$$F(z_0 + \Delta z_0) = \frac{a_{-k}}{(z - z_0)^k} + \cdots + \frac{a_{-1}}{z - z_0} + a_0 + \cdots$$

The portion of the Laurent series containing the negative powers of $z - z_0$ is called the *principal part* of $F(z)$ near z_0. When the singular point is a pole of order k at z_0, the principal part of $F(z)$ near z_0 contains k terms. When the principal part of $F(z)$ near z_0 has an infinite number of terms, the singular point is referred to as an *essential singularity*. Examples of functions with essential singular points at $z = 0$ are

$$e^{-1/z} = 1 + \frac{1}{z} + \frac{1}{2!z^2} + \frac{1}{3!z^3} + \cdots$$

$$\sin \frac{1}{z} = \frac{1}{z} - \frac{1}{3!z^3} + \frac{1}{5!z^5} - \cdots$$

Because each a_{-1} coefficient is 1, then the residue of each of these functions is 1. Application of Eq. (5.100) shows that

$$\int_C e^{-1/z} dz = 2\pi j \Sigma \text{ residues} = 2\pi j a_{-1} = 2\pi j$$

$$\int_C \sin 1/z \, dz = 2\pi j$$

The preceding discussion refers to the case in which the contour C encloses but one singular point. When the contour C encloses more than one singular point, the Laurent series contains all terms (i.e., none of the coefficients vanishes).

5.9 ANALYTIC CONTINUATION

A function $f(z)$ may be expressed as a power series in any region in which the function is analytic. In Fig. 5.22, suppose that the power series expansion for $f(z)$ about z_1 is $f_1(z)$. This expansion is valid in the region R_1. The power-series expansion $f(z)$ about z_2 is $f_2(z)$. This latter expansion is valid in the region R_2. The shaded area is the region common to R_1 and R_2. The function $f_2(z)$, which is analytic in R_2 and is equal to $f_1(z)$ in the common region, is called the *analytic continuation* of $f_1(z)$ into the region R_2.

Because an analytic function is uniquely determined throughout a region, if $f_2(z)$ is equal to $f_1(z)$ in the region common to R_1 and R_2, then $f_1(z)$ and $f_2(z)$ are merely different forms of the same function. For example, consider the function

$$f_1(z) = \frac{1}{1 + (z - 3)} = 1 - (z - 3) + (z - 3)^2 - \cdots$$

$$(5.101)$$

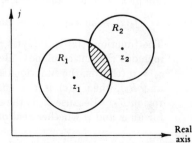

Fig. 5.22. Overlapping regions.

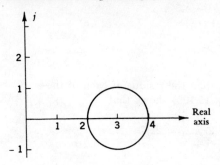

Fig. 5.23. Taylor series expansion about $z_0 = 3$.

The radius of convergence of $f_1(z)$ is $|z - 3| < 1$. The region R_1 for which $f_1(z)$ is analytic is the region within the unit circle about $z = 3$, as shown in Fig. 5.23. Next consider the function

$$f_2(z) = \frac{1}{z - 2} \tag{5.102}$$

This function is analytic everywhere except at $z = 2$. Because $f_2(z) = f_1(z)$ for any z in the region common to R_1 and R_2, then $f_2(z)$ is the analytic extension of $f_1(z)$ into the region R_2. This fact is readily verified by noting that Eq. (5.101) is merely another form of Eq. (5.102).

The concept of analytic extension may be used to determine whether or not a relationship for real variables is also applicable for complex variables. To illustrate, suppose that $f_1(z)$ is

$$f_1(z) = \sin(x + y) = \sin x \cos y + \cos x \sin y \quad \left\{ \begin{array}{l} 0 < x < \dfrac{\pi}{2} \\[2mm] 0 < y < \dfrac{\pi}{2} \end{array} \right. \tag{5.103}$$

Ordinarily in the derivation of the preceding expression the angles x and y do not exceed 90°. By replacing x by z_1 and y by z_2, it is an easy matter to determine whether or not this equation is valid for complex as well as real values. Hence

$$f_2(z) = \sin(z_1 + z_2) = \sin z_1 \cos z_2 + \cos z_1 \sin z_2 \tag{5.104}$$

The function $f_2(z)$ is analytic for all values of z_1 and z_2, so that R_2 includes the entire plane. Because $f_2(z) = f_1(z)$ when $z_1 = x$ and $z_2 = y$ (that is, the functions are equal in the region common to R_1 and R_2), then $f_2(z)$ is the analytic continuation of $f_1(z)$ into the region R_2. Because R_2 is the entire plane, then Eq. (5.103) is valid for all x and y whether real or complex.

PROBLEMS

5.1 Express each of the following complex quantities in polar form and in exponential form:

    (a) $3 + j4$ (b) $3 - j4$
    (c) $-3 + j4$ (d) $-3 - j4$

5.2 Express the answer to each of the following problems in exponential form:

    (a) $(3 + j4)(3 - j4)$ (b) $(3 + j4)(-3 - j4)$
    (c) $1/(3 + j4)$ (d) $(3 + j4)/(3 - j4)$

5.3 For each of the following functions, determine the values of z, if any, for which the Cauchy-Riemann equations are not satisfied:

    (a) z^3 (b) $z^2 + 3 + j4$
    (c) $1/(z + 1)$ (d) e^z

5.4 Determine the equation for the derivative $f'(z)$ for each of the functions given in Prob. 5.3. Evaluate each derivative at the point $z = 1 + j2$.

5.5 Show that, for $z = x + jy$:

    (a) $\sin z = \sin x \cosh y + j \cos x \sinh y$
    (b) $\sinh z = \sinh x \cos y + j \cosh x \sin y$
    (c) $e^z = e^x (\cos y + j \sin y)$

Evaluate each of the above for $z = \pi(1 + j2)$.

5.6 Evaluate

$$\int_C z \, dz$$

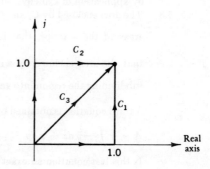

Fig. P5.6

along each of the three paths shown in Fig. P5.6, that is:

(a) Along the path C_1
(b) Along the path C_2
(c) Along the path C_3

Obtain the integral around the rectangle of Fig. P5.6 by subtracting the integral along path C_2 from that along path C_1. Verify this result by application of Cauchy's integral theorem.

5.7   Evaluate

$$\int_C z^2 \, dz$$

along each of the three paths shown in Fig. P5.7, that is:

(a) Along path C_1
(b) Along path C_2
(c) Along path C_3

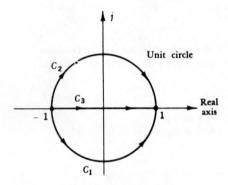

Fig. P5.7

Obtain the integral around the circle of Fig. P5.7 by subtracting the integral along path C_2 from that along path C_1. Verify this result by application of Cauchy's integral theorem.

5.8   The area enclosed by C_1 and C_2 in Fig. 5.10a is the summation of the areas of the n strips, that is, $\sum_{i=1}^{n} [(x_i)_{c_1} - (x_i)_{c_2}] \, \Delta y$. Thus, show that the area enclosed by a curve C is $A = \int_C x \, dy$. Similarly, by subdividing the region into vertical strips show that $A = - \int_C y \, dx$.

A third equation is obtained by adding the two preceding results.

$$A = \tfrac{1}{2} \int_C -y \, dx + x \, dy$$

Is this last equation an exact differential?

5.9   Evaluate

$$\frac{1}{2}\int_C -y\,dx + x\,dy$$

(a) Along the path $y = x$ from the point (0,0) to (1,1)
(b) Along the path $y = x^2$ from the point (0,0) to (1,1)
(c) Along the path $y^2 = x$ from the point (0,0) to (1,1)

5.10 Evaluate

$$\int_C y\,dx + x\,dy$$

(a) Along the path $y = x$ from the point (0,0) to (1,1)
(b) Along the path $y = x^2$ from the point (0,0) to (1,1)
(c) Along the path $y^2 = x$ from the point (0,0) to (1,1)

5.11 Evaluate

$$\int_C \frac{1}{z}\,dz$$

along each of the two paths shown in Fig. P5.11, that is:

(a) Along the path C_1
(b) Along the path C_2

Subtract the integral around C_2 from the integral around C_1 to obtain the integral along the closed contour. Verify this result by application of Cauchy's integral formula.

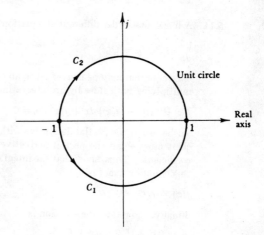

Fig. P5.11

215

5.12   Same as Prob. 5.11, but use the rectangular paths shown in Fig. P5.12.

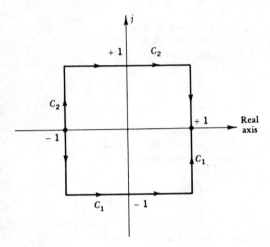

5.13   Determine the solution of the following differential equations:

(a) $2x(x + y)\, dx + (x^2 - 3y)\, dy = 0$
(b) $x(3x + 4y)\, dx + 2(x^2 + y)\, dy = 0$
(c) $(2xy + y \sin x + \sin y)\, dx + (x^2 + x \cos y - \cos x)\, dy = 0$

5.14   A linear first-order differential equation may be written in the form

$$\frac{dy}{dx} + P(x)y = Q(x)$$

This equation may be converted to an exact differential equation by multiplying both sides by an integrating factor $f(x)$.   That is,

$$f(x)\,[P(x)y - Q(x)]\, dx + f(x)\, dy = 0$$

To be exact, the partial derivative with respect to y of the dx coefficient must equal the partial derivative with respect to x of the dy coefficient.   Thus, show that the integrating factor $f(x)$ which makes this equation exact is

$$f(x) = e^{\int P(x)\, dx}$$

Finally, show that the solution is

$$y = \frac{\int Q(x)e^{\int P(x)\, dx}\, dx + C}{e^{\int P(x)\, dx}}$$

5.15 Use the result of Prob. 5.14 to determine the solution of each of the following linear first-order differential equations:

    (a) $y' = 2x - y$
    (b) $(2x^3y - 1)\,dx + x^4\,dy = 0$
    (c) $\sin x\,(dy/dx) + 2y \cos x = 1$

5.16 Use the method of residues to evaluate

$$\int_C \frac{z\,dz}{(z + 1)(z - 2)}$$

where C is a circle with center at the origin $(z_0 = 0)$ and radius:

    (a) $|z| < 1$
    (b) $1 < |z| < 2$
    (c) $|z| > 2$

5.17 Use the method of residues to evaluate

$$\int_C \frac{e^z}{z^2(z - 1)}\,dz$$

where C is a circle with center at $z_0 = 2$ and radius:

    (a) $|z - 2| < 1$
    (b) $1 < |z - 2| < 2$
    (c) $|z - 2| > 2$

5.18 Use the method of residues to evaluate

$$\int_C \frac{z + 1}{z(z + 2)}\,dz$$

where C is the contour shown in Fig. P5.18.

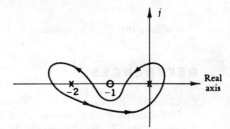

Fig. P5.18

5.19 Find the residues at the singular points of the following:

    (a) $\dfrac{z}{(z - 2)(z - 5)}$ (b) $\dfrac{1}{z(z^2 + 1)}$

    (c) $\dfrac{1}{z(z + 1)^2}$ (d) $\dfrac{1}{e^z - 1}$

217

5.20 Determine the residues for each of the following functions:

(a) $\dfrac{1}{\sin z}$ (b) $\dfrac{1}{z \cos z}$ (c) $\tan z$

5.21 Determine the Taylor series expansion for each of the following functions. The point about which the expansion is to be taken is shown after the function. Specify the radius of convergence:

(a) $\dfrac{1}{z(z-2)}, z = 1$ (b) $\dfrac{1}{z(z-2)}, z = 3$

(c) $\dfrac{z-1}{z(z-2)}, z = 1$ (d) $\dfrac{1}{z^2(z-1)}, z = 3$

(e) $\dfrac{1}{1-z}, z = 0$ (f) $\sin z, z = 0$

(g) $e^z, z = 0$ (h) $\ln(1+z), z = 0$

5.22 Determine all the Laurent expansions for the following function about the point $z = 2$ and about the point $z = 5$. Specify the region of convergence for each expansion.

$$f(z) = \frac{z}{(z-2)(z-5)}$$

Check that the a_{-1} coefficient for each expansion satisfies Eq. (5.100).

5.23 Determine all the Laurent expansions for the following function about the point $z = 0$ and about the point $z = 1$. Specify the region of convergence for each expansion.

$$f(z) = \frac{1}{z^2(z-1)}$$

Check that the a_{-1} coefficient for each expansion satisfies Eq. (5.100).

REFERENCES

1 Churchill, R. V.: "Complex Variables and Applications," 2d ed., McGraw-Hill Book Company, New York, 1960.
2 MacRobert, T. M.: "Functions of a Complex Variable," The Macmillan Company, New York, 1954.
3 Springer, G.: "Introduction to Riemann Surfaces," Addison-Wesley Publishing Company, Inc., Reading, Mass., 1957.
4 Titchmarsh, E. C.: "Theory of Functions," Oxford University Press, New York, 1939.
5 Whittaker, E. T., and G. N. Watson: "A Course of Modern Analysis," Cambridge University Press, New York, 1927.

chapter 6

Residues and contour integration

In solving linear differential equations with constant coefficients, the roots of the characteristic equation are either distinct or repeated. Consequently, only a few types of terms arise in the partial fraction expansion. A relatively small table of transform pairs suffices for solving such equations.

In the solution of partial differential equations, differential equations with variable coefficients, etc., the transform is frequently a rather complicated function of s. The complex inversion integral provides a good method for obtaining the inverse transformation of such functions.

6.1 THE COMPLEX INVERSION INTEGRAL

The line-integral representation of the complex inversion integral is developed as follows: Let the function $F(z)$ be analytic on and to the right of the line $x = a$ shown in Fig. 6.1 [that is, $F(z)$ is analytic for $x \geq a$]. In addition, $F(z)$ approaches zero uniformly as z becomes infinite.

$$\lim_{z \to \infty} |F(z)| \approx 0 \tag{6.1}$$

The contour C in Fig. 6.1 is composed of the portions C_1 and C_2. The straight-line section is C_1, and the semicircular section of radius b is C_2. Because $F(z)$ is analytic within and on C, then by Cauchy's integral formula it follows that for any point s within this contour

$$F(s) = \frac{1}{2\pi j} \int_{C_1} \frac{F(z)}{z - s} \, dz + \frac{1}{2\pi j} \int_{C_2} \frac{F(z)}{z - s} \, dz \tag{6.2}$$

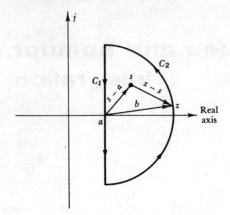

Fig. 6.1. Semicircular contour to the right of the line $x = a$.

By referring to Fig. 6.1, for values of z on the semicircle C_2

$$|z - s| + |s - a| \geq b$$
or
$$|z - s| \geq b - |s - a|$$

where $|s - a|$ is the distance from a to s and $|z - s|$ is the distance from s to any point z on C_2.

Consider the magnitude of the last integral on the right-hand side of Eq. (6.2); that is,

$$\left| \int_{C_2} \frac{F(z)}{z - s} \, dz \right| \leq \int_{C_2} \left| \frac{F(z)}{z - s} \right| |dz| \leq \int_{C_2} \frac{|F(z)|}{b - |s - a|} \, |dz| \qquad (6.3)$$

By letting M denote the maximum value of $|F(z)|$ for values of z on C_2,

$$\left| \int_{C_2} \frac{F(z)}{z - s} \, dz \right| \leq \frac{M}{b - |s - a|} \int_{C_2} |dz| = \frac{\pi b M}{b - |s - a|} \qquad (6.4)$$

where $\int_{C_2} |dz| = \pi b$ is the length of C_2. For any fixed point s, then $|s - a|$ is a constant, and

$$\lim_{b \to \infty} \frac{b}{b - |s - a|} \approx 1 \qquad (6.5)$$

As b becomes infinite, the contour C_2 becomes a semicircle of infinite radius. Because $|F(z)|$ approaches zero uniformly as z becomes infinite, M must also approach zero. Thus

$$\lim_{b \to \infty} M \approx 0 \qquad (6.6)$$

Utilizing the results of Eqs. (6.5) and (6.6) in taking the limit of Eq. (6.4) as b becomes infinite gives

$$\lim_{b \to \infty} \left| \int_{C_2} \frac{F(z)}{z-s} \, dz \right| = \lim_{b \to \infty} \pi M \frac{b}{b - |s-a|} \approx \lim_{b \to \infty} \pi M \approx 0 \tag{6.7}$$

Because the integral around C_2 vanishes as b becomes infinite, the limit of Eq. (6.2) is

$$F(s) = \lim_{b \to \infty} \frac{1}{2\pi j} \int_{a+jb}^{a-jb} \frac{F(z)}{z-s} \, dz = \frac{1}{2\pi j} \int_{a-j\infty}^{a+j\infty} \frac{F(z)}{s-z} \, dz \tag{6.8}$$

It is now desired to determine the function $f(t)$ which is the inverse of $F(s)$; that is,

$$f(t) = \mathcal{L}^{-1}[F(s)] = \mathcal{L}^{-1}\left[\frac{1}{2\pi j} \int_{a-j\infty}^{a+j\infty} \frac{F(z)}{s-z} \, dz \right] \tag{6.9}$$

The inverse transformation $\mathcal{L}^{-1}[F(s)]$ is taken with respect to the variable s and not with respect to the variable of integration z. Thus, taking the inverse operation inside the integral sign gives

$$f(t) = \mathcal{L}^{-1}[F(s)] = \frac{1}{2\pi j} \int_{a-j\infty}^{a+j\infty} F(z) \left[\mathcal{L}^{-1}\left(\frac{1}{s-z} \right) \right] dz \tag{6.10}$$

Because

$$\mathcal{L}^{-1}\left(\frac{1}{s-a} \right) = e^{at}$$

then

$$\mathcal{L}^{-1}\left(\frac{1}{s-z} \right) = e^{zt}$$

Substitution of the preceding result into Eq. (6.10) gives

$$f(t) = \mathcal{L}^{-1}[F(s)] = \frac{1}{2\pi j} \int_{a-j\infty}^{a+j\infty} F(z) e^{zt} \, dz \tag{6.11}$$

Because z is now the only variable which appears in the preceding integral, the variable z may be replaced by s. Thus,

$$f(t) = \mathcal{L}^{-1}[F(s)] = \frac{1}{2\pi j} \int_{a-j\infty}^{a+j\infty} F(s) e^{st} \, ds \tag{6.12}$$

Equation (6.12), which is identical to Eq. (1.55), is the line-integral representation of the complex inversion integral.

The symbols s and z are both used to represent a complex variable. The distinction is that z is ordinarily employed in the development of general mathematical formulas. The symbol s is usually reserved for more specialized equations such as in the Laplace transformation.

The preceding derivation shows that the inversion integral is a line integral taken to the right of all the singularities of $F(s)$, and in addition $F(s)$ must be such that

$$\lim_{s \to \infty} |F(s)| \approx 0 \tag{6.13}$$

This equation is automatically satisfied whenever the denominator of $F(s)$ is of a higher order than the numerator.

For most engineering work, it is much easier to evaluate the inversion integral by residues rather than by employing the line integral [Eq. (6.12)]. If the function $F(s)$ is common enough to be tabulated in a transform table, referring to the table is obviously the easiest way to obtain the inverse.

6.2 EVALUATION BY RESIDUES

In Fig. 6.2 is shown a contour $C = C_1 + C_2$, which encloses all the singularities of the function $F(s)$. The straight-line portion is C_1, and the semicircular portion of radius b is C_2. Because C encloses all the singularities of $F(s)$ and because e^{st} is analytic for all values of s,

$$\frac{1}{2\pi j} \int_{a-jb}^{a+jb} F(s)e^{st} \, ds + \frac{1}{2\pi j} \int_{C_2} F(s)e^{st} \, ds = \sum \text{residues of } F(s)e^{st}$$

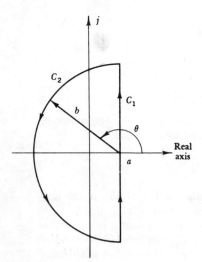

Fig. 6.2. Semicircular contour to the left of the line $x = a$.

where the summation is of residues within the contour C. In the following, it is shown that when $F(s)$ is such that Eq. (6.13) is satisfied, then the integral around C_2 vanishes. Thus, the complex inversion integral becomes

$$f(t) = \mathcal{L}^{-1}[F(s)] = \frac{1}{2\pi j} \int_{a-j\infty}^{a+j\infty} F(s)e^{st}\, ds$$

$$= \sum \text{residues of } F(s)e^{st} \qquad (6.14)$$

Because the contour encloses all the singularities of $F(s)$, then this is the summation of all the residues.

To prove that the integral around C_2 vanishes as b becomes infinite, first note that

$$\left| \int_{C_2} F(s)e^{st}\, ds \right| \leq \int_{C_2} |F(s)|\, |e^{st}|\, |ds| \qquad (6.15)$$

For values of s on C_2,

$$s = a + be^{j\theta} = (a + b\cos\theta) + jb\sin\theta \qquad (6.16)$$

Differentiation gives

$$ds = jbe^{j\theta}\, d\theta$$

Because $|e^{j\theta}| = 1$ and $|j| = 1$, then

$$|ds| = b\, d\theta \qquad (6.17)$$

From Eq. (6.16), it follows that the magnitude of s is

$$|s| = \sqrt{a^2 + 2ab\cos\theta + b^2} \leq \sqrt{a^2 + 2ab + b^2}$$

Hence

$$|s| \leq a + b \qquad (6.18)$$

Equation (6.18) could have been ascertained directly from the geometry of Fig. 6.2.

By utilizing Eq. (6.16), the magnitude of e^{st} may be expressed in the form

$$|e^{st}| = \left| e^{(a+b\cos\theta)t + j(b\sin\theta)t} \right| = \left| e^{(a+b\cos\theta)t} \right| \left| e^{j(b\sin\theta)t} \right|$$

Because $e^{jbt\sin\theta}$ is a unit vector (the angle of the vector is $bt\sin\theta$), its magnitude is 1. Therefore

$$|e^{st}| = e^{(a+b\cos\theta)t} \qquad (6.19)$$

223

Substitution of Eqs. (6.17) and (6.19) into Eq. (6.15) gives

$$\left| \int_{C_2} F(s)e^{st} \, ds \right| \leq b \int_{C_2} |F(s)| e^{(a+b\cos\theta)t} \, d\theta$$
$$\leq bMe^{at} \int_{\pi/2}^{3\pi/2} e^{bt\cos\theta} \, d\theta \qquad (6.20)$$

where M is the maximum value of $|F(s)|$ on C_2.

The preceding integration may be simplified by letting

$$\theta = \phi + \pi/2$$

in which case $\cos\theta = \cos(\phi + \pi/2) = -\sin\phi$. The new limits of integration are obtained by noting that, as θ varies from $\pi/2$ to $3\pi/2$, then ϕ varies from 0 to π. Thus, Eq. (6.20) becomes

$$\left| \int_{C_2} F(s)e^{st} \, ds \right| \leq b|F(s)| e^{at} \int_0^\pi e^{-bt\sin\phi} \, d\phi$$
$$\leq 2b|F(s)| e^{at} \int_0^{\pi/2} e^{-bt\sin\phi} \, d\phi \qquad (6.21)$$

As is illustrated in Fig. 6.3, for $0 \leq \phi \leq \pi/2$,

$$\sin\phi \geq \frac{2\phi}{\pi} \qquad 0 \leq \phi \leq \frac{\pi}{2} \qquad (6.22)$$

Utilizing the result of Eq. (6.22) in Eq. (6.21) gives

$$\left| \int_{C_2} F(s)e^{st} \, ds \right| \leq 2b|F(s)| e^{at} \int_0^{\pi/2} e^{-2bt\phi/\pi} \, d\phi$$
$$\leq 2b|F(s)| e^{at} \frac{1 - e^{-bt}}{2bt/\pi}$$
$$\leq \pi|F(s)| \frac{e^{at}}{t} (1 - e^{-bt}) \qquad (6.23)$$

As b becomes infinite, values of s on C_2 also become infinite. Thus

$$\lim_{b\to\infty} \left| \int_{C_2} F(s)e^{st} \, ds \right| \leq \frac{\pi e^{at}}{t} \lim_{b\to\infty} |F(s)| \approx 0 \qquad (6.24)$$

This verifies Eq. (6.14).

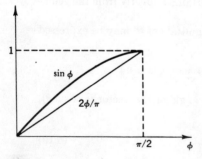

Fig. 6.3. A comparison of sin ϕ and the straight line $2\phi/\pi$.

Illustrative example Use the method of residues to obtain the inverse transform of Eq. (3.8).

Solution For this function

$$F(s) = \frac{F_0}{s(s+1)(s+2)}$$

The residues of $F(s)e^{st}$ are

$$R_1 = \lim_{s \to 0} s[F(s)e^{st}] = \frac{F_0 e^{st}}{(s+1)(s+2)} \Big|_{s=0} = \frac{F_0}{2}$$

$$R_2 = \lim_{s \to -1} (s+1)[F(s)e^{st}] = \frac{F_0 e^{st}}{s(s+2)} \Big|_{s=-1} = -F_0 e^{-t}$$

$$R_3 = \lim_{s \to -2} (s+2)[F(s)e^{st}] = \frac{F_0 e^{st}}{s(s+1)} \Big|_{s=-2} = \frac{F_0 e^{-2t}}{2}$$

Application of Eq. (6.14) gives

$$f(t) = (\tfrac{1}{2} - e^{-t} + \tfrac{1}{2}e^{-2t})F_0$$

The preceding result verifies the solution obtained in Eq. (3.14).

6.3 INVERSE TRANSFORMATIONS

In Sec. 3.2, it is shown how the time response (inverse transformation) is obtained by use of the partial-fraction-expansion technique. The present section illustrates the corresponding procedures for obtaining the time response by means of residues. The method of residues has the advantage that the response function is obtained directly in one operation. With partial fractions, it is necessary first to perform a partial fraction expansion and then obtain the inverse.

Because residues do not require a partial fraction expansion, it is not necessary that the denominator of $F(s)$ be expressed in factored form. This feature of the residue method proves invaluable for obtaining the inverse transform of partial differential equations, differential equations with variable coefficients, etc.

The general function given by Eq. (3.16) is

$$F(s) = \frac{A(s)}{B(s)}$$

$$= \frac{A(s)}{(s-r_1)(s-r_2) \cdots (s-r_i) \cdots (s-r_n)(s-r)^q}$$

In Chap. 3, the inverse transform of this function is obtained by the partial-fraction-expansion method. To provide a basis for comparison, it is now shown how the inverse transform is obtained by application of the method of residues.

Response due to distinct roots

A distinct root is a first-order pole of $F(s)$. In accordance with Eq. (6.14), the response term due to the pole at r_i is the residue of $F(s)e^{st}$ at $s = r_i$, that is,

$$R_i = \lim_{s \to r_i} (s - r_i)[F(s)e^{st}] = K_i e^{r_i t} \tag{6.25}$$

where

$$K_i = (s - r_i)F(s) \Big|_{s = r_i} = (s - r_i) \frac{A(s)}{B(s)} \Big|_{s = r_i}$$

This is the same result as given by Eq. (3.23), in which K_i is specified by Eq. (3.22).

Response due to complex conjugate roots

When $B(s)$ contains a pair of complex conjugate roots $(a + jb)$ and $(a - jb)$, the response due to each of these roots combines to form a sinusoidal response term. The residue at $a + jb$ is

$$R_{a+jb} = \lim_{s \to a+jb} (s - a - jb) \left[\frac{A(s)}{B(s)} e^{st} \right]$$

Because $B(s)$ contains the quadratic factor

$$(s - a - jb)(s - a + jb) = s^2 - 2as + a^2 + b^2$$

then the residue R_{a+jb} may be expressed in the form

$$\frac{\cancel{s - a - jb}}{\cancel{(s - a - jb)}(s - a + jb)} \frac{A(s)e^{st}}{B(s)/(s^2 - 2as + a^2 + b^2)} \Big|_{s = a+jb}$$

Evaluation at $s = a + jb$ yields

$$R_{a+jb} = \frac{1}{2jb} \frac{A(a + jb)}{B(a + jb)} e^{(a+jb)t} \tag{6.26}$$

The notation $A(a + jb)/B(a + jb)$ means that the quadratic term $s^2 - 2as + a^2 + b^2$ has been removed from $B(s)$ before evaluating $A(s)/B(s)$ at $s = a + jb$; that is,

$$\frac{A(a + jb)}{B(a + jb)} = \left[\frac{A(s)}{B(s)/(s^2 - 2as + a^2 + b^2)} \right]_{s = a+jb}$$

The residue at $a - jb$ is

$$\frac{\cancel{s - a + jb}}{(s - a - jb)\cancel{(s - a + jb)}} \frac{A(s)e^{st}}{B(s)/(s^2 - 2as + a^2 + b^2)} \Big|_{s=a-jb}$$

Evaluation at $s = a - jb$ yields

$$R_{a-jb} = -\frac{1}{2jb} \frac{A(a - jb)}{B(a - jb)} e^{(a-jb)t} \tag{6.27}$$

Because $A(a - jb)/B(a - jb)$ is the complex conjugate of $A(a + jb)/B(a + jb)$, each has the same magnitude but opposite angles. By letting α designate the angle of $A(a + jb)/B(a + jb)$, then

$$\frac{A(a + jb)}{B(a + jb)} = \left| \frac{A(a + jb)}{B(a + jb)} \right| e^{j\alpha}$$

$$\frac{A(a - jb)}{B(a - jb)} = \left| \frac{A(a + jb)}{B(a + jb)} \right| e^{-j\alpha}$$

Thus, the residue due to each complex conjugate root may be expressed in the form

$$R_{a+jb} = \frac{e^{at}}{b} \left| \frac{A(a + jb)}{B(a + jb)} \right| \frac{e^{jbt}e^{j\alpha}}{2j}$$

$$R_{a-jb} = -\frac{e^{at}}{b} \left| \frac{A(a + jb)}{B(a + jb)} \right| \frac{e^{-jbt}e^{-j\alpha}}{2j}$$

Adding these two residues gives the total response.

$$\frac{e^{at}}{b} \left| \frac{A(a + jb)}{B(a + jb)} \right| \frac{e^{j(bt+\alpha)} - e^{-j(bt+\alpha)}}{2j}$$

$$= \frac{e^{at}}{b} \left| \frac{A(a + jb)}{B(a + jb)} \right| \sin(bt + \alpha) \tag{6.28}$$

The preceding result verifies that of Eq. (3.38).

Response due to repeated roots

The response due to the repeated root r in Eq. (3.16) is the residue at this pole of order q. Hence, application of Eq. (5.77) gives

$$R = \frac{1}{(q - 1)!} \frac{d^{q-1}}{ds^{q-1}} \left[(s - r)^q \frac{A(s)}{B(s)} e^{st} \right]_{s=r} \tag{6.29}$$

To illustrate the application of Eq. (6.29), let it be desired to verify the solution of Eq. (3.33). The residue of the repeated root in Eq.

(3.33) is

$$R_1 = \frac{1}{1!}\frac{d}{ds}\left[(s+1.5)^2\frac{F_0 e^{st}}{s(s+1.5)^2}\right]_{s=-1.5} = \frac{d}{ds}\left[\frac{F_0 e^{st}}{s}\right]_{s=-1.5}$$

$$= F_0\left(\frac{ste^{st} - e^{st}}{s^2}\right)_{s=-1.5} = \frac{-F_0(1.5t+1)e^{-1.5t}}{2.25}$$

The residue of the distinct root in Eq. (3.33) is

$$R_2 = \lim_{s \to 0}\frac{F_0 e^{st}}{(s+1.5)^2} = \frac{F_0}{2.25}$$

Adding these residues verifies the result previously attained.

Equation (6.29) is in a more compact form than Eq. (3.32). However, the results obtained by either expression will, of course, be the same.

Illustrative example Use the method of residues to determine the inverse transform of the function

$$F(s) = \frac{1}{(s^2 - 2as + a^2 + b^2)^2} = \frac{1}{[s - (a+jb)]^2[s - (a-jb)]^2}$$

Solution This function has a pair of repeated complex conjugate roots. Application of Eq. (6.29) to obtain the residue at $a + jb$ gives

$$R_{a+jb} = \frac{1}{1!}\frac{d}{ds}\left\{\frac{e^{st}}{[s - (a-jb)]^2}\right\}_{s=a+jb}$$

$$= \left\{\frac{[s - (a-jb)]^2 te^{st} - 2[s - (a-jb)]e^{st}}{[s - (a-jb)]^4}\right\}_{s=a+jb}$$

Because $[s - (a-jb)]_{s=a+jb} = 2jb$, then R_{a+jb} becomes

$$R_{a+jb} = \frac{(-b^2 t - jb)e^{(a+jb)t}}{4b^4}$$

The residue R_{a-jb} is the complex conjugate of R_{a+jb}. The complex conjugate of a function is obtained by changing the sign of the imaginary portions (i.e., by substituting $-j$ for j). Thus

$$R_{a-jb} = \frac{-b^2 t + jb}{4b^4}e^{(a-jb)t}$$

Adding these residues gives

$$f(t) = \frac{-jbe^{at}}{4b^4}(e^{jbt} - e^{-jbt}) - \frac{b^2 te^{at}}{4b^4}(e^{jbt} + e^{-jbt})$$

Using Euler's equations to express the exponentials in trigonometric form gives

$$f(t) = e^{at} \left(\frac{1}{2b^3} \sin bt - \frac{t}{2b^2} \cos bt \right)$$

As is to be expected, the response contains the e^{at} factor.

6.4 CONTOUR INTEGRATION

By *contour integration* is meant the evaluation of an integral along a closed path C; that is,

$$\int_C F(z)\, dz$$

Simplicity in evaluating the contour integral is afforded by application of the residue theorem.

Certain contours lend themselves to a direct evaluation of several large and important classes of real definite integrals. The contour shown in Fig. 6.4 is well suited for evaluating integrals of the type

$$\lim_{r \to \infty} \int_{-r}^{r} F(x)\, dx = \int_{-\infty}^{\infty} F(x)\, dx \tag{6.30}$$

This contour is composed of the straight-line portion C_1, which runs along the real axis from $-r$ to $+r$, and the semicircular portion C_2, whose radius is r. If there are no poles of $F(z)$ on the real axis, application of the residue theorem gives

$$\int_{C_1} F(z)\, dz + \int_{C_2} F(z)\, dz = 2\pi j \sum \text{ residues of } F(z) \tag{6.31}$$

where the summation is of residues within the contour C.

In the first integral, it is to be noted that, for values of z on C_1 (that is, the real axis), $z = x$, $dz = dx$, and $F(z) = F(x)$; thus,

$$\int_{C_1} F(z)\, dz = \int_{-r}^{r} F(x)\, dx$$

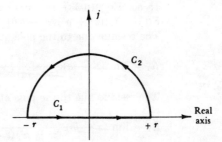

Fig. 6.4. A semicircular contour.

For values of z on the semicircular portion, $z = re^{j\theta}$ and

$$dz = jre^{j\theta}\, d\theta = jz\, d\theta$$

Thus, the integral along C_2 may be expressed in the form

$$\left| \int_{C_2} F(z)\, dz \right| = \left| \int_0^\pi jzF(z)\, d\theta \right| \le \int_0^\pi |zF(z)|\, d\theta \tag{6.32}$$

This integral around C_2 vanishes when $F(z)$ is such that $|zF(z)|$ converges uniformly to zero as r (and thus z) becomes infinite.

$$\lim_{z \to \infty} |zF(z)| \approx 0 \tag{6.33}$$

Equation (6.33) is automatically satisfied whenever the degree of the denominator of $F(z)$ exceeds the degree of the numerator by at least 2. Thus, if there are no poles of $F(z)$ on the real axis and if $F(z)$ satisfies Eq. (6.33), then Eq. (6.31) becomes

$$\int_{C_1} F(z)\, dz = \lim_{r \to \infty} \int_{-r}^r F(x)\, dx = 2\pi j \sum \text{ residues of } F(z) \tag{6.34}$$

For the case in which $F(x)$ is an even function, then the preceding expression becomes

$$\int_0^\infty F(x)\, dx = \frac{1}{2} \int_{-\infty}^\infty F(x)\, dx = \pi j \sum \text{ residues of } F(z) \tag{6.35}$$

The summation in Eqs. (6.34) and (6.35) is of residues in the upper half plane.

Illustrative example Evaluate the following definite integral:

$$\int_0^\infty \frac{dx}{(x^2 + a^2)(x^2 + b^2)} \tag{6.36}$$

Solution Because $F(x) = F(-x)$, then this is an even function. The degree of the denominator exceeds the numerator by 4; so Eq. (6.33) is satisfied. The poles of the function $F(z) = 1/[(z^2 + a^2)(z^2 + b^2)]$ are located at $\pm ja$ and $\pm jb$. The residue due to the pole at ja is

$$R_1 = \lim_{z \to ja} \underbrace{(z - ja)}\frac{1}{\underbrace{(z - ja)}(z + ja)(z^2 + b^2)} = \frac{1}{2ja(b^2 - a^2)}$$

The residue due to the pole at jb is

$$R_2 = \lim_{z \to jb} \underbrace{(z - jb)}\frac{1}{(z^2 + a^2)\underbrace{(z - jb)}(z + jb)} = \frac{-1}{2jb(b^2 - a^2)}$$

The sum of the residues due to poles in the upper half plane is

$$R_1 + R_2 = \frac{1}{2j(b^2 - a^2)} \left(\frac{1}{a} - \frac{1}{b}\right) = \frac{1}{2abj(a + b)}$$

Hence application of Eq. (6.35) gives

$$\int_0^\infty \frac{dx}{(x^2 + a^2)(x^2 + b^2)} = \frac{\pi}{2ab(a + b)} \tag{6.37}$$

Fourier integrals

These integrals have the form

$$\int_{-\infty}^\infty F(x) \cos sx \, dx \qquad \text{and} \qquad \int_{-\infty}^\infty F(x) \sin sx \, dx \tag{6.38}$$

The substitution of $F(z)e^{jsz}$ for $F(z)$, and hence the corresponding substitution of $F(x)e^{jsx}$ for $F(x)$ in Eq. (6.34), gives

$$\lim_{r \to \infty} \int_{-r}^r F(x)e^{jsx} \, dx = 2\pi j \sum \text{residues } [F(z)e^{jsz}] \tag{6.39}$$

The substitution of $e^{jsx} = \cos sx + j \sin sx$ gives

$$\int_{-\infty}^\infty F(x) \cos sx \, dx + j \int_{-\infty}^\infty F(x) \sin sx \, dx$$
$$= 2\pi j \sum \text{residues } [F(z)e^{jsz}] \tag{6.40}$$

The Σ residues $[F(z)e^{jsz}]$ may be broken up into real and imaginary parts.

$$\Sigma \text{ residues } [F(z)e^{jsz}] = \text{real part } \Sigma \text{ residues}$$
$$+ j \text{ imaginary part } \Sigma \text{ residues}$$

Thus, equating real and imaginary parts in Eq. (6.40) gives

$$\int_{-\infty}^\infty F(x) \cos sx \, dx = -2\pi \{\text{imag. part of } \Sigma \text{ residues } [F(z)e^{jsz}]\} \tag{6.41}$$

$$\int_{-\infty}^\infty F(x) \sin sx \, dx = 2\pi \{\text{real part of } \Sigma \text{ residues } [F(z)e^{jsz}]\} \tag{6.42}$$

In the preceding equations, the summation is of residues due to poles located in the upper half plane. Because Eqs. (6.41) and (6.42) are in effect an extension of Eq. (6.34), then the function $F(x)$ in Eqs. (6.41) and (6.42) must satisfy the same conditions of $F(x)$ in Eq. (6.34). In particular, the degree of the denominator of $F(x)$ must exceed that of the numerator by at least 2, and there can be no poles of $F(x)$ on the real axis.

Illustrative example   Evaluate the following definite integral:

$$\int_0^\infty \frac{\cos kx}{x^2 + a^2}\, dx \tag{6.43}$$

Solution   The terms $\cos kx$ and $1/(x^2 + a^2)$ are both even functions of x.   Because the product of even functions is an even function, the resulting integrand is an even function.   Hence

$$\int_0^\infty \frac{\cos kx}{x^2 + a^2}\, dx = \frac{1}{2} \int_{-\infty}^\infty \frac{\cos kx}{x^2 + a^2}\, dx$$

$$= -\pi \sum \text{ imaginary part of residues } [F(z)e^{jsz}]$$

For this case $s = k$, and $F(z) = 1/(z^2 + a^2)$.   The degree of the denominator of $F(z)$ exceeds the numerator by 2.   The only pole in the upper half plane is located at $z = ja$.

$$\text{Residue } [F(z)e^{jkz}] = \lim_{z \to ja} (z - ja) \frac{e^{jkz}}{(z + ja)(z - ja)}$$

$$= \frac{e^{-ak}}{2ja} = \frac{-je^{-ak}}{2a}$$

This residue is entirely imaginary; hence

$$\int_0^\infty \frac{\cos kx}{x^2 + a^2}\, dx = \frac{\pi e^{-ak}}{2a} \tag{6.44}$$

Integrals of the function $G(\cos \theta, \sin \theta)$

If $G(\cos \theta, \sin \theta)$ is a real rational function of $\cos \theta$ and $\sin \theta$, then

$$\int_0^{2\pi} G(\cos \theta, \sin \theta)\, d\theta = \int_C F(z) \frac{dz}{jz} = \frac{1}{j} \int_C \frac{F(z)}{z}\, dz \tag{6.45}$$

where the contour C is the unit circle shown in Fig. 6.5.   To verify Eq. (6.45), first note that, by restricting z to the unit circle, then $z = e^{j\theta}$ and $1/z = 1/e^{j\theta} = e^{-j\theta}$.   Thus, Euler's equations may be written in the form

$$\cos \theta = \frac{1}{2}\left(e^{j\theta} + e^{-j\theta}\right) = \frac{1}{2}\left(z + \frac{1}{z}\right)$$

$$\sin \theta = \frac{1}{2j}\left(e^{j\theta} - e^{-j\theta}\right) = \frac{1}{2j}\left(z - \frac{1}{z}\right)$$

These relationships show that the integrand $G(\cos \theta, \sin \theta)$ may be expressed as a rational function of z, which is designated $F(z)$.

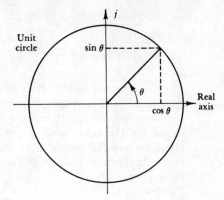

Fig. 6.5. Unit circle.

The derivative of z is

$$dz = je^{j\theta} \, d\theta = jz \, d\theta$$

or

$$d\theta = \frac{dz}{jz}$$

Thus, Eq. (6.45) has been verified. Application of the residue theorem to evaluate this integral gives

$$\int_0^{2\pi} G(\cos \theta, \sin \theta) \, d\theta = 2\pi \sum \text{residues} \frac{F(z)}{z} \tag{6.46}$$

where the summation is of residues within the unit circle.

Illustrative example Evaluate the following integral for the case in which $0 < \rho < 1$:

$$\int_0^{2\pi} \frac{d\theta}{1 - 2\rho \cos \theta + \rho^2} \tag{6.47}$$

Solution The function $F(z)$ is

$$F(z) = \frac{1}{1 - \rho(z + 1/z) + \rho^2} = \frac{z}{z - \rho(z^2 + 1) + \rho^2 z}$$

Hence

$$\frac{F(z)}{z} = \frac{1}{z - \rho(z^2 + 1) + \rho^2 z} = \frac{1}{(1 - \rho z)(z - \rho)}$$

The poles are at $z = 1/\rho$ and $z = \rho$. For $0 < \rho < 1$, only the pole at $z = \rho$ is within the unit circle; hence

$$\text{Residue} \frac{F(z)}{z} = \lim_{z \to \rho} \frac{1}{1 - \rho z} = \frac{1}{1 - \rho^2}$$

233

Application of Eq. (6.46) gives

$$\int_0^{2\pi} \frac{d\theta}{1 - 2\rho \cos \theta + \rho^2} = \frac{2\pi}{1 - \rho^2} \tag{6.48}$$

Contour integration is a powerful technique for evaluating definite integrals. After obtaining a little experience with different contours, one can ascertain quite readily the contour which is best suited for the solution of a particular problem. The following examples and the problems at the end of this chapter have been chosen to illustrate features of various contours employed in contour integration.

Illustrative example   Evaluate the integral

$$\int_0^\infty \frac{\sin x}{x} \, dx \tag{6.49}$$

Solution   In contour integration, it is generally much easier to work with exponentials whenever possible. Thus, by letting $F(z) = e^{jz}/z$,

$$\int_0^\infty \frac{e^{jz}}{z} \, dz = \int_0^\infty \frac{\cos z}{z} \, dz + j \int_0^\infty \frac{\sin z}{z} \, dz$$

The imaginary part of this integral is the desired result. The real part gives the value of the integral of $(\cos x)/x$. Because the function $F(z)$ has a pole on the real axis at $z = 0$, then the origin is excluded, as shown in the contour of Fig. 6.6. The C_1 portion is the small semicircle of radius ϵ, and the C_2 portion is the large semicircle of radius r. Application of Cauchy's inte-

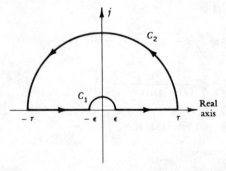

Fig. 6.6.   A semicircular contour which excludes the origin.

gral theorem gives

$$\int_{-r}^{-\epsilon} F(z)\, dz + \int_{C_1} F(z)\, dz + \int_{\epsilon}^{r} F(z)\, dz + \int_{C_2} F(z)\, dz = 0$$

$$(6.50)$$

For values of z on C_1, then $z = \epsilon e^{j\theta}$, $dz = j\epsilon e^{j\theta}\, d\theta$, and

$$dz/z = j\, d\theta$$

$$\int_{C_1} F(z)\, dz = \lim_{\epsilon \to 0} j \int_{\pi}^{0} e^{j\epsilon(\cos\theta + j\sin\theta)}\, d\theta = j \int_{\pi}^{0} d\theta = -j\pi \quad (6.51)$$

For values of z on C_2, then $z = re^{j\phi}$, $dz = jre^{j\phi}\, d\phi$, and

$$dz/z = j\, d\phi$$

$$\int_{C_2} F(z)\, dz = \lim_{r \to \infty} j \int_{0}^{\pi} e^{jr(\cos\phi + j\sin\phi)}\, d\phi$$

Because $e^{jr\cos\phi}$ is a unit vector, then

$$\left| \int_{C_2} F(z)\, dz \right| \le \lim_{r \to \infty} \int_{0}^{\pi} e^{-r\sin\phi}\, d\phi = \lim_{r \to \infty} 2 \int_{0}^{\pi/2} e^{-r\sin\phi}\, d\phi$$

Figure 6.3 illustrates the validity of Eq. (6.22); that is, for $0 \le \phi \le \pi/2$, then $\sin\phi \ge 2\phi/\pi$. Thus

$$\left| \int_{C_2} F(z)\, dz \right| \le \lim_{r \to \infty} 2 \int_{0}^{\pi/2} e^{-2r\phi/\pi}\, d\phi = \lim_{r \to \infty} \frac{e^{-2r\phi/\pi}}{r/\pi} \Big|_{\pi/2}^{0}$$

$$\le \lim_{r \to \infty} \frac{\pi}{r}(1 - e^{-r}) = 0$$

In the limit, as ϵ approaches zero and r infinity, Eq. (6.50) becomes

$$\int_{-\infty}^{\infty} F(z)\, dz = \int_{-\infty}^{\infty} \frac{e^{jx}}{x}\, dx = \int_{-\infty}^{\infty} \frac{\cos x}{x}\, dx + j \int_{-\infty}^{\infty} \frac{\sin x}{x}\, dx = j\pi$$

where for values of z on the real axis $z = x$, $F(z) = e^{jx}/x$, and $dz = dx$. Equating real and imaginary parts gives

$$\int_{-\infty}^{\infty} \frac{\sin x}{x}\, dx = \pi \qquad \text{and} \qquad \int_{-\infty}^{\infty} \frac{\cos x}{x}\, dx = 0$$

Because $(\sin x)/x$ is an even function, then

$$\int_{0}^{\infty} \frac{\sin x}{x}\, dx = \frac{\pi}{2} \tag{6.52}$$

Because $(\cos x)/x$ is an odd function, then, as proved above, this integral must vanish.

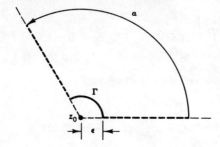

Fig. 6.7. A circular arc Γ of radius ϵ which subtends an angle α about z_0.

In contour integration, it is frequently necessary to construct a small circular arc around a pole, as is the case for the pole at the origin of Fig. 6.6. In the preceding example, the value of the integral around the small semicircular arc about the origin was obtained by direct integration [Eq. (6.51)]. It will now be shown how such an integral may be evaluated by application of the residue theorem.

Consider the function $f(z)/(z - z_0)$, which has a simple pole at z_0. As shown in Fig. 6.7, a circular arc Γ of radius ϵ subtends an angle α about z_0. For a simple pole at z_0, the Laurent expansion has the form

$$\frac{f(z)}{z - z_0} = \frac{a_{-1}}{z - z_0} + a_0 + a_1(z - z_0) + \cdots \qquad (6.53)$$

Integration of each term of this Laurent series along the circular arc Γ gives

$$\int_\Gamma \frac{f(z)}{z - z_0}\, dz = a_{-1} \int_\Gamma \frac{dz}{z - z_0} + a_0 \int_\Gamma dz$$
$$+ a_1 \int_\Gamma (z - z_0)\, dz + \cdots$$

For values of z on the circular arc, then $z - z_0 = \epsilon e^{j\theta}$, and $dz = j\epsilon e^{j\theta}\, d\theta$.

$$\int_\Gamma \frac{f(z)}{z - z_0}\, dz = ja_{-1} \int_0^\alpha d\theta + j\epsilon a_0 \int_0^\alpha e^{j\theta}\, d\theta$$
$$+ j\epsilon^2 a_1 \int_0^\alpha e^{2j\theta}\, d\theta + \cdots$$

In the limit as ϵ approaches zero, all the integrals in the preceding expression vanish except the first. Thus

$$\int_\Gamma \frac{f(z)}{z - z_0}\, dz = j\alpha a_{-1} = j\alpha \ (\text{residue at } z_0) \qquad (6.54)$$

Because this Laurent expansion is about the point z_0, then the a_{-1} coefficient is the residue at z_0.

Application of Eq. (6.54) to verify Eq. (6.51) gives

$$\int_{C_1} F(z)\,dz = \int_{C_1} \frac{e^{jz}}{z}\,dz = -j\pi \lim_{z \to 0} e^{jz} = -j\pi$$

The minus sign occurs because the semicircle C_1 goes about the pole $z = 0$ of Fig. 6.6 in the negative direction. Because C_1 is a semicircle, the subtended angle is π.

Illustrative example Show that, for $a > 0$,

$$\int_0^\infty \frac{\sin ax}{\sinh x}\,dx = \frac{\pi}{2}\tanh \frac{a\pi}{2} \tag{6.55}$$

Solution The function to be integrated is

$$F(z) = \frac{e^{jaz}}{\sinh z} = \frac{e^{jaz}}{(e^z - e^{-z})/2}$$

Because $\sinh z = -j \sin jz$, then the poles of $F(z)$ occur at $jz = \pm k\pi$ or $z = \mp jk\pi$. Thus, poles occur every π units along the imaginary axis, as shown in Fig. 6.8. The function $F(z)$ is analytic inside this contour, which excludes the pole at the origin and at $j\pi$. Application of Cauchy's integral theorem gives

$$\int_{-r}^{-\epsilon} F(z)\,dz + \int_{-\epsilon}^{+\epsilon} F(z)\,dz + \int_{\epsilon}^{r} F(z)\,dz + \int_{r}^{r+j\pi} F(z)\,dz$$
$$+ \int_{r+j\pi}^{\epsilon+j\pi} F(z)\,dz + \int_{\epsilon+j\pi}^{-\epsilon+j\pi} F(z)\,dz + \int_{-\epsilon+j\pi}^{-r+j\pi} F(z)\,dz$$
$$+ \int_{-r+j\pi}^{-r} F(z)\,dz = 0 \quad (6.56)$$

On the real axis, $z = x$. Thus

$$\int_{-r}^{-\epsilon} F(z)\,dz = \int_{-r}^{-\epsilon} \frac{(\cos ax + j \sin ax)\,dx}{\sinh x}$$

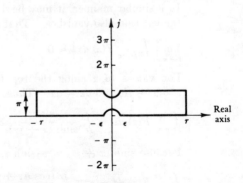

Fig. 6.8. A path of integration.

237

Similarly

$$\int_\epsilon^r F(z)\,dz = \int_\epsilon^r \frac{(\cos ax + j \sin ax)\,dx}{\sinh x}$$

Because $(\cos ax)/\sinh x$ is an odd function and $(\sin ax)/\sinh x$ is an even function, then adding the preceding integrals gives

$$\int_{-r}^{-\epsilon} F(z)\,dz + \int_\epsilon^r F(z)\,dz = 2j \int_\epsilon^r \frac{\sin ax}{\sinh x}\,dx \qquad (6.57)$$

Application of Eqs. (6.54) and (5.80) to evaluate the integral along the small semicircle about the origin gives

$$\int_{-\epsilon}^\epsilon F(z)\,dz = -j\pi \lim_{z\to 0} \frac{2e^{jaz}}{(d/dz)(e^z - e^{-z})} = -j\pi$$

In a similar manner, the integral along the small semicircle about $j\pi$ is found to be

$$\int_{\epsilon+j\pi}^{-\epsilon+j\pi} F(z)\,dz = -j\pi \lim_{z\to j\pi} \frac{2e^{jaz}}{(d/dz)(e^z - e^{-z})}$$

$$= -j\pi \frac{2e^{-a\pi}}{e^{j\pi} + e^{-j\pi}} = j\pi e^{-a\pi}$$

For values of z along the right side of the contour of Fig. 6.8, $z = r + jy$, and $dz = j\,dy$. Thus

$$\left| \int_r^{r+j\pi} F(z)\,dz \right| \le \int_0^\pi \frac{|2e^{ja(r+jy)}|\,|j\,dy|}{|e^{(r+jy)} + e^{-(r+jy)}|} \le \int_0^\pi \frac{2e^{-ay}\,dy}{e^r|e^{jy} + e^{-(2r+jy)}|}$$

In the limit as r becomes infinite, the preceding integral becomes

$$\lim_{r\to\infty} \left| \int_r^{r+j\pi} F(z)\,dz \right| \le \frac{2}{e^r} \int_0^\pi e^{-ay}\,dy \approx 0$$

In a similar manner, it may be shown that the integral along the left side also vanishes. That is,

$$\lim_{r\to\infty} \left| \int_{-r+j\pi}^{-r} F(z)\,dz \right| \approx 0$$

For values of z along the top from $r + j\pi$ to $\epsilon + j\pi$, then $z = x + j\pi$, and $dz = dx$. Thus

$$\int_{r+j\pi}^{\epsilon+j\pi} F(z)\,dz = \int_r^\epsilon \frac{e^{ja(x+j\pi)}}{\sinh (x + j\pi)}\,dx = e^{-a\pi} \int_r^\epsilon \frac{e^{jax}\,dx}{\sinh (x + j\pi)}$$

Because $\sinh (x + j\pi) = -\sinh x$, then

$$\int_{r+j\pi}^{\epsilon+j\pi} F(z)\,dz = e^{-a\pi} \int_\epsilon^r \frac{(\cos ax + j \sin ax)\,dx}{\sinh x}$$

Similarly, it may be shown that, for the region from $-\epsilon + j\pi$ to $-r + j\pi$,

$$\int_{-\epsilon+j\pi}^{-r+j\pi} F(z)\, dz = e^{-a\pi} \int_{-r}^{-\epsilon} \frac{(\cos ax + j \sin ax)\, dx}{\sinh x}$$

By noting as before that $(\cos ax)/\sinh x$ is an odd function and $(\sin ax)/\sinh x$ is an even function, then the addition of the two preceding integrals gives

$$\int_{-r}^{-\epsilon} F(z)\, dz + \int_{\epsilon}^{r} F(z)\, dz = 2je^{-a\pi} \int_{\epsilon}^{r} \frac{\sin ax\, dx}{\sinh x}$$

Finally, substituting the values of all the integrals into Eq. (6.56) and taking the limit as ϵ approaches zero and r infinity gives

$$2j(1 + e^{-a\pi}) \int_0^\infty \frac{\sin ax}{\sinh x}\, dx = j\pi(1 - e^{-a\pi})$$

Thus

$$\int_0^\infty \frac{\sin ax}{\sinh x}\, dx = \frac{\pi}{2} \frac{1 - e^{-a\pi}}{1 + e^{-a\pi}} = \frac{\pi}{2} \tanh \frac{a\pi}{2} \tag{6.58}$$

Illustrative example Show that

$$\int_0^\infty \cos x^2\, dx = \int_0^\infty \sin x^2\, dx = \frac{1}{2}\sqrt{\frac{\pi}{2}} \tag{6.59}$$

Solution The function to be integrated is

$$F(z) = e^{jz^2} = \cos z^2 + j \sin z^2$$

Because this is an entire function, application of Cauchy's theorem to the contour shown in Fig. 6.9 gives

$$\int_0^r F(z)\, dz + \int_{C_1} F(z)\, dz + \int_{C_2} F(z)\, dz = 0 \tag{6.60}$$

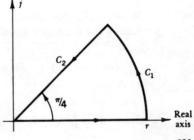

Fig. 6.9. A sector as a path of integration.

where C_1 and C_2 are the portions of the contour indicated in Fig. 6.9. For values of z on the real axis

$$\int_0^r F(z)\, dz = \int_0^r e^{jx^2}\, dx = \int_0^r (\cos x^2 + j \sin x^2)\, dx$$

For values of z on C_1, then $z = re^{j\theta}$, $z^2 = r^2 e^{2j\theta}$, and $dz = jre^{j\theta}\, d\theta$. Thus

$$\int_{C_1} F(z)\, dz = j \int_0^{\pi/4} re^{jr^2(\cos 2\theta + j \sin 2\theta)}\, e^{j\theta}\, d\theta$$

Because $e^{j\theta}$ and $e^{jr^2 \cos 2\theta}$ are unit vectors, then

$$\left| \int_{C_1} F(z)\, dz \right| \leq \int_0^{\pi/4} re^{-r^2 \sin 2\theta}\, d\theta$$

By making the change of variable $\phi = 2\theta$, then the new upper limit of integration at $\theta = \pi/4$ is $\phi = 2\theta = \pi/2$. Thus

$$\left| \int_{C_1} F(z)\, dz \right| \leq \frac{1}{2} \int_0^{\pi/2} re^{-r^2 \sin \phi}\, d\phi$$

Recalling from Eq. (6.22) that $\sin \phi \geq 2\phi/\pi$ for $0 \leq \phi \leq \pi/2$, then

$$\left| \int_{C_1} F(z)\, dz \right| \leq \frac{1}{2} \int_0^{\pi/2} re^{-2r^2\phi/\pi}\, d\phi = \frac{1 - e^{-r^2}}{2r/\pi}$$

In the limit as r becomes infinite, the preceding integral vanishes. For values of z on C_2, then $z = \rho e^{j\pi/4}$,

$$z^2 = \rho^2(\cos \pi/2 + j \sin \pi/2) = j\rho^2$$

and

$$dz = e^{j\pi/4}\, d\rho; \text{ thus,}$$

$$\int_{C_2} F(z)\, dz = e^{j\pi/4} \int_r^0 e^{-\rho^2}\, d\rho$$

Substituting the values of all the integrals into Eq. (6.60) and taking the limit as r approaches infinity gives

$$\int_0^\infty \cos x^2\, dx + j \int_0^\infty \sin x^2\, dx = e^{j\pi/4} \int_0^\infty e^{-\rho^2}\, d\rho = \frac{(1 + j)\sqrt{\pi}}{2\sqrt{2}}$$

where $e^{j\pi/4} = (1 + j)\sqrt{2}/2$, and the value of the last definite integral is $\int_0^\infty e^{-\rho^2}\, d\rho = \sqrt{\pi}/2$. The value of this integral is verified as follows:

$$\left(\int_{-\infty}^{\infty} e^{-\rho^2}\, d\rho\right)^2 = \int_{-\infty}^{\infty} e^{-x^2}\, dx \int_{-\infty}^{\infty} e^{-y^2}\, dy$$
$$= \int_{-\infty}^{\infty} \int_{-\infty}^{\infty} e^{-(x^2+y^2)}\, dx\, dy$$
$$= \int_0^{2\pi} \int_0^{\infty} e^{-r^2}\, r\, dr\, d\theta = \int_0^{2\pi} \left[\frac{e^{-r^2}}{2}\right]_{\infty}^0 d\theta$$
$$= \frac{1}{2} \int_0^{2\pi} d\theta = \pi$$

Hence

$$\int_0^{\infty} e^{-\rho^2}\, d\rho = \frac{1}{2} \int_{-\infty}^{\infty} e^{-\rho^2}\, d\rho = \sqrt{\frac{\pi}{2}} \tag{6.61}$$

6.5 MULTIVALUED FUNCTIONS

A single-valued function is one for which there is but one value of $F(z)$ for each value of z. For a multivalued function there is some value of z that yields more than one value of $F(z)$. In Fig. 6.10 is shown the vector $z = re^{j\theta}$. Rotating the vector by one or more complete revolutions (that is, 360°, 720°, etc.) does not change its value.

$$z = re^{j\theta} = re^{j(\theta + 2\pi n)} \qquad n = 0, \pm 1, \pm 2, \ldots \tag{6.62}$$

A single-valued function is distinguished by the fact that it has the same value regardless of the value of n. For example,

$$z^2 = r^2 e^{j(2\theta + 4\pi n)} = r^2 e^{j2\theta}$$
$$z^{-2} = r^{-2} e^{-j(2\theta + 4\pi n)} = r^{-2} e^{-j2\theta}$$
$$z^3 = r^3 e^{j(3\theta + 6\pi n)} = r^3 e^{j3\theta}$$
$$z^{-3} = r^{-3} e^{-j(3\theta + 6\pi n)} = r^{-3} e^{-j3\theta}$$

Because the value of each of the preceding functions is independent of n, then each is single-valued. It follows that any power series is single-valued in its ring of convergence.

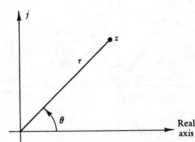

Fig. 6.10.   A vector $z = re^{j\theta}$.

Next consider the function

$$F(z) = \sqrt{z} = \sqrt{r}\, e^{j(\theta + 2\pi n)/2} \qquad n = 0, \pm 1, \pm 2, \ldots \qquad (6.63)$$

Regardless of the value of n, there are but two distinct values for $F(z)$; that is,

$$F(z) = \begin{cases} \sqrt{r}\, e^{j\theta/2} \\ \sqrt{r}\, e^{j(\theta/2+\pi)} \end{cases}$$

As is illustrated in Fig. 6.11, for a given value of z, then $\sqrt{z}$ may have either of two values. Such a multivalued function is said to be *double-valued*. In a similar way, it may be shown that $z^{\frac{1}{3}}$ is a triple-valued function. The principal value of $\sqrt{z}$ is the value of Eq. (6.63) for $n = 0$, $r > 0$, and $-\pi < \theta < \pi$. Thus, the principal value of $\sqrt{z}$ is

$$\sqrt{z} = \sqrt{r}\, e^{j\theta/2} \qquad r > 0,\ -\pi < \theta < \pi \qquad (6.64)$$

The domain $(r > 0,\ -\pi < \theta < \pi)$ is illustrated in Fig. 6.12a, in which the heavy line along the negative real axis is referred to as the *branch cut*. The significance of the branch cut is that, as values of z proceed along a path (as in contour integration), the path is not to cross the branch cut. In Fig. ·6.12a is shown a circular path of values of z that begins at a and proceeds to b to c to d to e. Points a and e lie on opposite sides of the cut. The corresponding plot of $\sqrt{z}$ is shown in Fig. 6.12b. If the path were to cross the cut in going from e to a, then the corresponding plot of $\sqrt{z}$ would be discontinuous because the value of $\sqrt{z}$ at e is

$$\sqrt{z} = \sqrt{r}\, e^{j(\pi-\epsilon)/2} \approx \sqrt{r}\, \epsilon^{j\pi/2} \qquad (6.65)$$

The value of $\sqrt{z}$ at a is

$$\sqrt{z} = \sqrt{r}\, e^{-j(\pi-\epsilon)/2} \approx \sqrt{r}\, e^{-j\pi/2} \qquad (6.66)$$

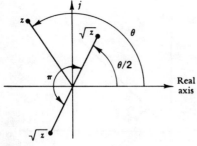

Fig. 6.11. The vector z and its two square roots.

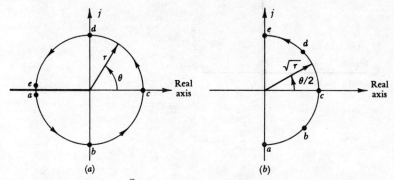

Fig. 6.12. Determination of $\sqrt{z}$ with branch cut along negative real axis. (a) Circular path of values of z (that is a, b, c, d, e); (b) corresponding plot of $\sqrt{z}$.

For values of z just above the branch cut, the angle of $\sqrt{z}$ is $\pi/2$, and just below the branch cut the angle is $-\pi/2$. As indicated by Eqs. (6.65) and (6.66) the function $\sqrt{z}$ is discontinuous along the branch cut. Because its derivative does not exist on the cut, it is not analytic along the cut. Multivalued functions may be regarded as single-valued functions in which the branch cut is a region for which the function is not analytic.

The branch cut could have been made along any radial line extending out from the origin. For example, if the angle of the ray were θ_0, then the domain of the principal value would be $(r > 0, \theta_0 < \theta < \theta_0 + 2\pi)$. The origin $(z = 0)$ of the branch cut is called the *branch point*. A branch point is the singular point common to all branch cuts. Branch points occur at poles and at zeros of multivalued functions.

Next consider the infinitely many-valued function

$$\ln z = \ln re^{j(\theta+2\pi n)} = \ln r + \ln e^{j(\theta+2\pi n)}$$
$$= \ln r + j(\theta + 2\pi n) \qquad n = 0, \pm 1, \pm 2, \ldots \qquad (6.67)$$

The principal value of $\ln z$ is the value of Eq. (6.67) for $n = 0, r > 0$, and $-\pi < \theta < \pi$.

$$\ln z = \ln r + j\theta \qquad r > 0, -\pi < \theta < \pi \qquad (6.68)$$

The preceding function is now a single-valued function in which the branch cut as shown in Fig. 6.13a is a region for which the function is not analytic. The branch point is located at the origin (i.e., at $z = 0$). As r varies from 0 to ∞ in Eq. (6.68), then $\ln r$ (the real part of the function) goes from $-\infty$ to $+\infty$. Similarly, as θ varies

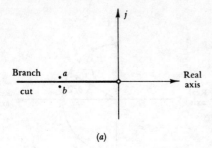

(a)

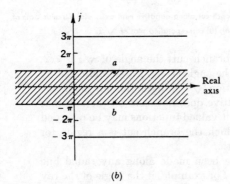

(b)

Fig. 6.13. Determination of ln z with branch cut along negative real axis. (a) Values of z (that is a and b); (b) corresponding plot of ln z.

from $-\pi$ to $+\pi$, then $j\theta$ (the imaginary part of the function) goes from $-j\pi$ to $+j\pi$. Thus, for $r > 0$, $-\pi < \theta < \pi$ and for $n = 0$, then ln z plots into the shaded strip shown in Fig. 6.13b. This is called the *principal strip*. For $n = 1$, ln z plots into the strip between π and 3π. Hence ln z may fall in an infinite number of strips, depending on the value of n.

In Fig. 6.13a is shown a point a just above the cut, and a point b just below the cut. The value of ln z at point a is

$$\ln z = \ln r + j(\pi - \epsilon) \approx \ln r + j\pi \qquad (6.69)$$

The value of ln z at point b is

$$\ln z = \ln r - j(\pi - \epsilon) \approx \ln r - j\pi \qquad (6.70)$$

The corresponding points a and b shown in Fig. 6.13b are discontinuous. The branch cut is thus seen to represent a region along which the function is not analytic.

In Fig. 6.14a is shown the branch cut for the function

$$\sqrt{z-1} = \sqrt{z_1} = \sqrt{r_1}\, e^{j(\theta_1 + 2\pi n)/2} \qquad n = 0,\ \pm 1,\ \pm 2,\ \dots$$
$$(6.71)$$

where $r_1 e^{j\theta_1} = z - 1$ is a vector with origin at $z = 1$, as shown in Fig. 6.14a. The principal value of this function is

$$\sqrt{z-1} = \sqrt{r_1}\, e^{j\theta_1/2} \qquad r_1 > 0,\ -\pi < \theta_1 < \pi \qquad (6.72)$$

In going across the branch cut the angle changes by 2π (that is, goes from π to $-\pi$), and thus the square root is discontinuous because it changes by the angle π. Similarly, the branch cut for $\sqrt{z-3}$ is shown in Fig. 6.14b. The branch cut for the product $\sqrt{(z-3)(z-1)}$ is shown in Fig. 6.14c. By superimposing Fig. 6.14a on Fig. 6.14b, it would initially be expected that the branch cut of Fig. 6.14c should continue on the real axis to the left of the point $z = 1$. In effect, the branch cuts cancel themselves out in this region. On the real axis to the left of $z = 1$ the angle of $z - 1 = r_1 e^{j\theta_1}$ changes by 2π and the angle of $z - 3 = r_2 e^{j\theta_2}$ also changes by 2π. Because the total change is 4π, the resultant square root changes by 2π. Thus, the square root is continuous in this region. When branch cuts coincide, it is necessary to check to determine whether or not they cancel. In Fig. 6.14d is shown the branch cut for the function $\sqrt{(z-3)(z-1)(z+2)}$. Figure 6.14d

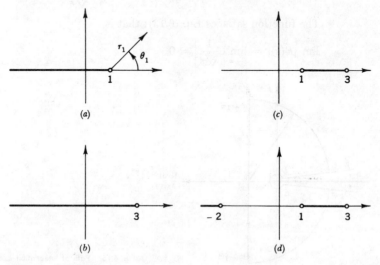

(a) (c) (b) (d)

Fig. 6.14. Branch cuts.

is obtained by superimposing the branch cut for the function $\sqrt{(z-3)(z-1)}$ (that is, Fig. 6.14c) and the branch cut for the function $\sqrt{z+2}$. The branch points of Fig. 6.14d are located at $z = -2, 1,$ and 3.

The following example illustrates how contour integration is applied to multivalued functions.

Illustrative example Find the inverse transform of the function $F(s) = 1/\sqrt{s}$.

Solution From Eq. (6.14)

$$f(t) = \mathcal{L}^{-1}[F(s)] = \frac{1}{2\pi j} \int_{a-j\infty}^{a+j\infty} \frac{e^{st}}{\sqrt{s}} \, ds \tag{6.73}$$

The function to be integrated has a singular point at the origin ($s = 0$) and a branch cut along the negative axis, as illustrated in Fig. 6.15. The closed contour which goes from A to B to C to D to E to F to A excludes all the singular points of the function. Because the function is analytic within this contour, application of Cauchy's theorem gives

$$\int_{a-j\infty}^{a+j\infty} \frac{e^{st}}{\sqrt{s}} \, ds + \int_{B}^{C} \frac{e^{st}}{\sqrt{s}} \, ds + \int_{C}^{D} \frac{e^{st}}{\sqrt{s}} \, ds + \int_{D}^{E} \frac{e^{st}}{\sqrt{s}} \, ds$$
$$+ \int_{E}^{F} \frac{e^{st}}{\sqrt{s}} \, ds + \int_{F}^{A} \frac{e^{st}}{\sqrt{s}} \, ds = 0 \quad (6.74)$$

The function satisfies Eq. (6.13); that is,

$$\lim_{s \to \infty} |F(s)| = \lim_{s \to \infty} \left| \frac{1}{\sqrt{s}} \right| \approx 0$$

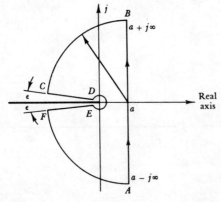

Fig. 6.15. Path of integration excluding the branch cut.

Thus, the integral of the function around each of the infinite arcs BC and FA vanishes.

For values of s on the section CD, then $s = re^{j(\pi-\epsilon)} \approx -r$, $\sqrt{s} = \sqrt{r}\, e^{j(\pi-\epsilon)/2} \approx \sqrt{r}\, e^{j\pi/2} = j\sqrt{r}$, and $ds = -dr$. Hence, the line integral along this section is

$$\int_C^D \frac{e^{st}}{\sqrt{s}}\, ds = -\int_\infty^0 \frac{e^{-rt}}{j\sqrt{r}}\, dr = \frac{1}{j}\int_0^\infty \frac{e^{-rt}}{\sqrt{r}}\, dr = \frac{1}{j}\sqrt{\frac{\pi}{t}} \tag{6.75}$$

The value of the preceding integral of $e^{-rt}/\sqrt{r}$ is obtained directly from a table of integrals.

For values of s on the line section EF, then

$$s = re^{-j(\pi-\epsilon)} \approx -r$$

$$\sqrt{s} = \sqrt{r}\, e^{-j(\pi-\epsilon)/2} \approx \sqrt{r}\, e^{-j\pi/2} = -j\sqrt{r}, \quad \text{and} \quad ds = -dr.$$

Hence, the line integral along this section is

$$\int_E^F \frac{e^{st}}{\sqrt{s}}\, ds = \int_0^\infty \frac{e^{-rt}}{j\sqrt{r}}\, dr = \frac{1}{j}\sqrt{\frac{\pi}{t}} \tag{6.76}$$

For the small circle of radius δ about the origin, $s = \delta e^{j\theta}$, $\sqrt{s} = \sqrt{\delta}\, e^{j\theta/2}$, $ds = j\delta e^{j\theta}\, d\theta$, and $e^{st} \approx 1$. Thus

$$\int_D^E \frac{e^{st}}{\sqrt{s}}\, ds = \lim_{\delta\to 0} j\sqrt{\delta} \int_{-\pi}^\pi e^{j\theta/2}\, d\theta = \lim_{\delta\to 0} 4\sqrt{\delta}\, j \approx 0 \tag{6.77}$$

Substitution of the preceding results into Eq. (6.74) gives

$$\int_{a-j\infty}^{a+j\infty} \frac{e^{st}}{\sqrt{s}}\, ds = \frac{-2}{j}\sqrt{\frac{\pi}{t}}$$

Thus, Eq. (6.73) may now be evaluated for the desired result

$$\mathcal{L}^{-1}\left(\frac{1}{\sqrt{s}}\right) = \frac{1}{2\pi j}\left(\frac{-2}{j}\sqrt{\frac{\pi}{t}}\right) = \frac{-1}{j^2\pi}\sqrt{\frac{\pi}{t}} = \frac{1}{\sqrt{\pi t}} \tag{6.78}$$

As illustrated in the preceding example, in applying contour integration to multivalued functions, the branch cut is regarded as a region for which the function is not analytic. Hence, the contour must exclude the branch cut.

Additional applications of the theory of complex variables to the solution of more advanced engineering problems are explained in the following chapters.

PROBLEMS

6.1 Use the method of residues to determine the inverse transform of each of the following functions:

(a) $\dfrac{1}{(s+a)(s+b)}$ (b) $\dfrac{s}{(s+a)(s+b)}$

(c) $\dfrac{s+1}{(s+a)(s+b)}$ (d) $\dfrac{1}{s^2-a^2}$

(e) $\dfrac{s}{s^2-a^2}$ (f) $\dfrac{s+k}{s^2-2as+a^2+b^2}$

6.2 Use the method of residues to determine the inverse transform of each of the following functions:

(a) $\dfrac{1}{s^2(s+a)}$ (b) $\dfrac{1}{s(s+a)^2}$

(c) $\dfrac{1}{s^4-\omega^4}$ (d) $\dfrac{s}{s^4-\omega^4}$

(e) $\dfrac{s}{(s^2-2as+a^2+b^2)^2}$ (f) $\dfrac{s+k}{(s^2-2as+a^2+b^2)^2}$

6.3 Use the method of residues to evaluate each of the following integrals:

(a) $\displaystyle\int_{-\infty}^{\infty}\dfrac{dx}{1+x^2}$ (b) $\displaystyle\int_{-\infty}^{\infty}\dfrac{x^2\,dx}{(x^2+a^2)(x^2+b^2)}$

(c) $\displaystyle\int_{-\infty}^{\infty}\dfrac{\cos x\,dx}{(x^2+a^2)(x^2+b^2)}$ (d) $\displaystyle\int_{-\infty}^{\infty}\dfrac{dx}{x^2-2ax+a^2+b^2}$

(e) $\displaystyle\int_{-\infty}^{\infty}\dfrac{\sin mx\,dx}{x^2-2ax+a^2+b^2}$ (f) $\displaystyle\int_{-\infty}^{\infty}\dfrac{\cos mx\,dx}{x^2-2ax+a^2+b^2}$

6.4 Use the method of residues to evaluate each of the following integrals:

(a) $\displaystyle\int_{0}^{2\pi}\dfrac{d\theta}{2+\cos\theta}$ (b) $\displaystyle\int_{0}^{2\pi}\dfrac{\cos\theta\,d\theta}{2+\cos\theta}$

(c) $\displaystyle\int_{0}^{2\pi}\dfrac{\cos\theta\,d\theta}{2+\sin\theta}$ (d) $\displaystyle\int_{0}^{2\pi}\dfrac{d\theta}{\sin\theta+2\cos\theta+3}$

(e) $\displaystyle\int_{0}^{2\pi}\dfrac{d\theta}{(a+b\cos\theta)^2}$ $\quad 0<b<a$

6.5 Show that

$$\int_{0}^{\infty}\frac{(\sin x)^2}{x^2}\,dx=\frac{1}{2}\int_{0}^{\infty}\frac{(1-\cos 2x)\,dx}{x^2}=\frac{\pi}{2}$$

Hint: Integrate $(1-e^{2iz})/z^2$ around the contour shown in Fig. 6.6 of the text.

6.6 Show that

$$\int_{-\infty}^{\infty}\frac{dx}{(x-1)(x^2+1)}=-\frac{\pi}{2}$$

Hint: Integrate $1/[(z-1)(z^2+1)]$ around the contour shown in Fig. P6.6.

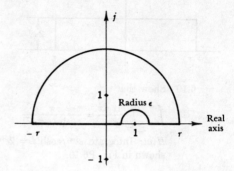

Fig. P6.6

6.7 Show that

$$\int_{-\infty}^{\infty} \frac{dx}{x^3 - 1} = -\frac{\pi}{\sqrt{3}}$$

Hint: Integrate $1/(z^3 - 1)$ around the contour shown in Fig. P6.6.

6.8 Show that

$$\int_0^{\infty} \frac{\cos x}{\sqrt{x}}\, dx = \int_0^{\infty} \frac{\sin x}{\sqrt{x}}\, dx = \sqrt{\frac{\pi}{2}}$$

Hint: Integrate $e^{iz}/\sqrt{z}$ around the contour shown in Fig. P6.8.

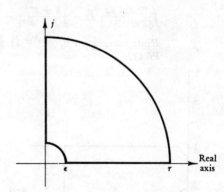

Fig. P6.8

6.9 Show that

$$\int_{-\infty}^{\infty} \frac{e^{ax}}{1 + e^x} = \frac{\pi}{\sin a\pi}$$

Hint: Integrate $e^{az}/(1 + e^z)$ along the contour shown in Fig. P6.9.

Fig. P6.9

6.10 Show that

$$\int_0^\infty \frac{\cos mx \, dx}{\cosh x} = \frac{\pi}{\cosh (m\pi/2)}$$

Hint: Integrate $e^{jmz}/\cosh z = 2e^{jmz}/(e^z + e^{-z})$ around the contour shown in Fig. P6.10.

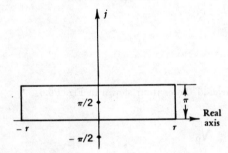

Fig. P6.10

6.11 Show that

$$\int_0^\infty \frac{\sin ax}{e^{2\pi x} - 1} \, dx = \frac{1 + e^{-a}}{4(1 - e^{-a})} - \frac{1}{2a}$$

Hint: Integrate $e^{jaz}/(e^{2\pi z} - 1)$ around the contour shown in Fig. P6.11.

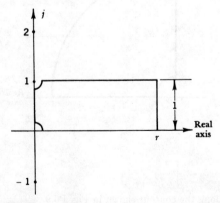

Fig. P6.11

6.12 For the function $\sqrt{(z-3)(z-1)(z+2)}$, determine the branch cuts for $r > 0$ and $0 < \theta < 2\pi$.

6.13 For the function $\sqrt{(z-3)(z-1)(z+2)(z+4)}$, determine the branch cuts for:

(a) $r > 0, \; -\pi < \theta < \pi$
(b) $r > 0, \; 0 < \theta < 2\pi$

6.14 For the function $\sqrt{z^2+1} = \sqrt{(z+j)(z-j)}$, determine the branch cuts for:

(a) $r > 0, \; -\pi/2 < \theta < 3\pi/2$
(b) $r > 0, \; -3\pi/2 < \theta < \pi/2$

6.15 For the function $\sqrt{(z+j)(z-j)(z-1)} = \sqrt{(r_1 e^{j\theta_1})(r_2 e^{j\theta_2})(r_3 e^{j\theta_3})}$ determine the branch cuts for $r_1 > 0, \; -\pi/2 < \theta_1 < 3\pi/2, \; r_2 > 0, \; -\pi/2 < \theta_2 < 3\pi/2$, and $r_3 > 0, \; -\pi < \theta_3 < \pi$.

6.16 Integrate the function $z^{-k}/(1+z)$ around the contour shown in Fig. P6.16 to show that

$$\int_0^\infty \frac{x^{-k}}{1+x}\,dx = \frac{\pi}{\sin k\pi} \qquad 0 < k < 1$$

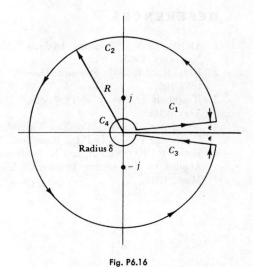

Fig. P6.16

6.17 Integrate the function $z^k/(1+z^2)$ around the contour shown in Fig. P6.16 to show that

$$\int_0^\infty \frac{x^k}{1+x^2}\,dx = \frac{\pi}{2}\frac{1}{\cos(k\pi/2)} \qquad -1 < k < 1$$

Because $z^k = e^{k \ln z} = e^{k(\ln r + j\theta)}$, it is necessary to construct a branch cut as shown in Fig. P6.16 in which $z = 0$ is the branch point and $0 < \theta < 2\pi$. (*Hint:* First show that the value of the integral along C_2 approaches zero as R becomes infinite and the value of the integral along C_4 approaches zero as δ approaches zero.)

6.18 Once a formula has been obtained, it is possible to obtain many more formulas by performing various manipulations. For example, the answer to Prob. 6.17 may be expressed in the form

$$\int_0^\infty \frac{e^{k \ln x}}{1 + x^2} \, dx = \frac{\pi}{2} \sec \frac{k\pi}{2}$$

(*a*) Differentiate this formula with respect to k, and then evaluate at $k = 0$ to show that

$$\int_0^\infty \frac{\ln x}{1 + x^2} \, dx = 0$$

(*b*) Differentiate the formula twice with respect to k, and then evaluate at $k = 0$ to show that

$$\int_0^\infty \frac{(\ln x)^2}{1 + x^2} \, dx = \frac{\pi^3}{8}$$

REFERENCES

1 Ahlfors, L. V.: "Complex Analysis," McGraw-Hill Book Company, New York, 1953.

2 Hille, E.: "Analytic Function Theory," Ginn and Company, Boston, 1959–1962.

3 Knopp, K.: "Theory of Functions," Dover Publications, Inc., New York, 1945.

4 McLachlan, N. W.: "Complex Variable and Operational Calculus," Cambridge University Press, New York, 1939.

5 Rothe, R., F. Ollendorf, and K. Pohlhausen: "Theory of Functions as Applied to Engineering Problems," The M.I.T. Press, Cambridge, Mass., 1933.

chapter 7

Partial differential equations

A major advantage of the transform method in the solution of partial differential equations is that the same general procedure is applicable for all problems. This procedure is similar to that employed for the solution of ordinary differential equations; that is,

1. Transform each term in the differential equation.
2. Insert the initial conditions (boundary conditions).
3. Solve for the transformed variable.
4. Invert to obtain the desired result.

Some partial differential equations which are applicable to numerous engineering phenomena are the wave equation,

$$\frac{\partial^2 u(x,t)}{\partial t^2} = c^2 \frac{\partial^2 u(x,t)}{\partial x^2} \tag{7.1}$$

the heat (or diffusion) equation,

$$\frac{\partial u(x,t)}{\partial t} = \alpha \frac{\partial^2 u(x,t)}{\partial x^2} \tag{7.2}$$

Laplace's equation,

$$\frac{\partial^2 u(x,y)}{\partial x^2} + \frac{\partial^2 u(x,y)}{\partial y^2} = 0 \tag{7.3}$$

and Poisson's equation,

$$\frac{\partial^2 u(x,y)}{\partial x^2} + \frac{\partial^2 u(x,y)}{\partial y^2} = f(x,y) \tag{7.4}$$

This chapter is primarily concerned with the wave equation and the heat (or diffusion) equation. Laplace's equation and Poisson's equation are considered in Chap. 8.

7.1 TRANSFORM OF FUNCTIONS OF TWO VARIABLES

To transform partial differential equations, it is necessary to transform functions of two variables. The transform of $y(x,t)$ with respect to t is

$$\mathcal{L}[y(x,t)] = \int_0^\infty y(x,t)e^{-st}\,dt = Y(x,s) \tag{7.5}$$

In transforming $y(x,t)$ with respect to t, the term x is regarded as a constant parameter. To be transformable with respect to t, the function $y(x,t)$ must be of exponential order and piecewise continuous with respect to t. The function is of exponential order if, for any value of x, there are a σ, M, and T such that

$$|y(x,t)e^{-st}| = |y(x,t)e^{-\sigma t}| \le Me^{-\epsilon t} \qquad (t > T)$$

where $\epsilon > 0$ is a very small number and $s = \sigma + j\omega$. The preceding conditions are essentially the same as those for the transform of a function $f(t)$ given in Chap. 1.

After performing the integration indicated in Eq. (7.5) and evaluating t between the limits of integration, the resulting expression is some function of x and s which is designated $Y(x,s)$.

The transform of the first partial derivative with respect to time of $y(x,t)$ is

$$\mathcal{L}\left[\frac{\partial}{\partial t}y(x,t)\right] = \int_0^\infty \frac{\partial}{\partial t}y(x,t)e^{-st}\,dt$$

By letting $u = e^{-st}$ and $dv = (\partial/\partial t)y(x,t)\,dt$, then integration by parts gives

$$\mathcal{L}\left[\frac{\partial}{\partial t}y(x,t)\right] = y(x,t)e^{-st}\Big|_0^\infty + s\int_0^\infty y(x,t)e^{-st}\,dt$$
$$= sY(x,s) - y(x,0) \tag{7.6}$$

Because $y(x,t)$ is of exponential order, then $y(x,t)e^{-st}$ vanishes when evaluated at $t = \infty$. At the lower limit $(t = 0)$, then $y(x,t)e^{-st}$ becomes $y(x,0)$.

The transform of the second partial derivative with respect to time is obtained by integrating by parts twice. Thus,

$$\mathcal{L}\left[\frac{\partial^2 y(x,t)}{\partial t^2}\right] = s^2 Y(x,s) - sy(x,0) - y'(x,0) \tag{7.7}$$

where

$$y'(x,0) = \frac{d}{dt}\, y(x,t)\,\Big|_{t=0}$$

The transform of the first partial derivative with respect to x is

$$\mathcal{L}\left[\frac{\partial}{\partial x}\, y(x,t)\right] = \int_0^\infty \frac{\partial}{\partial x}\, y(x,t)e^{-st}\, dt = \frac{\partial}{\partial x} \int_0^\infty y(x,t)e^{-st}\, dt$$

$$= \frac{\partial}{\partial x}\, Y(x,s) \tag{7.8}$$

Because the preceding integration is with respect to t, the partial differentiation with respect to x may be taken outside the integration. This is verified mathematically as follows:

$$\frac{\partial}{\partial x} \int_0^\infty y(x,t)e^{-st}\, dt = \frac{\partial}{\partial x}\, Y(x,s) = \lim_{\Delta x \to 0} \frac{Y(x + \Delta x,\, s) - Y(x,s)}{\Delta x}$$

$$= \lim_{\Delta x \to 0} \int_0^\infty \frac{y(x + \Delta x,\, t) - y(x,t)}{\Delta x}\, e^{-st}\, dt$$

$$= \int_0^\infty \frac{\partial}{\partial x}\, y(x,t)e^{-st}\, dt$$

Similarly the transform of the second partial derivative with respect to x is

$$\mathcal{L}\left[\frac{\partial^2}{\partial x^2}\, y(x,t)\right] = \frac{\partial^2}{\partial x^2}\, \mathcal{L}[y(x,t)] = \frac{\partial^2}{\partial x^2}\, Y(x,s) \tag{7.9}$$

The preceding expressions for the transform of $\partial y/\partial t$ and $\partial^2 y/\partial t^2$ [Eqs. (7.6) and (7.7)] and for the transform of $\partial y/\partial x$ and $\partial^2 y/\partial x^2$ are used to transform partial differential equations just as Eqs. (2.8) and (2.9) are used to transform ordinary differential equations.

7.2 THE WAVE EQUATION

The wave equation describes many physical situations. The motion of a stretched string is considered first because it provides a good geometric representation of the basic concepts of wave motion. An elastic (flexible) string is illustrated in Fig. 7.1a. The y coordinate

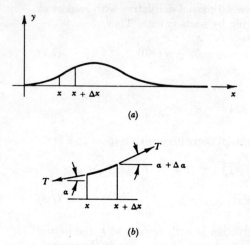

(a)

(b)

Fig. 7.1. An elastic string.

describes the vertical motion of the string. It is to be noted that $y = y(x,t)$ is a function not only of the position x but also of time t. A free-body diagram of the small segment of the string between positions x and $x + \Delta x$ is shown in Fig. 7.1b. The tangent to the string at position x is designated by the angle α, and the tangent at position $x + \Delta x$ is the angle $\alpha + \Delta\alpha$. For small motions of a perfectly flexible string, the tension T in the string is constant throughout its length. If ρ is the mass per unit length of the string, then the mass of the segment of the string shown in Fig. 7.1b is $\rho \, \Delta x$ (the angle α being kept small). The net vertical force acting on the segment of string shown in Fig. 7.1b is equal to the mass times the acceleration. The acceleration is measured at the midpoint $x + \Delta x/2$; hence

$$\rho \, \Delta x \, \frac{\partial^2 y(x + \Delta x/2, \, t)}{\partial t^2} = T \sin (\alpha + \Delta\alpha) - T \sin \alpha \qquad (7.10)$$

For small values of α,

$$\sin \alpha \approx \tan \alpha = \frac{\Delta y(x,t)}{\Delta x} \qquad (7.11)$$

and

$$\sin (\alpha + \Delta\alpha) \approx \tan (\alpha + \Delta\alpha) = \frac{\Delta y(x + \Delta x, \, t)}{\Delta x} \qquad (7.12)$$

The $\tan \alpha$ is the slope of the string at x, and $\tan (\alpha + \Delta\alpha)$ is the slope at position $x + \Delta x$.

Substituting the results of Eqs. (7.11) and (7.12) into Eq. (7.10) gives

$$\frac{\partial^2 y(x + \Delta x/2, t)}{\partial t^2} = \frac{T}{\rho} \left[\frac{\Delta y(x + \Delta x, t)}{\Delta x} - \frac{\Delta y(x,t)}{\Delta x} \right] \frac{1}{\Delta x} \qquad (7.13)$$

Because

$$\lim_{\Delta x \to 0} \left[\frac{\Delta y(x + \Delta x, t)}{\Delta x} - \frac{\Delta y(x,t)}{\Delta x} \right] \frac{1}{\Delta x} = \frac{\partial^2 y(x,t)}{\partial x^2}$$

then the limit of Eq. (7.13) as Δx approaches zero is

$$\frac{\partial^2 y(x,t)}{\partial t^2} = \frac{T}{\rho} \frac{\partial^2 y(x,t)}{\partial x^2} \qquad (7.14)$$

or

$$y_{tt}(x,t) = c^2 y_{xx}(x,t) \qquad (7.15)$$

where $c = \sqrt{T/\rho}$, $y_{tt} = \partial^2 y/\partial t^2$, and $y_{xx} = \partial^2 y/\partial x^2$.

Equation (7.15) is the partial differential equation (wave equation) which describes the motion of a stretched string as a function of the position x along the string and time t.

Illustrative example An infinitely long string is initially stretched along the x axis as shown in Fig. 7.2a. In addition,

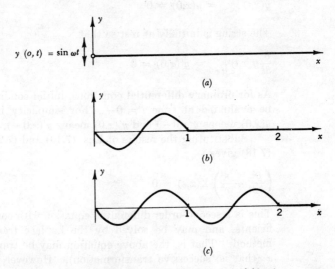

Fig. 7.2. Response of an infinitely long string to a sinusoidal input.

257

the string is initially at rest. The excitation for $t > 0$ is a sinusoidal motion at the end $x = 0$; that is,

$$y(x,t) \Big|_{x=0} = y(0,t) = \sin \omega t \qquad (7.16)$$

Determine the equation of motion for this string.

Solution   The wave equation is

$$\frac{\partial^2 y(x,t)}{\partial t^2} = c^2 \frac{\partial^2 y(x,t)}{\partial x^2} \qquad (7.17)$$

The transform of this wave equation is

$$s^2 Y(x,s) - sy(x,0) - y'(x,0) = c^2 \frac{\partial^2}{\partial x^2} Y(x,s)$$

or

$$\frac{\partial^2}{\partial x^2} Y(x,s) - \frac{s^2}{c^2} Y(x,s) = \frac{-sy(x,0) - y'(x,0)}{c^2} \qquad (7.18)$$

It is to be noted that the transformation process has eliminated the variable t so that the resultant transformed expression is a differential equation in x.

Because the string is stretched horizontally along the x axis at $t = 0$,

$$y(x,t) \Big|_{t=0} = y(x,0) = 0 \qquad (7.19)$$

The string is initially at rest so that

$$\frac{\partial}{\partial t} y(x,t) \Big|_{t=0} = y'(x,0) = 0 \qquad (7.20)$$

As for ordinary differential equations, initial conditions are to be evaluated at time $t = 0-$. For simplicity in notation, $y(x,0)$ means $y(x,0-)$ and $y'(x,0)$ means $y'(x,0-)$, etc.

Substituting the results of Eqs. (7.19) and (7.20) into Eq. (7.18) gives

$$\left(\frac{\partial^2}{\partial x^2} - \frac{s^2}{c^2} \right) Y(x,s) = 0 \qquad (7.21)$$

This is a second-order differential equation with constant coefficients, and may be solved by the Laplace transformation method. That is, the above equation may be transformed in x (that is, successive transformation). However, this differential equation is so simple that its general solution may be written directly.

$$Y(x,s) = C_1 e^{sx/c} + C_2 e^{-sx/c} \tag{7.22}$$

where C_1 and C_2 are constants which are functions of the parameter s. Two boundary conditions are required to evaluate C_1 and C_2. Because the string is infinitely long, an infinite length of time would be required for a wave to travel to the end at $x = \infty$; hence

$$y(\infty,t) = 0 \tag{7.23}$$

The other boundary condition is known by Eq. (7.16). Because the wave equation [Eq. (7.17)] contains a second derivative with respect to t and a second derivative with respect to x, then two boundary conditions in t [that is, Eqs. (7.19) and (7.20)] and two boundary conditions in x [that is, Eqs. (7.16) and (7.23)] are required.

Transforming Eqs. (7.16) and (7.23) gives

$$Y(0,s) = \mathcal{L}[y(0,t)] = \mathcal{L}(\sin \omega t) = \frac{\omega}{s^2 + \omega^2} \tag{7.24}$$

$$Y(\infty,s) = \mathcal{L}[y(\infty,t)] = 0 \tag{7.25}$$

Substituting $x = \infty$ into Eq. (7.22) and utilizing the result of Eq. (7.25) gives

$$Y(x,s)\Big|_{x=\infty} = C_1 e^{\infty} + C_2 e^{-\infty} = 0 \tag{7.26}$$

In order that Eq. (7.26) be satisfied [that is, $Y(\infty,s) = 0$], then C_1 must be zero, for otherwise the term $C_1 e^{\infty}$ would "blow up."

Substituting $x = 0$ into Eq. (7.22) and utilizing the result of Eq. (7.24) gives

$$Y(0,s) = C_2 e^0 = C_2 = \frac{\omega}{s^2 + \omega^2}$$

Substituting $C_1 = 0$ and the preceding value of C_2 into Eq. (7.22) gives

$$Y(x,s) = \frac{\omega}{s^2 + \omega^2} e^{-(x/c)s} \tag{7.27}$$

The term $e^{-(x/c)s}$ indicates that the time function is to be delayed by a time x/c. Hence, the inverse of Eq. (7.27) is

$$y(x,t) = \begin{cases} 0 & t < \dfrac{x}{c} \text{ or } x > ct \\[2ex] \sin \omega \left(t - \dfrac{x}{c} \right) & t \geq \dfrac{x}{c} \text{ or } x \leq ct \end{cases} \tag{7.28}$$

259

The physical meaning of this result is that the motion at the end $x = 0$ travels along the string at a velocity c. In Fig. 7.2b is shown the position of the string at time $t = 1/c$ for the case in which $\omega/c = 2\pi$. For this case, Eq. (7.28) becomes

$$y\left(x, \frac{1}{c}\right) = \begin{cases} 0 & x > 1 \\ \sin 2\pi(1 - x) & x \le 1 \end{cases}$$

Thus for $x > 1$ the string is still undisturbed. At $x = 1$ the string is just beginning to duplicate the motion of the input. Note that the motion will be the same as the input except that it is delayed by time $t = x/c = 1/c$.

In Fig. 7.2c is shown the position of the string at time $t = 2/c$. For this case, Eq. (7.28) becomes

$$y\left(x, \frac{2}{c}\right) = \begin{cases} 0 & x > 2 \\ \sin 2\pi(2 - x) & x \le 2 \end{cases}$$

The string is undisturbed for $x > 2$. The motion at $x = 2$ will be the same as the input except that it is delayed by time $t = x/c = 2/c$.

The solution given by Eq. (7.28) may be interpreted from two viewpoints:

1. The input motion at $x = 0$ travels along the string at a velocity c.
2. The motion at any position x is the same as the input except that it is delayed by time $t = x/c$.

Reflected waves

In Fig. 7.3 is shown a string of length L. After the wave travels down the length of the string to the fixed point at L, it is reflected. The nature of a reflected wave is brought out in the following example.

Illustrative example A string of length L is initially stretched along the x axis in an at-rest position.

$$y(x,0) = y'(x,0) = 0$$

The end at $x = L$ is fixed so that

$$y(L,t) = 0$$

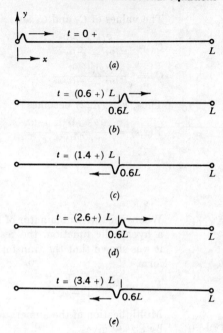

Fig. 7.3. Reflected wave.

A momentary excitation which begins at $t = 0$ is applied to the string at the end $x = 0$. The resulting position of the string for time slightly greater than zero ($t = 0+$) is shown in Fig. 7.3a. The equation for this input excitation is

$$y(0,t) = f(t)$$

The excitation $y(0,t)$ at $x = 0$ is a function of time only, that is, $f(t)$.

Determine the equation of motion for this string.

Solution The general transformed equation for a vibrating string is given by Eq. (7.18). For $y(x,0) = y'(x,0) = 0$, then $Y(x,s)$ is the same as that given by Eq. (7.22).

Transforming the two boundary conditions in x gives

$$\mathcal{L}[y(L,t)] = Y(L,s) = 0$$
$$\mathcal{L}[y(0,t)] = Y(0,s) = F(s)$$

Substituting these boundary conditions into Eq. (7.22) gives

$$Y(L,s) = C_1 e^{(L/c)s} + C_2 e^{-(L/c)s} = 0$$
$$Y(0,s) = C_1 + C_2 = F(s)$$

The values of C_1 and C_2 are

$$C_1 = \frac{-F(s)e^{-(L/c)s}}{e^{(L/c)s} - e^{-(L/c)s}}$$

$$C_2 = \frac{F(s)e^{(L/c)s}}{e^{(L/c)s} - e^{-(L/c)s}}$$

Thus, Eq. (7.22) becomes

$$Y(x,s) = \frac{e^{[(L-x)/c]s} - e^{-[(L-x)/c]s}}{e^{(L/c)s} - e^{-(L/c)s}} F(s)$$

$$= \frac{\sinh \dfrac{L-x}{c} s}{\sinh \dfrac{L}{c} s} F(s) \quad (7.29)$$

Whenever the denominator of a transformed equation contains a hyperbolic function, the response is periodic. In Sec. 2.4, it was shown that the transform of a periodic function has the form

$$F(s) = F_1(s) \frac{1}{1 - e^{-Ts}}$$

Multiplication of the numerator and denominator of Eq. (7.29) by $e^{-(L/c)s}$ gives

$$Y(x,s) = [e^{-(x/c)s} - e^{-[(2L-x)/c]s}] \frac{F(s)}{1 - e^{-(2L/c)s}} \quad (7.30)$$

The denominator term $1 - e^{-(2L/c)s}$ merely indicates that the motion is periodic. Each exponential term in the brackets of the preceding expression indicates a delayed response. The first exponential $e^{-(x/c)s}$ does not come into play until $t > x/c$. Thus, for $0 < t \le x/c$, the response is

$$y(x,t) = 0 \qquad 0 < ct \le x$$

For $x/c < t \le (2L - x)/c$, the response is that due to the first exponential term,

$$y(x,t) = f\left(t - \frac{x}{c}\right) \qquad x < ct \le 2L - x$$

The physical meaning of this response may be obtained by considering the motion at the station $x = 0.6L$. For ct slightly greater than $0.6L$ [that is, $ct = (0.6+)L$], the preceding expression applies.

$$y(x,t) = f\left(t - \frac{x}{c}\right) = f\left[\frac{(0.6+)L - 0.6L}{c}\right] = f(0+)$$

Thus, for $t = (0.6+)L/c$, the excitation motion is just beginning to be duplicated at the station $x = 0.6L$, as is illustrated in Fig. 7.3b.

For $(2L - x)/c < t \leq (2L + x)/c$, the response is that due to both delayed terms in Eq. (7.30); that is,

$$y(x,t) = f\left(t - \frac{x}{c}\right) - f\left(t - \frac{2L - x}{c}\right)$$

$$2L - x < ct \leq 2L + x$$

For ct slightly greater than $1.4L$ [that is, $ct = (1.4+)L$], the motion at the station $x = 0.6L$ is

$$y(x,t) = f\left[\frac{(1.4+)L - 0.6L}{c}\right] - f\left[\frac{(1.4+)L - 1.4L}{c}\right]$$

$$= f\left[\frac{(0.8+)L}{c}\right] - f(0+) = -f(0+)$$

The term

$$y(0,t) = f(t) = f[(0.8+)L/c]$$

is the motion at the origin at time $t = (0.8 +)L/c$. This motion is zero because the excitation is a momentary disturbance.

As shown in Fig. 7.3c, the negative of the excitation $-f(0+)$ has just reached the station $x = 0.6L$. The meaning of this is that, after the wave reaches the fixed end $x = L$, the wave is inverted (sign changed) as it goes back up the string. If the excitation were not momentary, the motion at $x = 0.6L$ would be the sum of that due to the inverted wave $[-f(0+)]$ plus that due to the motion $f[(0.8+)L/c]$ which is reaching this station for the first time.

The period is $t = 2L/c$ or $x = ct = 2L$. Thus, after the excitation arrives back at the starting point, $x = 2L = 0$, then the motion continues in a periodic fashion. As is illustrated in Fig. 7.3, a wave is inverted when it is reflected from a fixed point such as $x = L$ or from a point with prescribed motion such as the origin $x = 0$.

Transform variable

In transforming partial differential equations, any of the variables may be regarded as the transform variable. For example, if x is taken as the transform variable rather than t, then the basic Laplace transform equations are

$$\mathcal{L}[u(x,t)] = U(s,t)$$

$$\mathcal{L}\left[\frac{\partial}{\partial x} u(x,t)\right] = sU(s,t) - u(0,t)$$

$$\mathcal{L}\left[\frac{\partial^2}{\partial x^2} u(x,t)\right] = s^2 U(s,t) - su(0,t) - u'(0,t)$$

$$\mathcal{L}\left[\frac{\partial}{\partial t} u(x,t)\right] = \frac{\partial}{\partial t} \mathcal{L}[u(x,t)] = \frac{\partial}{\partial t} U(s,t)$$

$$\mathcal{L}\left[\frac{\partial^2}{\partial t^2} u(x,t)\right] = \frac{\partial^2}{\partial t^2} \mathcal{L}[u(x,t)] = \frac{\partial^2}{\partial t^2} U(s,t)$$

Thus, the transform of the wave equation is

$$\frac{\partial^2}{\partial t^2} U(s,t) = c^2[s^2 U(s,t) - su(0,t) - u'(0,t)]$$

When x is the transform variable, the initial condition terms are $u(0,t)$ and $u'(0,t)$. For the case in which t is the transform variable, the initial condition terms are $u(x,0)$ and $u'(x,0)$. Computational effort is usually saved when the initial-condition terms in the transformed equation are known boundary conditions. Thus, if $u(0,t)$ and $u'(0,t)$ are two of the given boundary conditions, direct substitution is afforded by taking x as the transform variable. Similarly, if $u(x,0)$ and $u'(x,0)$ are two of the given boundary conditions, direct substitution is afforded by using t as the transform variable. It should be pointed out that it is not necessary that the initial-condition terms in the transformed equation be known boundary conditions.

Transverse motion of a beam

The transverse motion of a beam is also described by the wave equation. In Fig. 7.4a is shown a beam of uniform cross section. Consider a cross section through the beam at position x and also a cross section at position $x + \Delta x$. The term $y(x,t)$ is the strain at the x cross section due to the load. Similarly $y(x + \Delta x, t)$ is the strain at the $x + \Delta x$ cross section (strain can be measured experimentally by means of a strain gauge). The stretch of the bar between these two sections is $y(x + \Delta x, t) - y(x,t)$.

The unit strain ϵ is the stretch per unit length; hence

$$\epsilon = \lim_{\Delta x \to 0} \frac{y(x + \Delta x, t) - y(x,t)}{\Delta x} \approx \frac{\partial}{\partial x} y(x,t) \tag{7.31}$$

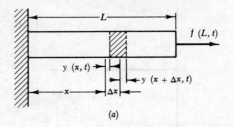

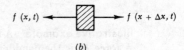

Fig. 7.4. A beam of uniform cross section.

(b)

From Hooke's law

$$f(x,t) = AE\epsilon = AE \frac{\partial}{\partial x} y(x,t) \tag{7.32}$$

where A is the cross-sectional area and E is the modulus of elasticity.

In Fig. 7.4b is a free-body diagram of the shaded portion of the beam. The summation of the forces acting on this section is equal to the mass times the acceleration

$$f(x + \Delta x, t) - f(x,t) = \rho A \Delta x \frac{\partial^2 y(x + \Delta x/2, t)}{\partial t^2} \tag{7.33}$$

where ρ is the mass per unit volume, A is the cross-sectional area, and thus $\rho A \Delta x$ is the mass of the volume under consideration. Dividing both sides of Eq. (7.33) by Δx and taking the limit as Δx approaches zero gives

$$\frac{\partial f(x,t)}{\partial x} = \rho A \frac{\partial^2 y(x,t)}{\partial t^2} \tag{7.34}$$

where

$$\frac{\partial f(x,t)}{\partial x} = \lim_{\Delta x \to 0} \frac{f(x + \Delta x, t) - f(x,t)}{\Delta x}$$

Substituting $f(x,t)$ from Eq. (7.32) into Eq. (7.34) and solving for $\partial^2/\partial t^2[y(x,t)]$ gives

$$\frac{\partial^2 y(x,t)}{\partial t^2} = c^2 \frac{\partial^2 y(x,t)}{\partial x^2} \tag{7.35}$$

where $c = \sqrt{E/\rho}$.

Equation (7.35) shows that the beam motion is governed by the wave equation.

265

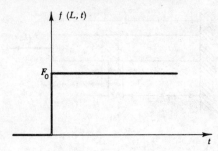

Fig. 7.5. A step change in force.

Illustrative example A beam is initially at rest and unstrained. A force F_0 is then suddenly applied at the end $x = L$. Thus, as shown in Fig. 7.5, the force at the end of the beam is

$$f(L,t) = \begin{cases} 0 & t < 0 \\ F_0 & t > 0 \end{cases}$$

Determine the equation for the strain $y(x,t)$ in the bar.

Solution Because the bar is initially at rest and under no strain,

$$y(x,0) = 0$$
$$y'(x,0) = 0$$

The transform of Eq. (7.35) is

$$s^2 Y(x,s) - sy(x,0) - y'(x,0) = c^2 \frac{\partial^2 Y(x,s)}{\partial x^2}$$

Hence

$$\left(\frac{d^2}{dx^2} - \frac{s^2}{c^2} \right) Y(x,s) = 0$$

The solution of this differential equation is

$$Y(x,s) = C_1 e^{(x/c)s} + C_2 e^{-(x/c)s} \qquad (7.36)$$

To determine C_1 and C_2, it is necessary to establish the two boundary conditions associated with x.

Because there can be no strain at the wall, then at $x = 0$ and for any time t

$$y(0,t) = 0$$

Transforming gives

$$Y(0,s) = 0 \qquad (7.37)$$

The second boundary condition is obtained from the known loading. From Eq. (7.32), it follows that

$$\frac{\partial y(x,t)}{\partial x} = \frac{1}{AE} f(x,t)$$

For $x = L$,

$$\frac{\partial y(L,t)}{\partial x} = \frac{1}{AE} f(L,t)$$

Transforming gives

$$\frac{\partial}{\partial x} Y(L,s) = \frac{1}{AE} \frac{F_0}{s} \qquad (7.38)$$

The transform of the load $f(L,t)$ shown in Fig. 7.5 is F_0/s. Because $Y(0,s) = 0$, then for $x = 0$ Eq. (7.36) becomes

$$Y(0,s) = C_1 + C_2 = 0$$

or $C_1 = -C_2$.

Differentiating Eq. (7.36) with respect to x,

$$\frac{\partial Y(x,s)}{\partial x} = \frac{sC_1}{c} e^{(x/c)s} - \frac{sC_2}{c} e^{-(x/c)s}$$

For $x = L$,

$$\frac{\partial Y(L,s)}{\partial x} = \frac{sC_1}{c} \left(e^{(L/c)s} + e^{-(L/c)s} \right) = \frac{F_0}{AEs}$$

Solving for C_1 gives

$$C_1 = \frac{cF_0}{AEs^2(e^{(L/c)s} + e^{-(L/c)s})}$$

Thus, the general transformed equation is

$$Y(x,s) = \frac{cF_0}{AEs^2} \frac{e^{(x/c)s} - e^{-(x/c)s}}{e^{(L/c)s} + e^{-(L/c)s}} = \frac{cF_0}{AEs^2} \frac{\sinh (x/c)s}{\cosh (L/c)s} \qquad (7.39)$$

Multiplication of the numerator and denominator of Eq. (7.39) by $e^{-(L/c)s} - e^{-(3L/c)s}$ gives

$$Y(x,s) = \frac{cF_0}{AEs^2} \{ e^{-[(L-x)/c]s} - e^{-[(L+x)/c]s} - e^{-[(3L-x)/c]s}$$
$$+ e^{-[(3L+x)/c]s} \} \frac{1}{1 - e^{-(4L/c)s}} \qquad (7.40)$$

The denominator term $1 - e^{-(4L/c)s}$ means that the solution is periodic with a period of $4L/c$. Each exponential term merely indicates a delayed response effect. Thus, for $0 < t \le (L - x)/c$ none of the delayed terms in Eq. (7.40) has come into play; hence

$$y(x,t) = 0 \qquad 0 < ct \le L - x$$

For $(L - x)/c < t \le (L + x)/c$, the response is that due to the first exponential term; hence

$$y(x,t) = \frac{cF_0}{AE}\left(t - \frac{L - x}{c}\right) \qquad L - x < ct \le L + x$$

For $(L + x)/c < t \le (3L - x)/c$ the response is the sum of the first two exponential terms.

$$y(x,t) = \frac{cF_0}{AE}\left[\left(t - \frac{L - x}{c}\right) - \left(t - \frac{L + x}{c}\right)\right]$$

$$= \frac{cF_0}{AE}\left(\frac{2x}{c}\right) \qquad L + x < ct \le 3L - x$$

For $(3L - x)/c < t \le (3L + x)/c$, the effect of the third delayed term must be added to the preceding result.

$$y(x,t) = \frac{cF_0}{AE}\left[\frac{2x}{c} - \left(t - \frac{3L - x}{c}\right)\right]$$

$$= \frac{cF_0}{AE}\left(-t + \frac{3L + x}{c}\right) \qquad 3L - x < ct \le 3L + x$$

Finally, for $(3L + x)/c < t \le 4L/c$, the response is

$$y(x,t) = \frac{cF_0}{AE}\left[\left(-t + \frac{3L + x}{c}\right) + \left(t - \frac{3L + x}{c}\right)\right] = 0$$

$$3L + x < ct \le 4L$$

The resulting plot of the strain at some station x is shown in Fig. 7.6a.

Stress σ is related to strain ϵ by the relationship

$$\sigma = E\epsilon = E\frac{\partial}{\partial x}y(x,t)$$

Thus, differentiating the preceding strain equations with respect to x and multiplying by E gives

$$\sigma = 0 \qquad\qquad 0 < ct \le L - x$$

$$\sigma = \frac{F_0}{A} \qquad L - x < ct \le L + x$$

$$\sigma = \frac{2F_0}{A} \qquad L + x < ct \le 3L - x$$

$$\sigma = \frac{F_0}{A} \qquad 3L - x < ct \leq 3L + x$$

$$\sigma = 0 \qquad 3L + x < ct \leq 4L$$

The corresponding stress plot is shown in Fig. 7.6b. This figure shows that initially there is no stress in the bar at station x. The stress due to the suddenly applied force F_0 travels down the bar at a velocity c and thus reaches position x at time $t = (L - x)/c$. A step change in stress to a value F_0/A then occurs. The stress wave continues down the bar to the wall and then bounces back up the bar. The time required to travel down the length of the bar from x and back up to x is $2x/c$. The total time from start is

$$t = (2x/c) + (L - x)/c = (L + x)/c$$

As shown in Fig. 7.6b the stress is now $2F_0/A$. The stress wave then proceeds out to the end of the bar, whence it bounces back down the bar toward x again. The time it reaches x is $t = 2(L - x)/c + (L + x)/c = (3L - x)/c$. At this time, the stress is reduced to the F_0/A value. The stress wave then hits the wall again, and when it gets to station x at time

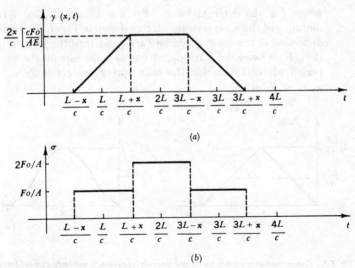

(a)

(b)

Fig. 7.6. Response of a beam to a step change in force. (a) Strain at station x; (b) stress at station x.

269

$t = (3L + x)/c$, the stress is reduced to zero. This process is then repeated with period $4L/c$. At the end $x = L$, the stress is always zero. Hence, this end acts in the same manner as a fixed point for a vibrating string (i.e., the wave is inverted as it is reflected). At the wall ($x = 0$), the stress is not zero, nor is its value prescribed. Thus, the stress wave is merely reflected without being inverted at the wall. This is somewhat analogous to the fact that water waves slapping against a pier (wall) are merely reflected without being inverted, because the water level at the pier is not prescribed.

The strain at the end $x = L$ is a periodic triangular wave, as shown in Fig. 7.7a. The static strain at $x = L$ due to a load F_0 would be $L\epsilon = LF_0/AE$. Thus, the peak strain $2(LF_0/AE)$, as shown in Fig. 7.7a, is twice the value of the static strain. It is interesting to compare this motion at the end $x = L$ with the corresponding motion of a mass-spring system to a suddenly applied force F_0. The differential equation of operation for a mass-spring system is

$$M\ddot{x} + Kx = f$$

or

$$\ddot{x} + \omega_n{}^2 x = \frac{f}{M}$$

where f is the external force. For a suddenly applied force, the motion y of the mass is sinusoidal, as shown in Fig. 7.7b. The static deflection of the mass would be F_0/K, and thus the peak amplitude $(2F_0/K)$ is twice the static deflection, as is the case for the bar. The period of oscillation for the mass-spring system is $2\pi/\omega_n$, where $\omega_n = \sqrt{K/M}$.

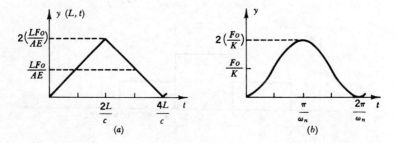

Fig. 7.7. Comparison of a uniform beam and mass-spring system to suddenly applied force F_0. (a) Strain at $x = L$ for a uniform beam; (b) motion of the mass of a mass-spring system.

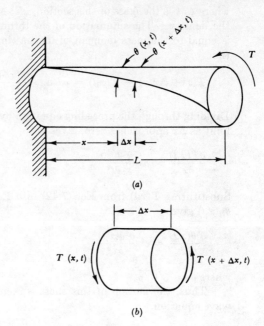

Fig. 7.8. A uniform shaft.

Torsion of a shaft

In Fig. 7.8a is shown a uniform shaft. The twist in the shaft at station x is designated by $\theta(x,t)$, and the twist at $x + \Delta x$ is designated by $\theta(x + \Delta x, t)$. The unit twist is

$$\frac{\theta(x + \Delta x, t) - \theta(x,t)}{\Delta x} \approx \frac{\partial \theta(x,t)}{\partial x} \tag{7.41}$$

Torque T is related to unit twist by the basic relationship

$$T(x,t) = JG \frac{\partial \theta(x,t)}{\partial x} \tag{7.42}$$

where $J = \pi R^4/2$ is the polar moment of inertia of the cross section and G is the modulus of elasticity in shear. In Fig. 7.8b is shown a free-body diagram of the cylindrical section of width Δx. The mass moment of inertia is

$$\frac{\Delta M r^2}{2} = \frac{\rho(\pi R^2 \, \Delta x)R^2}{2} = \rho J \, \Delta x$$

where ΔM is the mass of this section, $\pi R^2 \Delta x$ is the volume, and ρ is the density. The summation of the torques acting on this section is equal to the mass moment of inertia times the angular acceleration.

$$T(x + \Delta x,\, t) - T(x,t) = \rho J \,\Delta x \,\frac{\partial^2 \theta(x + \Delta x/2,\, t)}{\partial t^2} \tag{7.43}$$

Dividing through the preceding equation by Δx, and then taking the limit as Δx approaches zero, gives

$$\frac{\partial T(x,t)}{\partial x} = \rho J \,\frac{\partial^2 \theta(x,t)}{\partial t^2} \tag{7.44}$$

Substituting $T(x,t)$ from Eq. (7.42) into Eq. (7.44) and solving for $\theta_{tt}(x,t)$ gives

$$\frac{\partial^2 \theta(x,t)}{\partial t^2} = c^2 \,\frac{\partial^2 \theta(x,t)}{\partial x^2} \tag{7.45}$$

where $c = \sqrt{G/\rho}$.

Thus, the motion of this shaft is seen to be governed by the wave equation.

Illustrative example Determine the equation of motion for the shaft shown in Fig. 7.9a. The shaft is initially subjected to a uniform twist such that the angular deflection at the end L is θ_0. Thus, at any station x, the initial twist is $(x/L)\theta_0$; that is,

$$\theta(x,0) = \frac{x}{L}\,\theta_0 \tag{7.46}$$

The shaft is initially at rest, so that

$$\frac{\partial \theta(x,0)}{\partial t} = 0 \tag{7.47}$$

The torque which maintains the shaft in this initial position is released at time $t = 0$ so that the shaft is free to oscillate. Because $T(L,t) = 0$, then from Eq. (7.42)

$$T(L,t) = JG \,\frac{\partial \theta(L,t)}{\partial x} = 0 \tag{7.48}$$

Equations (7.46) and (7.47) are the two boundary conditions in time. Equation (7.48) is one boundary condition in x. The

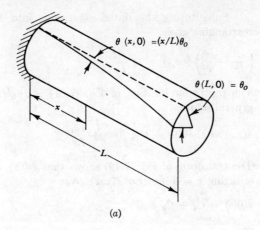

$$\theta(x,0) = (x/L)\theta_O$$

$$\theta(L,0) = \theta_O$$

(a)

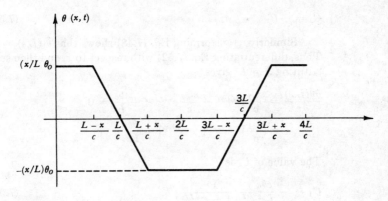

(b)

Fig. 7.9. (a) Shaft under an initial twist. (b) Motion at station x of shaft.

remaining boundary condition is obtained by noting that there can never be any twist of the shaft at the wall.

$$\theta(0,t) = 0 \tag{7.49}$$

Solution Transforming Eq. (7.45) gives

$$s^2\bar{\theta}(x,s) - s\theta(x,0) - \theta'(x,0) = c^2 \frac{\partial^2 \bar{\theta}(x,s)}{\partial x^2} \tag{7.50}$$

The bar above θ is used to designate the transformed function in Eq. (7.50).

Substituting the initial conditions into Eq. (7.50) and rearranging gives

$$\left(\frac{d^2}{dx^2} - \frac{s^2}{c^2}\right)\bar{\theta}(x,s) = -\frac{xs\theta_0}{Lc^2} \tag{7.51}$$

The particular solution of Eq. (7.51) is $x\theta_0/Ls$. Hence, the general solution is

$$\bar{\theta}(x,s) = C_1 e^{(x/c)s} + C_2 e^{-(x/c)s} + \frac{x\theta_0}{Ls} \tag{7.52}$$

The transform of Eq. (7.49) shows that $\bar{\theta}(0,s) = 0$; hence substituting $x = 0$ into Eq. (7.52) gives

$$\bar{\theta}(0,s) = C_1 + C_2 = 0$$

or

$$C_1 = -C_2 \tag{7.53}$$

Similarly, transforming Eq. (7.48) shows that $\bar{\theta}_x(L,s) = 0$. Thus, differentiating Eq. (7.52) with respect to x, and then substituting $x = L$, gives

$$\frac{\partial \bar{\theta}(L,s)}{\partial x} = \frac{sC_1}{c}e^{(L/c)s} - \frac{sC_2}{c}e^{-(L/c)s} + \frac{\theta_0}{Ls} = 0$$

$$= \frac{sC_1}{c}\left[e^{(L/c)s} + e^{-(L/c)s}\right] + \frac{\theta_0}{Ls}$$

The value of C_1 is

$$C_1 = -\frac{c\theta_0}{Ls^2}\frac{1}{e^{(L/c)s} + e^{-(L/c)s}} \tag{7.54}$$

Utilizing the results of Eqs. (7.53) and (7.54) in Eq. (7.52) gives

$$\bar{\theta}(x,s) = \frac{x\theta_0}{Ls} - \frac{c\theta_0}{Ls^2}\frac{e^{(x/c)s} - e^{-(x/c)s}}{e^{(L/c)s} + e^{-(L/c)s}} = \frac{x\theta_0}{Ls} - \frac{c\theta_0}{Ls^2}\frac{\sinh(x/c)s}{\cosh(L/c)s} \tag{7.55}$$

The right-hand side of Eq. (7.55) has the same form as Eq. (7.39); hence multiplication of numerator and denominator by $e^{-(L/c)s} - e^{-(3L/c)s}$ gives

$$\bar{\theta}(x,s) = \frac{x}{L}\frac{\theta_0}{s} - \frac{c}{L}\frac{\theta_0}{s^2}\{e^{-[(L-x)/c]s} - e^{-[(L+x)/c]s} - e^{-[(3L-x)/c]s}$$

$$+ e^{-[(3L+x)/c]s}\}\frac{1}{1 - e^{-(4L/c)s}} \tag{7.56}$$

Inverting Eq. (7.56) yields the following response expressions for the various time intervals:

$$\theta(x,t) = \begin{cases} \dfrac{x}{L}\,\theta_0 & 0 < ct \leq L - x \\[2mm] \dfrac{x}{L}\,\theta_0 - \dfrac{c\theta_0}{L}\left(t - \dfrac{L - x}{c}\right) & L - x < ct \leq L + x \\[2mm] -\dfrac{x}{L}\,\theta_0 & L + x < ct \leq 3L - x \\[2mm] \dfrac{x}{L}\,\theta_0 + \dfrac{c\theta_0}{L}\left(t - \dfrac{3L + x}{c}\right) & 3L - x < ct \leq 3L + x \\[2mm] \dfrac{x}{L}\,\theta_0 & 3L + x < ct \leq 4L \end{cases}$$

The corresponding motion $\theta(x,t)$ is plotted in Fig. 7.9b. The motion $\theta(L,t)$ at the end $x = L$ is the periodic triangular wave (solid line) of Fig. 7.10b.

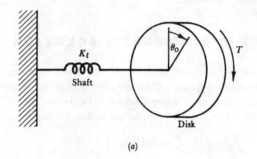

(a)

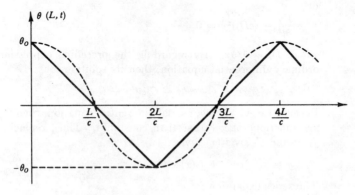

Fig. 7.10. (a) Disk and spring. (b) Solid line, motion at station L of shaft in Fig. 7.9a; dotted line, motion of disk in (a).

275

A schematic diagram of a shaft of torsional spring rate K_t and a disk of inertia J is shown in Fig. 7.10a. The summation of torques acting on the disk is the external torque T minus the shaft torque $K_t\theta$, which opposes the motion of the disk. Thus,

$$T - K_t\theta = J\ddot{\theta}$$

or

$$\ddot{\theta} + \omega_n{}^2\theta = \frac{T}{J}$$

where $\omega_n = \sqrt{K_t/J}$. If the disk is rotated through the angle θ_0 and then released at time $t = 0$, the resulting motion is the sinusoid shown by the dotted curve of Fig. 7.10b. The amplitude is θ_0, and the period of oscillation is $2\pi/\omega_n$ rather than $4L$. As illustrated in Figs. 7.7 and 7.10b, there are many similarities between the behavior of systems governed by partial differential equations and the behavior of corresponding systems described by ordinary differential equations.

7.3 D'ALEMBERT'S SOLUTION

The D'Alembert solution of the wave equation is a fundamental classical method. This solution yields information concerning the basic nature of wave motion. To derive the D'Alembert solution, first write the wave equation in the form

$$\frac{\partial^2 y}{\partial t^2} = c^2 \frac{\partial^2 y}{\partial x^2} = (cD_x)^2 y$$

or

$$\frac{\partial^2 y}{\partial t^2} - (cD_x)^2 y = 0 \tag{7.57}$$

where $D_x = \partial/\partial x$. By regarding the preceding expression as an ordinary differential equation, then its solution is

$$y = C_1 e^{cD_x t} + C_2 e^{-cD_x t} = C_1 e^{ctD_x} + C_2 e^{-ctD_x}$$

Because Eq. (7.57) is a partial differential equation, then C_1 and C_2 are functions of x rather than constants. Thus, replacing C_1 by $f_1(x)$ and C_2 by $f_2(x)$ gives

$$y = f_1(x)e^{ctD_x} + f_2(x)e^{-ctD_x} \tag{7.58}$$

The series expansion for e^x is

$$e^x = 1 + x + \frac{x^2}{2!} + \frac{x^3}{3!} + \cdots$$

Thus, $e^{ctD_x}f_1(x)$ may be written in the form

$$e^{ctD_x}f_1(x) = \left[1 + ctD_x + \frac{(ctD_x)^2}{2!} + \cdots \right] f_1(x)$$

$$= f_1(x) + ctf_1'(x) + \frac{(ct)^2 f_1''(x)}{2!} + \cdots \tag{7.59}$$

Comparison of Eqs. (7.59) and (5.87) shows that $e^{ctD_x}f_1(x)$ is the Taylor series expansion for $f_1(x)$; that is,

$$e^{ctD_x}f_1(x) = f_1(x + ct)$$

In a similar manner, it may be shown that

$$e^{-ctD_x}f_2(x) = f_2(x - ct)$$

Hence, the general solution of the wave equation as given by Eq. (7.58) is

$$y(x,t) = f_1(x + ct) + f_2(x - ct) \tag{7.60}$$

To verify this solution, note that

$$\frac{\partial y}{\partial t} = cf_1'(x + ct) - cf_2'(x - ct)$$

$$\frac{\partial^2 y}{\partial t^2} = c^2 f_1''(x + ct) + c^2 f_2''(x - ct)$$

$$\frac{\partial y}{\partial x} = f_1'(x + ct) + f_2'(x - ct)$$

$$\frac{\partial^2 y}{\partial x^2} = f_1''(x + ct) + f_2''(x - ct)$$

Thus, the general solution given by Eq. (7.60) satisfies the wave equation

$$\frac{\partial^2 y}{\partial t^2} = c^2 \frac{\partial^2 y}{\partial x^2} = c^2 f_1''(x + ct) + c^2 f_2''(x - ct)$$

In the general solution, the term $f_1(x + ct)$ represents a wave traveling to the left, and the term $f_2(x - ct)$ represents a wave traveling to the right.

To illustrate the application of D'Alembert's solution, consider the problem of the string of infinite length, which was solved by transform methods in the preceding section. Because the string is initially at rest and stretched along the x axis, the resulting motion is that due to the wave originating at the origin. For this string the wave motion is to the right, so that $f_1(x + ct) = 0$. Thus, evaluating Eq. (7.60) at $x = 0$ gives

$$y(0,t) = f_2(-ct) = y_0 \sin \omega t \tag{7.61}$$

277

where $y_0 \sin \omega t$ is the given excitation at $x = 0$. The desired result $y(x,t) = f_2(x - ct)$ is now obtained by replacing $-ct$ in Eq. (7.61) by $x - ct$. Dividing $-ct$ by $-c$ shows that t is to be replaced by $-(x - ct)/c = t - x/c$. Thus,

$$y(x,t) = f_2(x - ct) = y_0 \sin \omega \left(t - \frac{x}{c} \right)$$

As a second example, consider the bar of length L subjected to a suddenly applied force F_0, which was analyzed by the transform method in the preceding section. Because the bar is initially at rest and unstretched, the wave motion is that due to the suddenly applied load. At time t the stress has traveled a distance ct down the beam. The deflection at a station x is zero before the first stress wave arrives; hence, for $ct \leq L - x$,

$$y(x,t) = 0 \qquad 0 < ct \leq L - x$$

After the first stress wave passes the x station ($L - x < ct$), but before the wave hits the wall ($ct \leq L$), then the distance from the stress wave to the x station is $ct - (L - x)$. Thus, the strain is

$$y(x,t) = \frac{F_0}{AE} [ct - (L - x)] = \frac{cF_0}{AE} \left(t - \frac{L - x}{c} \right)$$
$$L - x < ct \leq L$$

After the stress wave is reflected from the wall, the stress is doubled in the region between the stress wave and the wall. The distance from the stress wave to the wall is $ct - L$, and the distance from the x station to the stress wave is $(L + x) - ct$. The deflection is now

$$y(x,t) = \frac{2F_0}{AE} (ct - L) + \frac{F_0}{AE} [(L + x) - ct]$$
$$= \frac{cF_0}{AE} \left(t - \frac{L - x}{c} \right) \qquad L < ct \leq L + x$$

Because the two preceding equations are the same, this result is valid for the region ($L - x < ct \leq L + x$). After this reflected (second) stress wave passes x (on its way to the end of the bar), there is no change in the stress between the wall and x until the third stress wave, which is reflected from the end $x = L$, comes back. The substitution of $ct = L + x$ into the preceding equation gives the value of the deflection for this interval.

$$y(x,t) = \frac{cF_0}{AE} \left(\frac{2x}{c} \right) \qquad L + x < ct \leq 3L - x$$

Thus, by proceeding in a similar manner the previously obtained solution for this problem is verified. It is to be noted that the stress wave is inverted when it bounces back from the end $x = L$. Thus, the force in the beam between the stress wave and end is reduced from $2F_0$ to F_0.

7.4 THE HEAT, OR DIFFUSION, EQUATION

In Fig. 7.11 is shown a portion of a solid body. For unidirectional flow of heat the temperature $u(x,t)$ is a function of the x coordinate and time t. The rate of heat flowing across the x plane is

$$q_x = -kA \frac{\partial u(x,t)}{\partial x} \tag{7.62}$$

where k is the thermal conductivity of the material, A is the area of section through which heat flows, and $\partial u(x,t)/\partial x$ is the rate of change of temperature with respect to distance (i.e., temperature gradient in x direction).

The rate of heat flowing across the $x + \Delta x$ plane is

$$q_{x+\Delta x} = -kA \frac{\partial u(x + \Delta x,\, t)}{\partial x}$$

The rate at which heat is being accumulated between the two planes is

$$q_x - q_{x+\Delta x} = kA \left[\frac{\partial u(x + \Delta x,\, t)}{\partial x} - \frac{\partial u(x,t)}{\partial x} \right] \tag{7.63}$$

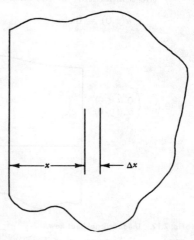

Fig. 7.11. A semi-infinite solid body.

The coefficient c of heat capacity per unit volume is the quantity of heat required to raise the temperature of a unit volume of material by $1°$. Thus, a measure of the amount of heat contained in a volume of width Δx is $cA\ \Delta xu(x + \Delta x/2, t)$. Differentiation with respect to time yields the rate at which heat is being accumulated in the elemental volume.

$$q_x - q_{x+\Delta x} = cA\ \Delta x\ \frac{\partial}{\partial t}\ u\left(x + \frac{\Delta x}{2}, t\right) \tag{7.64}$$

From Eqs. (7.63) and (7.64), it follows that

$$cA\ \frac{\partial}{\partial t}\ u\left(x + \frac{\Delta x}{2}, t\right) = kA\left[\frac{\partial u(x + \Delta x, t)}{\partial x} - \frac{\partial u(x,t)}{\partial x}\right]\frac{1}{\Delta x} \tag{7.65}$$

The limit as Δx approaches zero is

$$\frac{\partial u(x,t)}{\partial t} = \alpha\frac{\partial^2 u(x,t)}{\partial x^2} \tag{7.66}$$

where $\alpha = k/c$ is the thermal diffusivity of the material. Equation (7.66) is called the *heat*, or *diffusion, equation*. This equation is not limited to the case in which heat is being diffused; rather, it is applicable to many other diffusion processes.

Temperatures in semi-infinite solids

In Fig. 7.12 is shown a semi-infinite solid. Suppose initially that the temperature of the entire body is zero.

$$u(x,0) = 0 \tag{7.67}$$

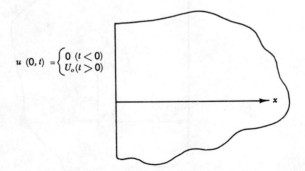

$$u\ (0, t) = \begin{cases} 0 \ (t < 0) \\ U_o(t > 0) \end{cases}$$

Fig. 7.12. Unidirectional heat flow.

The temperature at the left-hand wall is

$$u(0,t) = \begin{cases} 0 & t < 0 \\ U_0 & t > 0 \end{cases} \tag{7.68}$$

Because time is required for heat to flow across the body, the temperature at $x = \infty$ remains at its initial value.

$$u(\infty,t) = 0 \tag{7.69}$$

Equation (7.66) contains a first-order partial derivative with respect to time and a second-order partial derivative with respect to x; hence, one boundary condition in t [Eq. (7.67)] and two boundary conditions in x [Eqs. (7.68) and (7.69)] are required. To determine the equation for the temperature distribution $u(x,t)$, first transform Eq. (7.66). That is,

$$sU(x,s) - u(x,0) = \alpha \frac{\partial^2 U(x,s)}{\partial x^2}$$

Because $u(x,0) = 0$, then

$$\left(\frac{\partial^2}{\partial x^2} - \frac{s}{\alpha} \right) U(x,s) = 0 \tag{7.70}$$

The general solution of this differential equation is

$$U(x,s) = C_1 e^{x\sqrt{s/\alpha}} + C_2 e^{-x\sqrt{s/\alpha}} \tag{7.71}$$

The transform of Eqs. (7.68) and (7.69) is

$$U(0,s) = \frac{U_0}{s}$$

$$U(\infty,s) = 0$$

Because $U(\infty,s) = 0$, the substitution of $x = \infty$ into Eq. (7.71) gives

$$U(\infty,s) = C_1 e^{\infty} + C_2 e^{-\infty} = C_1 e^{\infty} = 0$$

In order that this equation be satisfied, it is necessary that $C_1 = 0$.

Because $U(0,s) = U_0/s$, the substitution of $x = 0$ and $C_1 = 0$ in Eq. (7.71) gives

$$U(0,s) = C_2 = \frac{U_0}{s}$$

Thus, for $C_1 = 0$ and $C_2 = U_0/s$, then Eq. (7.71) becomes

$$U(x,s) = \frac{U_0}{s} e^{-(x/\sqrt{\alpha})\sqrt{s}} \tag{7.72}$$

The inverse transform of Eq. (7.72) is the complementary error function. The error function and the complementary error function occur frequently in the solution of the heat, or diffusion, equation, especially with regard to bodies of infinite length. The transforms of these functions are derived by first considering the function

$$f(t) = \frac{e^{-k^2/4t}}{t^{\frac{3}{2}}} \tag{7.73}$$

The transform of Eq. (7.73) is

$$F(s) = \int_0^\infty \frac{e^{-k^2/4t}e^{-st} \, dt}{t^{\frac{3}{2}}} \tag{7.74}$$

By letting $\lambda = k/(2\sqrt{t})$ and hence $d\lambda = -(k/4t^{\frac{3}{2}}) \, dt$ in Eq. (7.74), then

$$F(s) = -\frac{4}{k} \int_\infty^0 e^{-[\lambda^2 + (k/2\lambda)^2 s]} \, d\lambda = \frac{4}{k} e^{-k\sqrt{s}} \int_0^\infty e^{-[\lambda - (k/2\lambda)\sqrt{s}]^2} \, d\lambda$$

Another representation for $F(s)$ may be obtained by letting $\lambda^2 = st$ and hence $2\lambda \, d\lambda = s \, dt$ in Eq. (7.74); thus,

$$F(s) = \int_0^\infty \frac{2\sqrt{s}}{\lambda^2} e^{-\lambda^2} e^{-(k/2\lambda)^2 s} \, d\lambda$$

$$= \frac{4}{k} e^{-k\sqrt{s}} \int_0^\infty \frac{k\sqrt{s}}{2\lambda^2} e^{-[\lambda - (k/2\lambda)\sqrt{s}]^2} \, d\lambda$$

Adding the two preceding expressions for $F(s)$ gives

$$2F(s) = \frac{4}{k} e^{-k\sqrt{s}} \int_0^\infty \left(1 + \frac{k\sqrt{s}}{2\lambda^2}\right) e^{-[\lambda - (k/2\lambda)\sqrt{s}]^2} \, d\lambda$$

The substitution of $x = \lambda - k\sqrt{s}/2\lambda$ and $dx = (1 + k\sqrt{s}/2\lambda^2) \, d\lambda$ yields (note that, for $\lambda = 0$, the new lower limit is $x = -\infty$)

$$2F(s) = \frac{4}{k} e^{-k\sqrt{s}} \int_{-\infty}^\infty e^{-x^2} \, dx$$

Equation (6.61) shows that the value of the integral from 0 to ∞ is $\sqrt{\pi}/2$, and thus the value of the preceding definite integral is $\sqrt{\pi}$.

$$2F(s) = \frac{4\sqrt{\pi}}{k} e^{-k\sqrt{s}}$$

Thus

$$F(s) = \mathcal{L}\left(\frac{e^{-k^2/4t}}{t^{\frac{3}{2}}}\right) = \frac{2\sqrt{\pi}}{k} e^{-k\sqrt{s}}$$

Taking the inverse transform of the preceding expression gives

$$\mathcal{L}^{-1}(e^{-k\sqrt{s}}) = \frac{k}{2\sqrt{\pi}} \frac{e^{-k^2/4t}}{t^{\frac{3}{2}}} \tag{7.75}$$

Other transform pairs may be obtained by the application of transform theorems to Eq. (7.75). For example, application of Eq. (2.27) for obtaining the transform of $tf(t)$ yields the following transform pair:

$$\mathcal{L}^{-1}\left(\frac{e^{-k\sqrt{s}}}{\sqrt{s}}\right) = \frac{e^{-k^2/4t}}{\sqrt{\pi t}} \tag{7.76}$$

Similarly, the application of Eq. (2.20) to Eq. (7.75) yields the transform pair

$$\mathcal{L}^{-1}\left(\frac{e^{-k\sqrt{s}}}{s}\right) = \frac{k}{2\sqrt{\pi}} \int_0^t \frac{e^{-k^2/4t}}{t^{\frac{3}{2}}} dt$$

By letting $\lambda = k/(2\sqrt{t})$, $d\lambda = -k/(4t^{\frac{3}{2}})\,dt$ and noting that the lower limit of integration when $t = 0$ is $\lambda = \infty$, and similarly the upper limit when $t = t$ is $\lambda = k/(2\sqrt{t})$, gives

$$\mathcal{L}^{-1}\left(\frac{e^{-k\sqrt{s}}}{s}\right) = \frac{2}{\sqrt{\pi}} \int_{k/2\sqrt{t}}^{\infty} e^{-\lambda^2} d\lambda$$

$$= \frac{2}{\sqrt{\pi}} \int_0^{\infty} e^{-\lambda^2} d\lambda - \frac{2}{\sqrt{\pi}} \int_0^{k/2\sqrt{t}} e^{-\lambda^2} d\lambda$$

$$= 1 - \frac{2}{\sqrt{\pi}} \int_0^{k/2\sqrt{t}} e^{-\lambda^2} d\lambda = \mathrm{erfc}\left(\frac{k}{2\sqrt{t}}\right) \tag{7.77}$$

The preceding result is referred to as the *complementary error function* (erfc). The integral term is called the *error function* (erf); that is,

$$\mathrm{erf}\left(\frac{k}{2\sqrt{t}}\right) = \frac{2}{\sqrt{\pi}} \int_0^{k/2\sqrt{t}} e^{-\lambda^2} d\lambda \tag{7.78}$$

This error function is given in many tables in the same manner as trigonometric functions. A plot of the error function is shown in Fig. 7.13.

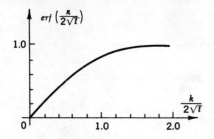

Fig. 7.13. The error function.

From Eqs. (7.77) and (7.78) it follows that

$$\text{erfc}\left(\frac{k}{2\sqrt{t}}\right) = 1 - \text{erf}\left(\frac{k}{2\sqrt{t}}\right) \tag{7.79}$$

Hence, the complementary error function is merely 1 minus the error function.

The inverse transform of Eq. (7.72) is obtained by replacing k in Eq. (7.77) by $x/\sqrt{\alpha}$; hence

$$u(x,t) = U_0\left(1 - \frac{2}{\sqrt{\pi}}\int_0^{\frac{x/\sqrt{\alpha}}{2\sqrt{t}}} e^{-\lambda^2}\,d\lambda\right)$$

$$= U_0\left[\text{erfc}\left(\frac{x/\sqrt{\alpha}}{2\sqrt{t}}\right)\right] = U_0\left[1 - \text{erf}\left(\frac{x/\sqrt{\alpha}}{2\sqrt{t}}\right)\right] \tag{7.80}$$

Thus, with the aid of an error-function table, the temperature at any station x and at any time t may be readily determined.

(A short table of additional Laplace transforms is given in Appendix A.)

Temperature in finite solids

In Fig. 7.14 is shown a rectangular solid of length L. The top and bottom are insulated. Initially the temperature of the entire body is zero; then, at time $t = 0$, there is a step change in the temperature at the left face ($x = 0$). That is,

$$u(0,t) = \begin{cases} 0 & t < 0 \\ U_0 & t > 0 \end{cases}$$

The temperature at the right end is maintained at zero, so that

$$u(L,t) = 0$$

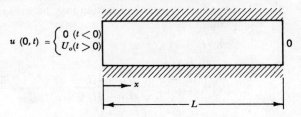

$$u\,(0,\,t)\;=\;\begin{cases}0\;(t<0)\\U_o(t>0)\end{cases}$$

Fig. 7.14. A rectangular solid insulated at top and bottom.

Because $u(x,0) = 0$, the transform for $U(x,s)$ is the same as that given by Eq. (7.71). That is,

$$U(x,s) \;=\; C_1 e^{x\sqrt{s/\alpha}} + C_2 e^{-x\sqrt{s/\alpha}}$$

Transforming the boundary conditions gives

$$U(0,s) \;=\; \frac{U_0}{s}$$
$$U(L,s) \;=\; 0$$

For $x = 0$, Eq. (7.71) becomes

$$U(0,s) \;=\; C_1 + C_2 \;=\; \frac{U_0}{s}$$

Similarly, for $x = L$,

$$U(L,s) \;=\; C_1 e^{L\sqrt{s/\alpha}} + C_2 e^{-L\sqrt{s/\alpha}} \;=\; 0$$

Solving these equations for C_1 and C_2 gives

$$C_1 \;=\; \frac{-\,U_0 e^{-L\sqrt{s/\alpha}}}{s(e^{L\sqrt{s/\alpha}} - e^{-L\sqrt{s/\alpha}})}$$
$$C_2 \;=\; \frac{U_0 e^{L\sqrt{s/\alpha}}}{s(e^{L\sqrt{s/\alpha}} - e^{-L\sqrt{s/\alpha}})}$$

Substituting these constants into Eq. (7.71) gives

$$\begin{aligned}
U(x,s) &\;=\; \frac{U_0}{s}\,\frac{e^{(L-x)\sqrt{s/\alpha}} - e^{-(L-x)\sqrt{s/\alpha}}}{e^{L\sqrt{s/\alpha}} - e^{-L\sqrt{s/\alpha}}}\\[2mm]
&\;=\; \frac{U_0}{s}\,\frac{\sinh\,(L-x)\,\sqrt{s/\alpha}}{\sinh L\,\sqrt{s/\alpha}}
\end{aligned} \qquad (7.81)$$

The preceding hyperbolic function may be written in a power series expansion as follows:

$$\frac{\sinh{(L-x)}\sqrt{s/\alpha}}{\sinh L \sqrt{s/\alpha}}$$

$$= \frac{(L-x)\sqrt{s/\alpha} + [(L-x)\sqrt{s/\alpha}]^3/3! + \cdots}{L\sqrt{s/\alpha} + (L\sqrt{s/\alpha})^3/3! + \cdots}$$

$$= \frac{\dfrac{L-x}{\sqrt{\alpha}} + \left(\dfrac{L-x}{\sqrt{\alpha}}\right)^3 \dfrac{s}{3!} + \cdots}{\dfrac{L}{\sqrt{\alpha}} + \left(\dfrac{L}{\sqrt{\alpha}}\right)^3 \dfrac{s}{3!} + \cdots} \tag{7.82}$$

Because the limit as s approaches 0 of Eq. (7.82) is $(L-x)/L$, then $s = 0$ is not a singular point of Eq. (7.82). The singular points are the other values of s such that

$$\sinh L \sqrt{\frac{s}{\alpha}} = -j \sin \; jL \sqrt{\frac{s}{\alpha}} = 0$$

The sine is zero for angles of $\pm \pi n$. For $jL \sqrt{s/\alpha} = \pm \pi n$, then

$$s = -\frac{\alpha \pi^2 n^2}{L^2} \qquad n = 1, 2, 3, \ldots \tag{7.83}$$

The singular points of Eq. (7.81) are those given by Eq. (7.83) plus the one at the origin ($s = 0$) due to the lone s term in the denominator of Eq. (7.81). The residue of $U(x,s)e^{st}$ due to the pole at the origin is

$$\lim_{s \to 0} \frac{U_0 e^{st} \sinh{(L-x)}\sqrt{s/\alpha}}{\sinh L \sqrt{s/\alpha}} = \frac{L-x}{L} U_0$$

The residue of $U(x,s)e^{st}$ due to the singular point at $s = -\alpha \pi^2 n^2/L^2$ is obtained by application of Eq. (5.80), in which

$$P(s) = \frac{U_0 e^{st}}{s} \sinh{(L-x)}\sqrt{\frac{s}{\alpha}}$$

$$Q(s) = \sinh L \sqrt{\frac{s}{\alpha}}$$

$$Q'(s) = \frac{L}{2\sqrt{s\alpha}} \cosh L \sqrt{\frac{s}{\alpha}}$$

Thus,

$$\lim_{s \to -\alpha \pi^2 n^2/L^2} P(s) = -\frac{U_0 L^2 e^{-\alpha(\pi n/L)^2 t}}{\alpha \pi^2 n^2} \sinh \frac{j\pi n(L-x)}{L}$$

$$= \frac{U_0 L^2 e^{-\alpha(\pi n/L)^2 t}}{j\alpha \pi^2 n^2} \sin \frac{\pi n(L-x)}{L}$$

$$\lim_{s \to -\alpha \pi^2 n^2/L^2} Q'(s) = \frac{L^2}{j2\alpha \pi n} \cosh j\pi n = \frac{L^2}{j2\alpha \pi n} \cos \pi n = \frac{(-1)^n L^2}{j2\alpha \pi n}$$

The corresponding residue is

$$\lim_{s \to -\alpha\pi^2 n^2/L^2} \frac{P(s)}{Q'(s)} = \frac{2U_0 e^{-\alpha(\pi n/L)^2 t}}{(-1)^n \pi n} \sin \frac{\pi n(L-x)}{L}$$

Summing all the residues yields the inverse transform of Eq. (7.81); that is,

$$u(x,t) = U_0 \left[\frac{L-x}{L} + \frac{2}{\pi} \sum_{n=1}^{\infty} \frac{e^{-\alpha(\pi n/L)^2 t}}{(-1)^n n} \sin \frac{\pi n(L-x)}{L} \right] \quad (7.84)$$

In the next section, it is shown how the response of a system to an arbitrarily varying excitation $f(t)$ may be ascertained by knowing the response of the system to a unit impulse or a unit step function. In Sec. 3.8, it was shown how to determine the response of an ordinary differential equation to an arbitrary input. The next section is essentially an extension of this method to partial differential equations.

7.5 RESPONSE TO AN ARBITRARY INPUT

When the excitation is an arbitrary function $f(t)$, the transformed equation may be written in the form

$$U(x,s) = F(s)W(x,s) \quad (7.85)$$

where $F(s) = \mathcal{L}[f(t)]$ is the transform of the excitation and $W(x,s)$ is the transform of the differential equation when the excitation is a unit impulse. Application of the convolution (superposition) integral [Eq. (3.87)] gives

$$u(x,t) = \int_0^t f(t-\tau)w(x,\tau) \, d\tau$$
$$= \int_0^t f(\tau)w(x, t-\tau) \, d\tau \quad (7.86)$$

Thus, by knowing the impulse response $w(x,\tau)$, then the response to an arbitrary excitation may be obtained by application of either form of the superposition integral given in Eq. (7.86). The superposition integral is also referred to as the *Faltung integral*.

To illustrate this method, it is to be noted that Eq. (7.81) is the transform for the case in which the excitation is a step change U_0. Dividing Eq. (7.81) by U_0/s gives the transform $W(x,s)$ for a unit impulse excitation.

$$W(x,s) = \frac{\sinh(L-x)\sqrt{s/\alpha}}{\sinh L \sqrt{s/\alpha}}$$

Inverting yields $w(x,t)$, whence the response is obtained by application of Eq. (7.86). The impulse response $w(x,t)$ yields information regarding the basic behavior or nature of a system.

The response to an arbitrary input $f(t)$ may also be determined from the unit step-function response. The transformed equation for a unit step function is $G(x,s) = W(x,s)/s$. Hence, Eq. (7.85) may be expressed in the form

$$U(x,s) = sF(s)G(x,s) \tag{7.87}$$

Because $\mathcal{L}[f'(t)] = sF(s) - f(0)$, then

$$\mathcal{L}^{-1}[sF(s)] = f'(t) + f(0)u_1(t)$$

Application of the convolution integral to Eq. (7.87) gives

$$
\begin{aligned}
u(x,t) &= \int_0^t f'(\tau)g(x, t - \tau)\, d\tau + f(0) \int_0^t u_1(\tau)g(x, t - \tau)\, d\tau \\
&= \int_0^t f'(\tau)g(x, t - \tau)\, d\tau + f(0)g(x,t) \int_0^t u_1(\tau)\, d\tau \\
&= \int_0^t f'(\tau)g(x, t - \tau)\, d\tau + f(0)g(x,t) \tag{7.88}
\end{aligned}
$$

The term $u_1(\tau)$ is a unit impulse which occurs at $\tau = 0$.

For $\tau \approx 0$, the function $g(x,t)$ remains essentially unchanged and may be factored from the integral as indicated in the preceding development. The term $\int_0^t u_1(\tau)\, d\tau$ is the area of the impulse which is unity.

By grouping the s factor in Eq. (7.87) with $G(x,s)$ rather than $F(s)$, then the response is expressed in the form

$$u(x,t) = \int_0^t f(t - \tau)g'(x,\tau)\, d\tau + f(t)g(x,0) \tag{7.89}$$

By expressing the integral portions of Eqs. (7.88) and (7.89) in the alternate convolution form, these equations become

$$u(x,t) = \int_0^t f'(t - \tau)g(x,\tau)\, d\tau + f(0)g(x,t)$$
$$u(x,t) = \int_0^t f(\tau)g'(x, t - \tau)\, d\tau + f(t)g(x,0)$$

The four preceding integral equations are called *Duhamel's integrals*.

For the case of the preceding example, the response to a step U_0 is given by Eq. (7.84). Thus, $g(x,t)$ is obtained by dividing Eq. (7.84) by U_0.

7.6 MULTIDIMENSIONAL EQUATIONS

The heat, or diffusion, equation as given by Eq. (7.2) is for flow in one dimension. For two-dimensional flow this equation becomes

$$\frac{\partial u}{\partial t} = \alpha \left(\frac{\partial^2 u}{\partial x^2} + \frac{\partial^2 u}{\partial y^2} \right) \tag{7.90}$$

To solve a two-dimensional equation by the transform method, it is necessary to transform twice. For example, a first transformation with respect to t would eliminate the t variable; then a second transformation with respect to x would eliminate the x variable. The resulting expression is a differential equation in y. This differential equation is solved, then inverted with respect to x, and finally inverted with respect to t to yield the desired answer.

For three dimensions, the diffusion equation is

$$\frac{\partial u}{\partial t} = \alpha \left(\frac{\partial^2 u}{\partial x^2} + \frac{\partial^2 u}{\partial y^2} + \frac{\partial^2 u}{\partial z^2} \right) = \alpha \nabla^2 u$$

where

$$\nabla^2 = \frac{\partial^2}{\partial x^2} + \frac{\partial^2}{\partial y^2} + \frac{\partial^2}{\partial z^2}$$

Similarly for three dimensions the wave equation is

$$\frac{\partial^2 u}{\partial t^2} = c^2 \left(\frac{\partial^2 u}{\partial x^2} + \frac{\partial^2 u}{\partial y^2} + \frac{\partial^2 u}{\partial z^2} \right) = c^2 \nabla^2 u$$

It is necessary to apply three successive transformations to solve a three-dimensional equation.

Steady flow exists when the temperature at each point (x,y,z) in a body maintains some equilibrium, or steady-state value. Because the temperature is independent of time, $\partial u / \partial t = 0$. Thus, for steady flow the heat equation reduces to Laplace's equation

$$\frac{\partial^2 u}{\partial x^2} + \frac{\partial^2 u}{\partial y^2} + \frac{\partial^2 u}{\partial z^2} = \nabla^2 u = 0$$

Another partial differential equation which has considerable application to engineering problems is the Poisson equation

$$\frac{\partial^2 u}{\partial x^2} + \frac{\partial^2 u}{\partial y^2} + \frac{\partial^2 u}{\partial z^2} = \nabla^2 u = f(x,y,z)$$

An indication of the wide range of applications of the wave equation, diffusion equation, Laplace's equation, and Poisson's equation is given in Table 7.1.

Table 7.1†

Laplace Equation

$$\nabla^2 u = 0$$

Static deflection of stretched membrane in regions without lateral force
Stream function for ideal incompressible fluid with no vorticity
Velocity potential for ideal incompressible irrotational fluid in regions without sources or sinks
Steady-state temperature in homogeneous body without heat sources or sinks
Steady-state neutron flux in homogeneous matter without neutron sources
Voltage in homogeneous conductor
Voltage in region containing no space charge
Scalar magnetic potential in region free of currents
Steady-state pressure of incompressible fluid in homogeneous permeable field
Elastic volume expansion in plane strain
Angle of rigid body rotation in plane strain
Stress function for beam in elastic bending
Conjugate harmonic stress functions in plane elasticity
Gravitational potential in empty space

Poisson Equation

$$\nabla^2 u = f(x,y,z)$$

Static deflection of stretched string or membrane with lateral force
Stream function for ideal incompressible fluid with vorticity
Velocity potential for ideal incompressible irrotational fluid in regions with sources or sinks
Steady-state temperature in body, nonhomogeneous and/or with heat sources or sinks
Steady-state neutron flux in matter, nonhomogeneous and/or with neutron sources
Voltage in conductor, nonhomogeneous and/or with distributed current sources
Voltage in region containing space charge
Stress function for elastic torsion
Gravitational potential in regions with matter

Wave Equation

$$c^2 \nabla^2 u = \frac{\partial^2 u}{\partial t^2}$$

Deflection of vibrating string or membrane
Current and voltage in transmission line without leakage and resistance
Scalar and vector potentials in nonconducting homogeneous regions free of charges and currents
Velocity in surface (tidal) waves of ideal fluid

Table 7.1 (continued)

Displacement in longitudinally vibrating bar
Displacement for sound waves in a pipe
Angle of twist in torsional vibration of circular shaft
Displacement components for elastic waves of distortion or dilatation

Diffusion Equation

$$\alpha \nabla^2 u = \frac{\partial u}{\partial t}$$

Temperature in a homogeneous body
Neutron flux in homogeneous matter
Current and voltage in transmission line without leakage and inductance
Pressure of incompressible fluid in homogeneous permeable field
Concentration of material in homogeneous diffusive medium
Vorticity diffusing through fluid

---

† P. L. Balise, Mathematical Analogies in Engineering, *J. Eng. Educ.*, June, 1963, p. 783.

If x is regarded as the transform variable, then the Laplace transform of the first term in Laplace's and Poisson's equations is

$$\mathcal{L}\left[\frac{\partial^2 u(x,y,z)}{\partial x^2}\right] = s^2 U(s,y,z) - su(0,y,z) - u'(0,y,z)$$

Often in problems involving Laplace's or Poisson's equation, the boundary conditions are in the form $u(0,y,z)$ and $u(L,y,z)$. The initial-condition terms that occur in Fourier sine transforms are $u(0,y,z)$ and $u(L,y,z)$. Thus, direct substitution is possible by using the Fourier sine transform rather than the Laplace transform. For other problems, direct substitution is afforded by the Fourier cosine transform.

Additional features of partial differential equations are discussed in greater detail in the following chapters. Particular consideration is given to the wave equation, diffusion equation, Laplace's equation, and Poisson's equation.

PROBLEMS

7.1 Use transform methods to solve each of the following partial differential equations:

(a) $\dfrac{\partial y(x,t)}{\partial t} + 2y(x,t) = 0 \qquad y(x,0) = x$

(b) $\dfrac{\partial^2 y(x,t)}{\partial t^2} + y(x,t) = 0 \qquad y(x,0) = 0; y'(x,0) = x$

(c) $\dfrac{\partial y(x,t)}{\partial t} + \dfrac{\partial y(x,t)}{\partial x} = 2$ $y(x,0) = x; y(0,t) = t$

(d) $\dfrac{\partial y(x,t)}{\partial x} + \dfrac{\partial y(x,t)}{\partial t} = 1$ $y(0,t) = y(x,0) = 0$

(e) $\dfrac{\partial^2 y(x,t)}{\partial x^2} + \dfrac{\partial^2 y(x,t)}{\partial x \partial t} = 2(t+1)$ $y(x,0) = y(0,t) = y'(0,t) = 0$

(f) $\dfrac{\partial y(x,t)}{\partial x} + \dfrac{\partial y(x,t)}{\partial t} = 0$ $y(x,0) = e^x; y(0,t) = e^{-t}$

(g) $\dfrac{\partial^2 y(x,t)}{\partial t^2} + y(x,t) = 0$ $y(x,0) = 0; y'(x,0) = e^x$

(h) $y(x,t) - \dfrac{x}{2}\dfrac{\partial}{\partial x} y(x,t) = 0$ $y(1,t) = t$

(i) $y(x,t) + x\dfrac{\partial y(x,t)}{\partial x} = 0$ $y(1,t) = e^{-t}$

7.2   In the method of successive transformations, the first and second transforms are distinguished as follows:

$\mathcal{L}[y(x,t)] = Y(x,s)$
$\mathcal{L}[Y(x,s)] = \bar{Y}(z,s)$

Note that s is the transform variable corresponding to t and z is the transform variable corresponding to x. The bar is used to distinguish the second transform from the first transform. For the partial differential equation,

$$\dfrac{\partial y(x,t)}{\partial t} + \dfrac{\partial y(x,t)}{\partial x} = x + t \qquad y(x,0) = y(0,t) = 0$$

Show that

$$\bar{Y}(z,s) = \dfrac{1}{(zs)^2}$$

Invert $\bar{Y}(z,s)$ with respect to z to determine $Y(x,s)$, and then invert $Y(x,s)$ with respect to s to determine the desired solution $y(x,t)$.

7.3   Verify the solution of Prob. 7.2 by transforming first with respect to x and then with respect to t. Use the notation

$\mathcal{L}[y(x,t)] = Y(z,t)$
$\mathcal{L}[Y(z,t)] = \bar{Y}(z,s)$

Note that the same result for $\bar{Y}(z,s)$ is obtained regardless of the order of transforming. Also, the inverse may be taken in any order.

7.4   The transform of Eq. (7.17) with respect to t is given by Eq. (7.21). Show that the transform of Eq. (7.21) with respect to x is

$$z^2 \bar{Y}(z,s) - zY(0,s) - Y'(0,s) - \dfrac{s^2}{c^2}\bar{Y}(z,s) = 0$$

or

$$\bar{Y}(z,s) = \dfrac{zY(0,s)}{z^2 - (s/c)^2} + \dfrac{Y'(0,s)}{z^2 - (s/c)^2}$$

Utilize the known initial condition [Eq. (7.16)], and invert to show that

$$Y(x,s) = \frac{\omega}{s^2 + \omega^2} \cosh \frac{s}{c} x + \frac{Y'(0,s)}{s/c} \sinh \frac{s}{c} x$$

$$Y(x,s) = \left[\frac{\omega}{s^2 + \omega^2} + \frac{Y'(0,s)}{s/c} \right] \frac{e^{(s/c)x}}{2} + \left[\frac{\omega}{s^2 + \omega^2} - \frac{Y'(0,s)}{s/c} \right] \frac{e^{-(s/c)x}}{2}$$

To satisfy the remaining initial condition that $Y(\infty,s) = 0$, it is apparent that $Y'(0,s) = -s/c[\omega/(s^2 + \omega^2)]$. Thus, show that the preceding equation reduces to Eq. (7.27), whence the solution given by Eq. (7.28) follows.

7.5 In Fig. 7.2c is shown the position of a string for the case in which $ct = 2$ and $\omega/c = 2\pi$. Plot the position of the string when $ct = 2$ for each of the following cases: (a) $\omega/c = \pi$; (b) $\omega/c = 3\pi$; (c) $\omega/c = 4\pi$.

7.6 Use the transform method to determine the motion $y(x,t)$ of the string shown in Fig. 7.2a for the case in which $y(x,0) = y'(x,0) = y(\infty,0) = 0$ and $y(0,t) = 1$.

7.7 The equation for sound traveling through a pipe of constant area is

$$\frac{\partial^2 \phi(x,t)}{\partial x^2} = \frac{1}{c^2} \frac{\partial^2 \phi(x,t)}{\partial t^2}$$

where c is the velocity of sound and $\phi(x,t)$ is the pressure. Determine the equation for the pressure wave $\phi(x,t)$ traveling through the pipe when the boundary conditions are

$$\phi(0,t) = f(t) \qquad \phi(x,0) = 0$$
$$\phi(\infty,t) = 0 \qquad \phi'(x,0) = 0$$

7.8 The bar shown in Fig. 7.4 has an initial load such that at the end $x = L$ the strain is $y(L,0) = y_0$. The initial strain at a station x is $y(x,0) = (x/L)y_0$. The bar is initially at rest so that $y(x,0) = 0$. The load is then released at time $t = 0$. Determine:

(a) The equation for the subsequent strain $y(x,t)$
(b) The equation for the subsequent stress $\sigma(x,t)$

7.9 The shear stress due to torsion in a shaft is given by the equation $\sigma = rG(\partial\theta/\partial x)$, where r is the distance from the center of the shaft to the point where the stress is to be determined. The stress on the outer surface is

$$\sigma = RG \left(\frac{\partial\theta}{\partial x} \right)$$

where R is the outside radius of the shaft. In Fig. 7.9b is shown a plot of the twist $\theta(x,t)$ for the uniform shaft of Fig. 7.9a. Make a plot of the corresponding stress $\sigma(x,t)$ on the outer surface.

7.10 A uniform shaft is initially at rest and under no load so that $\theta(x,0) = (\partial/\partial t)\theta(x,0) = 0$. Determine the equation for the subsequent motion $\theta(x,t)$ for the case in which a suddenly applied torque T_0 is applied at the end $x = L$.

7.11 The Taylor series expansion for $f(x)$ about a point x_0 is

$$e^{\Delta x D_x}f(x_0) = f(x_0 + \Delta x) = f(x_0) + \frac{f'(x_0)\,\Delta x}{1!} + \frac{f''(x_0)\,\Delta x^2}{2!} + \cdots$$

If y_0 is regarded as a constant parameter,

$$e^{\Delta x D_x}f(x_0,y_0) = f(x_0 + \Delta x,\, y_0) = f(x,y_0)$$

Multiplication of both sides of the preceding expression by $e^{\Delta y D_y}$ gives the Taylor series expansion for a function of two variables, $f(x,y)$. That is,

$$e^{\Delta x D_x + \Delta y D_y}f(x_0,y_0) = f(x_0 + \Delta x,\, y_0 + \Delta y) = f(x,y)$$

Show that

$$f(x,y) = f(x_0,y_0) + \Delta x\,\frac{\partial f}{\partial x} + \Delta y\,\frac{\partial f}{\partial y}$$

$$+ \frac{1}{2!}\left(\Delta x^2\,\frac{\partial^2 f}{\partial x^2} + 2\Delta x\,\Delta y\,\frac{\partial^2 f}{\partial x\partial y} + \Delta y^2\,\frac{\partial^2 f}{\partial y^2}\right) + \cdots$$

7.12 A bar of infinite length is insulated on the top and bottom. Initially, the temperature is $0°$ throughout. Determine the resulting temperature distribution for the case in which the temperature at the end $x = 0$ is

$$u(0,t) = \begin{cases} 0 & t < 0 \\ U_0 & t > 0 \end{cases}$$

7.13 In Prob. 7.12, let the temperature at the end $x = 0$ be

$$u(0,t) = \begin{cases} 0 & t < 0 \\ f(t) & t > 0 \end{cases}$$

Solve this problem by:

(a) Applying the convolution integral in such a manner as to obtain the answer in terms of the response to a unit step function. The unit step response is the answer to Prob. 7.12 divided by U_0.

(b) Using the appropriate Duhamel equation.

7.14 Solve the heat, or diffusion, equation for the semi-infinite solid shown in Fig. 7.11 when the boundary conditions are $u(x,0) = u(\infty,t) = U_0$ and $u(0,t) = 0$.

7.15 Solve the heat, or diffusion, equation for the semi-infinite solid shown in Fig. 7.11 when the boundary conditions are $u(x,0) = u(\infty,t) = 0$ and $u(0,t) = f(x)$.

7.16 A bar of infinite length is in effect insulated at the end $x = \infty$ because no heat flows beyond this point. Thus, $\partial u(\infty,t)/\partial x = 0$. Determine the temperature distribution $u(x,t)$ for the case in which $u(x,0) = 1$ and $u(0,t) = 0$.

7.17 For the bar shown in Fig. 7.14 the temperature at each end is maintained at $0°$ [that is, $u(0,t) = u(L,t) = 0$]. Determine the resulting temperature $u(x,t)$ when the initial temperature distribution is:

    (a) $u(x,0) = k \sin (\pi x/L)$
    (b) $u(x,0) = kx$

REFERENCES

1 Balise, P. L.: Mathematical Analogies in Engineering, *J. Eng. Educ.*, June, 1963, pp. 780–786.

2 Carslaw, H. S., and J. C. Jaeger: "Operational Methods in Applied Mathematics," 2d ed., Oxford University Press, New York, 1948.

3 Churchill, R. V.: "Operational Mathematics," 2d ed., McGraw-Hill Book Company, New York, 1958.

4 Doetsch, Gustav: "Guide to the Applications of Laplace Transforms," D. Van Nostrand Company, Inc., Princeton, N.J., 1961.

5 McLachlan, N. W.: "Modern Operational Calculus," The Macmillan Company, New York, 1948.

6 Miller, Fredric H.: "Partial Differential Equations," John Wiley & Sons, Inc., New York, 1941.

7 Sneddon, I. N.: "Elements of Partial Differential Equations," McGraw-Hill Book Company, New York, 1957.

chapter 8

Boundary-value problems

One partial differential equation may describe numerous physical situations (e.g., see Table 7.1). Thus, there are many solutions which satisfy the same partial differential equation. The form of each solution is determined by the boundary conditions for the particular problem. For example, the form of the solution of the wave equation for a vibrating beam depends upon whether the ends are fixed, pinned, free, etc. Similarly, the form of the solution of the heat equation depends upon whether the surface is insulated, maintained at a specified temperature, free to radiate heat, etc. It is to be noted that boundary conditions are a permanent type of constraint. The values of the constants which appear in a given solution are determined by specific values of initial conditions such as initial temperatures, forces, positions, etc.

Problems in which the form of the solution depends upon the boundary conditions, or constraints, are referred to as *boundary-value problems*. Thus, partial differential equations such as the wave equation, heat equation, Laplace's equation, and Poisson's equation constitute boundary-value problems.

In the method of separation of variables, which is discussed in Sec. 8.4, the partial differential equation is reduced to several ordinary differential equations. In order that these differential equations have solutions which satisfy the boundary conditions, it is usually found that a parameter λ must satisfy a certain equation. This equation is referred to as the *characteristic equation*. The values of λ which satisfy the characteristic equation are called the *characteristic values*, or *eigenvalues*, for the problem. Ordinarily, there are an infinite number of eigenvalues, λ, and thus the solution is obtained

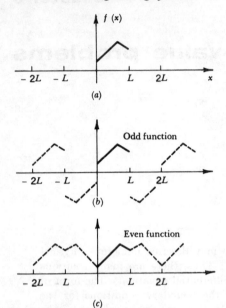

Fig. 8.1. Fourier extensions. (a) A function $f(x)$ defined for $0 \leq x \leq L$; (b) extension of $f(x)$ to an odd function; (c) extension of $f(x)$ to an even function.

in the form of an infinite series. The final step in the solution is to determine the constants which arise in this infinite series such that this solution satisfies the initial conditions. When problems are solved by transformation methods, these essential features, or characteristics, of the solution tend to be disguised in the transformation process. However, transformation methods usually yield the solution in a more direct manner than does the method of separation of variables. Also, transformation methods are applicable to a wider range of problems.

The finite sine and cosine transformations are better suited to the solution of many boundary-value problems than the Laplace transformation. For example, many problems involving the wave equation or the heat equation are solved most directly by use of the finite sine transformation or the finite cosine transformation. For Laplace's and Poisson's differential equations, the boundary conditions are usually such that finite sine or cosine transformations provide the best method of solution. For problems which require successive transformations, it may be best to use Laplace transformations only, sine or cosine transformations only, or a combination in which one or more transformations are Laplace and the others are sine or cosine transformations. In Chap. 11, it is shown that Hankel and Legendre transformations are particularly well suited for the solution of partial differential equations with variable coefficients.

In Fig. 8.1a is shown a function $f(x)$ which is defined over the finite interval $0 \leq x \leq L$. This function may be regarded as a periodic function. The dotted lines in Fig. 8.1b show the extension when $f(x)$ is considered to be an odd function; that is,

$$f(x) = -f(-x)$$

The dotted lines in Fig. 8.1c show the extension when $f(x)$ is considered to be an even function; i.e.,

$$f(x) = f(-x)$$

8.1 FINITE FOURIER SINE TRANSFORMATION

The Fourier series for an odd function is given by Eq. (1.19). When $f(x)$ is regarded as an odd function, the Fourier series for $f(x)$ is obtained by substituting x for t in Eq. (1.19). That is,

$$f(x) = \sum_{n=1}^{\infty} B_n \sin n\omega_0 x = \sum_{n=1}^{\infty} B_n \sin \frac{2\pi nx}{T} \tag{8.1}$$

where $\omega_0 = 2\pi/T$. The coefficient B_n is given by Eq. (1.17). Replacing t by x and ω_0 by $2\pi/T$ gives

$$B_n = \frac{4}{T} \int_0^{T/2} f(x) \sin \frac{2\pi nx}{T} dx \qquad n = 1, 2, 3, \ldots$$

As shown in Fig. 8.1b, the period for the odd function $f(x)$ extends from $-L$ to $+L$ (that is, $T = 2L$). Thus

$$B_n = \frac{2}{L} \int_0^L f(x) \sin \frac{\pi nx}{L} dx = \frac{2}{L} F(n)$$

The preceding integration converts the function $f(x)$ to some function of n [that is, $F(n)$]. This is called the *finite Fourier sine transform* $S[f(x)]$; that is,

$$S[f(x)] = \int_0^L f(x) \sin \frac{\pi nx}{L} dx = F(n) \qquad n = 1, 2, \ldots \tag{8.2}$$

299

The inverse transform of this finite Fourier sine transform is obtained by substituting $B_n = (2/L)F(n)$ into Eq. (8.1). Thus,

$$f(x) = \frac{2}{L} \sum_{n=1}^{\infty} F(n) \sin \frac{\pi n x}{L} \tag{8.3}$$

Equation (8.2) is the direct finite Fourier sine transform which transforms a function $f(x)$ to another function $F(n)$. Equation (8.3) is the inverse finite Fourier sine transform which inverts $F(n)$ back to $f(x)$.

Because the finite Fourier sine transform is based upon the Fourier series representation for an odd function, the only restriction on $f(x)$ is that it satisfy the Dirichlet conditions over the interval 0 to L.

The finite Fourier sine transform has operational properties similar to the Laplace transform. A brief listing of finite Fourier sine transform pairs is given in Appendix B. The derivation of some of these pairs follows.

Consider first the transform for the second derivative $f''(x)$. Integration by parts in which $u = \sin(\pi n x/L)$ and $v = f'(x)$ gives

$$S[f''(x)] = \int_0^L f''(x) \sin \frac{\pi n x}{L} \, dx = f'(x) \sin \frac{\pi n x}{L} \Big|_0^L$$
$$- \frac{\pi n}{L} \int_0^L f'(x) \cos \frac{\pi n x}{L} \, dx$$

The first term on the right-hand side of the preceding equation vanishes at both the upper limit $x = L$ and the lower limit $x = 0$. Application of integration by parts to the last term by letting $u = \cos(\pi n x/L)$ and $v = f(x)$ gives

$$S[f''(x)] = -\frac{\pi n}{L} f(x) \cos \frac{\pi n x}{L} \Big|_0^L - \left(\frac{\pi n}{L}\right)^2 \int_0^L f(x) \sin \frac{\pi n x}{L} \, dx$$
$$= -\frac{\pi n}{L} \left[\frac{\pi n}{L} F(n) + (-1)^n f(L) - f(0) \right] \tag{8.4}$$

where $\cos \pi n = (-1)^n$. The boundary conditions are specified at positions $x = 0$ and L.

Application of Eq. (8.2) to obtain the finite sine transform of $f(x) = 1$ for $0 \le x \le L$ gives

$$S(1) = \int_0^L \sin \frac{\pi n x}{L} \, dx = -\frac{L}{\pi n} \cos \frac{\pi n x}{L} \Big|_0^L = \frac{L}{\pi n} [1 - (-1)^n] \tag{8.5}$$

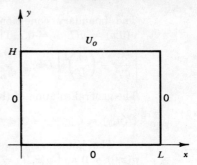

Fig. 8.2. A rectangular plate.

Application of Eq. (8.2) to obtain the finite sine transform of
$f(x) = x$ for $0 \leq x \leq L$ gives

$$S(x) = \int_0^L x \sin \frac{\pi n x}{L} \, dx = \frac{L}{\pi n} \left[\frac{L}{\pi n} \sin \frac{\pi n x}{L} - x \cos \frac{\pi n x}{L} \right]_0^L$$

$$= \frac{L^2}{\pi n} (-1)^{n+1} \tag{8.6}$$

By regarding x as the transform variable, then for a function of two
variables $u(x,y)$ the basic finite-sine-transform relationships are

$$S[u(x,y)] = U(n,y)$$

$$S\left[\frac{\partial^2}{\partial x^2} u(x,y) \right] = -\frac{\pi n}{L} \left[\frac{\pi n}{L} U(n,y) + (-1)^n u(L,y) - u(0,y) \right]$$

$$S\left[\frac{\partial^2}{\partial y^2} u(x,y) \right] = \frac{\partial^2}{\partial y^2} S[u(x,y)] = \frac{\partial^2}{\partial y^2} U(n,y)$$

Illustrative example   In Fig. 8.2 is shown a rectangular plate.
The temperature on the left, right, and bottom sides is zero,
and the temperature on the top is maintained at U_0. For
steady flow, the temperature at each point (x,y) achieves a
constant value such that $\partial u/\partial t = 0$. Thus, the heat equation
[Eq. (7.90)] reduces to Laplace's equation. That is,

$$\frac{\partial^2 u}{\partial x^2} + \frac{\partial^2 u}{\partial y^2} = 0 \tag{8.7}$$

Determine the equation for the temperature distribution
$u(x,y)$ throughout this plate.

Solution   By regarding x as the transform variable, the finite
sine transform of Laplace's equation is

$$-\frac{\pi n}{L} \left[\frac{\pi n}{L} U(n,y) + (-1)^n u(L,y) - u(0,y) \right] + \frac{\partial^2}{\partial y^2} U(n,y) = 0$$

The boundary conditions at the left and right sides are $u(0,y) = u(L,y) = 0$. Thus

$$\left[\frac{\partial^2}{\partial y^2} - \left(\frac{\pi n}{L}\right)^2\right] U(n,y) = 0 \tag{8.8}$$

The general solution of this differential equation is

$$U(n,y) = C_1 e^{\pi n y/L} + C_2 e^{-\pi n y/L} \tag{8.9}$$

The boundary conditions at the bottom and top are $u(x,0) = 0$ and $u(x,H) = U_0$. The transform of each boundary condition is

$$S[u(x,0)] = U(n,0) = 0$$

$$S[u(x,H)] = U(n,H) = \frac{U_0 L[1 - (-1)^n]}{\pi n}$$

Evaluating Eq. (8.9) at $y = 0$ gives

$$U(n,0) = C_1 + C_2 = 0$$

Similarly, evaluating Eq. (8.9) at $y = H$ gives

$$U(n,H) = C_1 e^{\pi n H/L} + C_2 e^{-\pi n H/L} = \frac{U_0 L[1 - (-1)^n]}{\pi n}$$

From the two preceding equations, the values of C_1 and C_2 are found to be

$$C_1 = -C_2 = \frac{U_0 L[1 - (-1)^n]}{\pi n(e^{\pi n H/L} - e^{-\pi n H/L})}$$

Thus, the general solution, Eq. (8.9), becomes

$$U(n,y) = \frac{e^{\pi n y/L} - e^{-\pi n y/L}}{e^{\pi n H/L} - e^{-\pi n H/L}} \frac{U_0 L[1 - (-1)^n]}{\pi n}$$

$$= \frac{\sinh{(\pi n y/L)}}{\sinh{(\pi n H/L)}} \frac{U_0 L[1 - (-1)^n]}{\pi n} \tag{8.10}$$

Application of Eq. (8.3) to invert the preceding equation gives for the desired result

$$u(x,y) = \frac{2U_0}{\pi} \sum_{n=1}^{\infty} \left[\frac{\sinh{(\pi n y/L)}}{\sinh{(\pi n H/L)}} \frac{1 - (-1)^n}{n}\right] \sin{\frac{\pi n x}{L}} \tag{8.11}$$

If, in Eq. (8.10), $U_0 L[1 - (-1)^n]/\pi n$, which is the transform of U_0, is replaced by $F(n)$, which is the transform of $f(x)$, then the transformed equation for the case in which the temperature on the top surface is some arbitrary function $f(x)$ rather than U_0 is obtained. That is,

$$U(n,y) = \frac{\sinh\,(\pi ny/L)}{\sinh\,(\pi nH/L)}\,F(n)$$

Application of Eq. (8.3) to obtain the inverse gives

$$u(x,y) = \frac{2}{L} \sum_{n=1}^{\infty} \frac{\sinh\,(\pi ny/L)}{\sinh\,(\pi nH/L)}\,F(n)\,\sin\frac{\pi nx}{L} \tag{8.12}$$

where from Eq. (8.2)

$$F(n) = \int_0^L f(x)\,\sin\frac{\pi nx}{L}\,dx \tag{8.13}$$

Thus, the solution is given by Eq. (8.12), in which the function $F(n)$ is known from Eq. (8.13).

The sine transform is converted from the rectangular form of Eq. (8.2) to the polar form by making the change of variable $\theta = (\pi/L)x$. Thus,

$$S_1[f(\theta)] = \frac{L}{\pi} \int_0^{\pi} f(\theta)\,\sin n\theta\,d\theta \qquad n = 1, 2, 3, \ldots$$

The corresponding inversion formula is

$$F(\theta) = \frac{2}{L} \sum_{n=1}^{\infty} S_1[f(\theta)]\,\sin n\theta$$

The more common representation for this polar form is obtained by multiplying $S_1[f(\theta)]$ by the constant π/L. Thus, letting $S[f(\theta)]$ equal $(\pi/L)S_1[f(\theta)]$ gives

$$S[f(\theta)] = \int_0^{\pi} f(\theta)\,\sin n\theta\,d\theta = F(n) \qquad n = 1, 2, 3$$

and

$$f(\theta) = \frac{2}{\pi} \sum_{n=1}^{\infty} S[f(\theta)]\,\sin n\theta = \frac{2}{\pi} \sum_{n=1}^{\infty} F(n)\,\sin n\theta$$

The application of successive transformations for solving the heat equation for flow in two dimensions will now be explained.

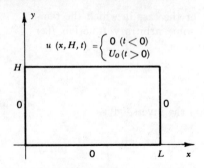

Fig. 8.3. A rectangular plate.

Successive transformations

In Fig. 8.3 is shown a rectangular plate for which the boundary conditions on the four sides are

$$u(0,y,t) = 0$$
$$u(L,y,t) = 0$$
$$u(x,0,t) = 0$$
$$u(x,H,t) = \begin{cases} 0 & t < 0 \\ U_0 & t > 0 \end{cases} \tag{8.14}$$

In addition, the temperature of the plate is initially zero throughout. That is,

$$u(x,y,0) = 0$$

The heat equation is

$$\frac{\partial u}{\partial t} = \alpha \left(\frac{\partial^2 u}{\partial x^2} + \frac{\partial^2 u}{\partial y^2} \right)$$

By regarding t as the transform variable, the Laplace transform is

$$sU(x,y,s) - u(x,y,0) = \alpha \left(\frac{\partial^2}{\partial x^2} + \frac{\partial^2}{\partial y^2} \right) U(x,y,s) \tag{8.15}$$

where $u(x,y,0) = 0$. By regarding x as the transform variable, the finite sine transform of the preceding equation is

$$\frac{s}{\alpha} \bar{U}(n,y,s) = -\frac{\pi n}{L} \left[\frac{\pi n}{L} \bar{U}(n,y,s) + (-1)^n U(L,y,s) - U(0,y,s) \right]$$
$$+ \frac{\partial^2}{\partial y^2} \bar{U}(n,y,s) \tag{8.16}$$

The symbol $\bar{U}(n,y,s)$ is used to distinguish the second transform. Because $U(L,y,s) = \mathcal{L}[u(L,y,t)] = 0$ and $U(0,y,s) = \mathcal{L}[u(0,y,t)] = 0$, the preceding equation becomes

$$\left[\frac{\partial^2}{\partial y^2} - \left(\frac{s}{\alpha} + \frac{\pi^2 n^2}{L^2}\right)\right] \bar{U}(n,y,s) = 0 \tag{8.17}$$

The general solution of this differential equation is

$$\bar{U}(n,y,s) = C_1 e^{ay} + C_2 e^{-ay} \tag{8.18}$$

where $a^2 = s/\alpha + \pi^2 n^2/L^2$.

Transforming the last two boundary conditions of Eq. (8.14) with respect to s gives

$$U(x,0,s) = 0$$
$$U(x,H,s) = \frac{U_0}{s}$$

The transforms of the preceding equations with respect to n are

$$\bar{U}(n,0,s) = 0$$
$$\bar{U}(n,H,s) = \frac{U_0 L[1 - (-1)^n]}{\pi n s}$$

Evaluating Eq. (8.18) at $y = 0$ and at $y = H$ gives

$$\bar{U}(n,0,s) = C_1 + C_2 = 0 \ .$$
$$\bar{U}(n,H,s) = C_1 e^{aH} + C_2 e^{-aH} = \frac{U_0 L[1 - (-1)^n]}{\pi n s}$$

The constants C_1 and C_2 are

$$C_1 = -C_2 = \frac{U_0 L[1 - (-1)^n]}{\pi n s(e^{aH} - e^{-aH})}$$

Thus, the general solution becomes

$$\bar{U}(n,y,s) = \frac{e^{ay} - e^{-ay}}{e^{aH} - e^{-aH}} \frac{U_0 L[1 - (-1)^n]}{\pi n s}$$
$$= \frac{\sinh ay}{\sinh aH} \frac{U_0 L[1 - (-1)^n]}{\pi n s} \tag{8.19}$$

This equation for the second transform $\bar{U}(n,y,s)$ is the same regardless of whether the partial differential equation is first transformed with respect to x and then t or whether, as in this example, it is first transformed with respect to t and then x. The work involved is about the same regardless of the order of transforming.

Application of Eq. (8.3) to obtain the inverse with respect to n gives

$$U(x,y,s) = \frac{2U_0}{\pi} \sum_{n=1}^{\infty} \frac{\sinh ay}{s \sinh aH} \frac{1 - (-1)^n}{n} \sin \frac{\pi n x}{L} \tag{8.20}$$

To obtain the inverse Laplace transform of $\sinh ay/(s \sinh aH)$, first note $\sinh aH = 0$ when

$$aH = \pm jm\pi \qquad m = 1, 2, 3, \ldots$$

Thus,

$$a^2H^2 = \left(\frac{s}{\alpha} + \frac{\pi^2 n^2}{L^2}\right) H^2 = -m^2\pi^2$$

Solving for the value of s which makes $\sinh aH = 0$ gives

$$s = -\alpha\pi^2 \left(\frac{m^2}{H^2} + \frac{n^2}{L^2}\right) = -K_{mn} \qquad m = 1, 2, 3, \ldots$$

The poles of $\sinh ay/(s \sinh aH)$ occur at $s = 0$ and $s = -K_{mn}$. The residue at $s = 0$ is

$$\lim_{s \to 0} \frac{\sinh ay}{\sinh aH} e^{st} = \frac{\sinh (\pi ny/L)}{\sinh (\pi nH/L)}$$

The residue at $s = -K_{mn}$ is obtained by letting

$$P(s) = (\sinh ay)e^{st}/s$$

and

$$Q(s) = \sinh aH.$$

When $s = -K_{mn}$, then $a = j\pi m/H$. Thus,

$$P(s)\Big|_{s = -K_{mn}} = \frac{\sinh (j\pi my/H)e^{-K_{mn}t}}{-K_{mn}} = \frac{j \sin (\pi my/H)e^{-K_{mn}t}}{-K_{mn}}$$

$$Q'(s)\Big|_{s = -K_{mn}} = \frac{H^2 \cosh j\pi m}{j2\pi\alpha m} = \frac{H^2 \cos \pi m}{j2\pi\alpha m}$$

$$\frac{P(s)}{Q'(s)}\Big|_{s = -K_{mn}} = \frac{2\pi\alpha m \sin (\pi my/H)e^{-K_{mn}t}}{(-1)^m H^2 K_{mn}}$$

where $\cos \pi m = (-1)^m$.

Summing all the residues gives

$$u(x,y,t) = \frac{2U_0}{\pi} \sum_{n=1}^{\infty} \left[\frac{\sinh (\pi ny/L)}{\sinh (\pi nH/L)} \right.$$

$$\left. - \sum_{m=1}^{\infty} \frac{2\pi\alpha m \sin (\pi my/H)e^{-K_{mn}t}}{(-1)^m H^2 K_{mn}} \right] \frac{1 - (-1)^n}{n} \sin \frac{\pi nx}{L} \qquad (8.21)$$

Next consider the case in which the temperature on the top surface for $t > 0$ is some arbitrary function $f(x)$ rather than U_0; that is,

$$u(x,H,t) = \begin{cases} 0 & t < 0 \\ f(x) & t > 0 \end{cases}$$

The Laplace transform is

$$U(x,H,s) = \mathcal{L}[(1)f(x)] = f(x)\mathcal{L}(1) = \frac{f(x)}{s}$$

Next, taking the finite sine transform gives

$$\bar{U}(n,H,s,) = \frac{F(n)}{s}$$

Replacing $U_0L[1 - (-1)^n]/\pi ns$ in Eq. (8.19) by $F(n)/s$ gives

$$\bar{U}(n,y,s) = \frac{\sin ay}{\sinh aH} \frac{F(n)}{s} \tag{8.22}$$

The inverse with respect to n is

$$U(x,y,s) = \frac{2}{L} \sum_{n=1}^{\infty} \frac{\sinh ay}{s \sinh aH} F(n) \sin \frac{\pi nx}{L} \tag{8.23}$$

The inverse of $\sinh ay/(s \sinh aH)$ has previously been determined. Hence, the desired result is

$$u(x,y,t) = \frac{2}{L} \sum_{n=1}^{\infty} \left[\frac{\sinh (\pi ny/L)}{\sinh (\pi nH/L)} \right.$$
$$\left. - \sum_{m=1}^{\infty} \frac{2\pi \alpha m \sin (\pi my/H)e^{-K_{mn}t}}{(-1)^m H^2 K_{mn}} \right] F(n) \sin \frac{\pi nx}{L} \tag{8.24}$$

In the limit as t approaches infinity, the summation from $m = 1$ to ∞ in the preceding expression vanishes (that is, $e^{-K_{mn}t}$ approaches zero). For this case, Eq. (8.24) reduces to Eq. (8.12), which is the steady-state solution.

8.2 FINITE FOURIER COSINE TRANSFORMATION

The Fourier series for an even function is given by Eq. (1.15). When $f(x)$ is regarded as an even function, the Fourier series for $f(x)$ is obtained by substituting x for t in Eq. (1.15). That is,

$$f(x) = K + \sum_{n=1}^{\infty} A_n \cos n\omega_0 x = K + \sum_{n=1}^{\infty} A_n \cos \frac{2\pi nx}{T} \tag{8.25}$$

where $\omega_0 = 2\pi/T$. The coefficients K and A_n are given by Eqs. (1.10) and (1.11). As illustrated in Fig. 8.1c, when $f(x)$ is regarded as an even function, the corresponding period is $T = 2L$. Thus, replacing t by x, T by $2L$, and ω_0 by $2\pi/T = \pi/L$ in Eqs. (1.10) and (1.11) gives

$$K = \frac{1}{L} \int_0^L f(x)\, dx$$

$$A_n = \frac{2}{L} \int_0^L f(x) \cos \frac{\pi n x}{L}\, dx = \frac{2}{L} F(n)$$

The last integral yields a function of n which is called the *finite Fourier cosine transform* $C[f(x)]$. That is,

$$C[f(x)] = \int_0^L f(x) \cos \frac{\pi n x}{L}\, dx = F(n) \tag{8.26}$$

For $n = 0$, then $F(0) = \int_0^L f(x)\, dx = LK$. Thus, replacing K by $F(0)/L$ and A_n by $(2/L)F(n)$ in Eq. (8.25) gives

$$f(x) = \frac{F(0)}{L} + \frac{2}{L} \sum_{n=1}^{\infty} F(n) \cos \frac{\pi n x}{L} \tag{8.27}$$

Equation (8.26) is the direct finite Fourier cosine transform, and Eq. (8.27) is the inverse finite Fourier cosine transform. A brief list of cosine-transform pairs is given in Appendix C.

As is the case for the finite Fourier sine transform, the finite Fourier cosine transform exists for any function $f(x)$ which satisfies the Dirichlet conditions over the interval 0 to L.

The cosine transform for the second derivative is obtained as follows: Integration by parts in which $u = \cos(\pi n x/L)$ and $v = f'(x)$ gives

$$C[f''(x)] = \int_0^L f''(x) \cos \frac{\pi n x}{L}\, dx$$

$$= f'(x) \cos \frac{\pi n x}{L} \Big|_0^L + \frac{\pi n}{L} \int_0^L f'(x) \sin \frac{\pi n x}{L}\, dx$$

By now letting $u = \sin(\pi n x/L)$ and $v = f(x)$, application of integration by parts to the last integral gives

$$\int_0^L f'(x) \sin \frac{\pi n x}{L}\, dx = f(x) \sin \frac{\pi n x}{L} \Big|_0^L - \frac{\pi n}{L} \int_0^L f(x) \cos \frac{\pi n x}{L}\, dx$$

$$= -\frac{\pi n}{L} C[f(x)] = -\frac{\pi n}{L} F(n)$$

Thus, the finite cosine transform for the second derivative is

$$C[f''(x)] = -\frac{\pi^2 n^2}{L^2} F(n) + (-1)^n f'(L) - f'(0) \tag{8.28}$$

For the finite cosine transform the boundary conditions are expressed in terms of the derivatives at $x = 0$ and $x = L$.

The finite cosine transform of $f(x) = x$ for $0 \leq x \leq L$ is

$$C(x) = \int_0^L x \cos \frac{\pi n x}{L} \, dx = \frac{L}{\pi n} \left[\frac{L}{\pi n} \cos \frac{\pi n}{L} x + x \sin \frac{\pi n}{L} x \right]_0^L$$

$$= \frac{L^2}{(\pi n)^2} [(-1)^n - 1] = \begin{cases} \dfrac{-2L^2}{(\pi n)^2} & n \text{ odd} \\ 0 & n \text{ even} \end{cases} \quad (8.29)$$

An alternate technique which may be employed to evaluate sine or cosine transforms is obtained by noting that

$$\int_0^L f(x) e^{jsx} \, dx = \int_0^L f(x) \cos sx \, dx + j \int_0^L f(x) \sin sx \, dx$$

The substitution of $s = \pi n / L$ gives

$$\int_0^L f(x) e^{j\pi n x/L} \, dx = \int_0^L f(x) \cos \frac{\pi n x}{L} \, dx + j \int_0^L f(x) \sin \frac{\pi n x}{L} \, dx$$

$$= C[f(x)] + jS[f(x)] \quad (8.30)$$

The real part of Eq. (8.30) is the finite cosine transform, and the imaginary part is the finite sine transform. To illustrate this method, let it be desired to determine both the sine and cosine transforms of $f(x) = x$.

$$\int_0^L x e^{jsx} \, dx = \frac{1}{(js)^2} [e^{jsx}(jsx - 1)]_0^L$$

$$= \frac{-1}{s^2} [e^{jsL}(jsL - 1) - (-1)]$$

Substituting $s = \pi n / L$ and noting that $e^{jsL} = e^{j\pi n} = (-1)^n$ gives

$$\int_0^L x e^{jsx} \, dx = \frac{-L^2}{(\pi n)^2} [(-1)^n(j\pi n - 1) + 1]$$

$$= \frac{L^2}{(\pi n)^2} [(-1)^n - 1] - j \left[\frac{L^2}{\pi n} (-1)^n \right] \quad (8.31)$$

The real part of the preceding expression is the finite cosine transform of x, and the imaginary part is the finite sine transform.

8.3 FOURIER SINE AND COSINE TRANSFORMATIONS

Taking the limit as L approaches infinity of the finite Fourier sine transform [Eq. (8.2)] yields the Fourier sine transform. As n changes by one unit, the function $\pi n / L$ changes by π / L units [that is,

$\pi(n + 1)/L - \pi n/L = \pi/L]$. In the limit as L becomes very large, the change π/L becomes very small. Because $\pi n/L$ approaches a continuous function as L becomes infinite, then replacing $\pi n/L$ by ω in Eq. (8.2) gives

$$\int_0^\infty f(x) \sin (\omega x) \, dx = F(\omega) \tag{8.32}$$

where $F(\omega)$ is the Fourier sine transform of $f(x)$. Note that $F(n)$ in Eq. (8.2) becomes $F(\omega)$ in the preceding equation.

The inverse Fourier sine transform is obtained from Eq. (8.3) as follows: The change in the function $\omega = \pi n/L$ as n changes by one unit is $\Delta\omega = \pi/L$. Thus, replacing $1/L$ by $\Delta\omega/\pi$, $\pi n/L$ by ω, and $F(n)$ by $F(\omega)$ in Eq. (8.3) gives

$$f(x) = \frac{2}{\pi} \sum_{\omega=0}^\infty F(\omega) \sin (\omega x) \, \Delta\omega = \frac{2}{\pi} \int_0^\infty F(\omega) \sin (\omega x) \, d\omega \tag{8.33}$$

The direct and inverse Fourier sine transforms are usually expressed in the following manner, in which the constant $2/\pi$ is shared between the two expressions. That is,

$$F(\omega) = \sqrt{\frac{2}{\pi}} \int_0^\infty f(x) \sin (\omega x) \, dx \tag{8.34}$$

$$F(x) = \sqrt{\frac{2}{\pi}} \int_0^\infty F(\omega) \sin (\omega x) \, d\omega \tag{8.35}$$

By proceeding in a similar manner, then in the limit as L approaches infinity, the finite Fourier cosine transform becomes the Fourier cosine transform. The equations for the direct and inverse Fourier cosine transforms are

$$F(\omega) = \sqrt{\frac{2}{\pi}} \int_0^\infty f(x) \cos (\omega x) \, dx \tag{8.36}$$

$$f(x) = \sqrt{\frac{2}{\pi}} \int_0^\infty F(\omega) \cos (\omega x) \, d\omega \tag{8.37}$$

Because of the infinite range of integration, there are relatively few functions $f(x)$ for which the Fourier sine and cosine transforms converge. Thus, these transforms, as is the case with the Fourier transform [Eq. (1.40)], have limited application for solving differential equations. However, it should be pointed out that the Fourier sine, Fourier cosine, and Fourier transforms are very useful for investigating certain engineering phenomena.

For example, the Fourier sine and cosine transforms form the basis for an excellent technique for determining the transfer function of a system from the transient response, and vice versa. A significant application of the Fourier transformation arises in the design of systems which are described by discrete or random variables. In particular, the Fourier transform of the autocorrelation function yields the power-density spectrum of the system.

8.4 SEPARATION OF VARIABLES

The method of separation of variables is a versatile classical method for solving partial differential equations. This method is applicable to a wide range of equations such as the wave equation, heat equation, Laplace's equation, Poisson's equation, etc. The application of the method of separation of variables to the wave equation is considered first.

Wave equation

This partial differential equation is

$$\frac{\partial^2 y(x,t)}{\partial t^2} = c^2 \frac{\partial^2 y(x,t)}{\partial x^2}$$

The solution $y(x,t)$ is a function of x and t. In the method of separation of variables, it is assumed that the solution is a product of a function of t alone [that is, $T(t)$] and a function of x alone [that is, $X(x)$]. Thus, the solution is assumed to have the form

$$y(x,t) = T(t)X(x) \tag{8.38}$$

The second partial derivative with respect to t is

$$\frac{\partial^2 y(x,t)}{\partial t^2} = T''(t)X(x)$$

The second partial derivative with respect to x is

$$\frac{\partial^2 y(x,t)}{\partial x^2} = T(t)X''(x)$$

Substitution of these results into the wave equation gives

$$\frac{T''(t)}{T(t)} = c^2 \frac{X''(x)}{X(x)} \tag{8.39}$$

Because $T''(t)/T(t)$ is independent of x and similarly $X''(x)/X(x)$ is independent of t, both expressions must equal a constant k.

$$\frac{T''(t)}{T(t)} = c^2 \frac{X''(x)}{X(x)} = k \tag{8.40}$$

The preceding expression may be written as two separate differential equations.

$$T'' - kT = 0$$
$$X'' - \frac{k}{c^2} X = 0$$

Thus, the partial differential equation has been reduced to a pair of ordinary differential equations. The solution of these differential equations depends on the value of k. It is necessary to investigate the three cases

$$k > 0$$
$$k = 0$$
$$k < 0$$

For $k = \lambda^2 > 0$, the general solution for each of the differential equations is

$$T(t) = C_1 e^{\lambda t} + C_2 e^{-\lambda t}$$
$$X(x) = C_3 e^{\lambda x/c} + C_4 e^{-\lambda x/c}$$

For this case

$$y(x,t) = (C_1 e^{\lambda t} + C_2 e^{-\lambda t})(C_3 e^{\lambda x/c} + C_4 e^{-\lambda x/c})$$

Because this is an exponential type of response, it cannot describe the undamped periodic motion of the wave equation. Thus, the solution for $k > 0$ must be rejected.

For $k = 0$, then $T(t) = C_1 t + C_2$ and $X(x) = C_3 x + C_4$. For this case the solution is

$$y(x,t) = (C_1 t + C_2)(C_3 x + C_4)$$

This solution must also be rejected because it is not periodic.

The acceptable solution occurs for $k = -\lambda^2 < 0$. For this case

$$T(t) = C_1 e^{j\lambda t} + C_2 e^{-j\lambda t}$$
$$X(x) = C_3 e^{j\lambda x/c} + C_4 e^{-j\lambda x/c}$$

By use of Euler's equations, $T(t)$ may be written in the form

$$T(t) = C_1(\cos \lambda t + j \sin \lambda t) + C_2(\cos \lambda t - j \sin \lambda t)$$
$$= A \cos \lambda t + B \sin \lambda t$$

where $A = C_1 + C_2$ and $B = j(C_1 - C_2)$.

Similarly, $X(x)$ may be written in the form

$$X(x) = C \cos \frac{\lambda x}{c} + D \sin \frac{\lambda x}{c}$$

where $C = C_3 + C_4$ and $D = j(C_3 - C_4)$. Thus, the solution is

$$y(x,t) = (A \cos \lambda t + B \sin \lambda t)\left(C \cos \frac{\lambda x}{c} + D \sin \frac{\lambda x}{c} \right) \qquad (8.41)$$

The constants A, B, C, D, and λ are evaluated from the boundary conditions for a particular problem.

Illustrative example An elastic string of length L is fixed at both ends so that

$$y(0,t) = y(L,t) = 0$$

Because the string is initially at rest,

$$y'(x,0) = 0$$

As shown in Fig. 8.4, the initial position is

$$y(x,0) = f(x)$$

Determine the equation of motion $y(x,t)$.

Solution Substitution of the first boundary condition that $y(0,t) = 0$ into Eq. (8.41) gives

$$y(0,t) = (A \cos \lambda t + B \sin \lambda t)C = 0 \qquad (8.42)$$

If $A = B = 0$, the trivial solution $y(x,t) = 0$ results. Thus, the preceding equation must be satisfied by having $C = 0$.

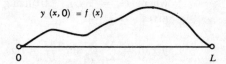

$y\,(x,0) = f\,(x)$

Fig. 8.4. An elastic string of length L.   0 L

To utilize the boundary condition $y'(x,0) = 0$, let $C = 0$ in Eq. (8.41), and then differentiate with respect to t.

$$y'(x,t) = (-\lambda A \sin \lambda t + \lambda B \cos \lambda t)D \sin \frac{\lambda x}{c}$$

For $t = 0$,

$$y'(x,0) = \lambda BD \sin \frac{\lambda x}{c} = 0 \tag{8.43}$$

If $C = D = 0$, the trivial solution $y(x,t) = 0$ results. Thus, the preceding expression must be satisfied by having $B = 0$. For $B = C = 0$, Eq. (8.41) becomes

$$y(x,t) = AD \cos \lambda t \sin \frac{\lambda x}{c} \tag{8.44}$$

Substituting the boundary condition $y(L,t) = 0$ into Eq. (8.44) gives

$$y(L,t) = AD \cos \lambda t \sin \frac{\lambda L}{c}$$

This equation is satisfied when

$$\sin \frac{\lambda L}{c} = 0 \quad \text{or} \quad \frac{\lambda L}{c} = \pi n \tag{8.45}$$

The values of n which satisfy the preceding expression and for which $k = -\lambda^2 < 0$ are $n = \pm 1, \pm 2, \pm 3, \ldots$. Employing the negative values of n yields the same result as the positive values. Hence, it suffices to use the positive values.

$$\lambda_n = \frac{\pi n c}{L} \quad n = 1, 2, 3, \ldots \tag{8.46}$$

The portion of the solution $y_n(x,t)$ due to the nth value, λ_n, is

$$y_n(x,t) = (AD)_n \cos \lambda_n t \sin \frac{\lambda_n x}{c} \tag{8.47}$$

The total solution now has the form

$$y(x,t) = \sum_{n=1}^{\infty} C_n \cos \lambda_n t \sin \frac{\lambda_n x}{c} \tag{8.48}$$

where $C_n = (AD)_n$. Utilizing the last boundary condition gives

$$y(x,0) = \sum_{n=1}^{\infty} C_n \sin \frac{\pi n x}{L} = f(x)$$

This is recognized as the Fourier series expansion for an odd function. The coefficient C_n is

$$C_n = \frac{2}{L} \int_0^L f(x) \sin \frac{\pi n x}{L} \, dx \tag{8.49}$$

Thus, the desired result is given by Eq. (8.48), in which the coefficient C_n is known from Eq. (8.49).

It should be pointed out that the method of separation of variables is limited to problems in which the answer may be expressed as the product of a time function [for example, $\cos \lambda_n t$ in Eq. (8.48)] and an x function [for example, $\sin (\lambda_n x/c)$ in Eq. (8.48)]. This method is not applicable to the illustrative examples on the wave equation in Chap. 7, for the solutions of these examples were not of a product form. The transform method has no such restriction regarding the form of the solution.

Heat equation

Consider now the application of the method of separation of variables to the heat equation

$$\frac{\partial u(x,t)}{\partial t} = \alpha \frac{\partial^2 u(x,t)}{\partial x^2}$$

Because $u(x,t)$ is a function of x and t, the assumed product solution has the form

$$u(x,t) = T(t)X(x)$$

The partial derivative with respect to t is

$$\frac{\partial u}{\partial t} = T'(t)X(x)$$

The second partial derivative with respect to x is

$$\frac{\partial^2 u}{\partial x^2} = T(t)X''(x)$$

Substitution of these results into the heat equation gives

$$\frac{T'(t)}{T(t)} = \alpha \frac{X''(x)}{X(x)} = k \tag{8.50}$$

Because T'/T is independent of x and X''/X is independent of t, each of these terms must be equal to a constant k. The preceding equation may be expressed as two differential equations,

$$T' - kT = 0$$

$$X'' - \frac{k}{\alpha} X = 0$$

The solution of the differential equation in T is

$$T = Ce^{kt} \tag{8.51}$$

For positive k, then T becomes infinite as t approaches infinity. Hence, $k > 0$ must be rejected. For $k = 0$, then $T = C$. The solution of the differential equation in X is $X = C_1 x + C_2$. Thus, the general solution is

$$u(x,t) = TX = C(C_1 x + C_2)$$

This indicates that u becomes infinite as x becomes infinite, and thus $k = 0$ must be rejected.

The desired solution occurs for $k < 0$. Because α is a positive constant, then for $k < 0$ let $k/\alpha = -\lambda^2 < 0$. The solution of the differential equation in X is

$$X = C_1 e^{j\lambda x} + C_2 e^{-j\lambda x} = A \cos \lambda x + B \sin \lambda x \tag{8.52}$$

where $A = C_1 + C_2$ and $B = j(C_1 - C_2)$. The general solution of the heat equation may now be expressed in the form

$$u(x,t) = Ce^{-\alpha\lambda^2 t}(A \cos \lambda x + B \sin \lambda x) \tag{8.53}$$

where $-\alpha\lambda^2 = k$. The constants A, B, C, and λ are evaluated from the boundary conditions for a particular problem.

For the illustrative example pertaining to Fig. 7.14, the solution is given by Eq. (7.84). The infinite summation is in the form of the product of a t function $(e^{-\alpha(\pi n/L)^2 t})$ and an x function [$\sin \pi n(L - x)/L$]. However, the entire solution is not in a product form, and thus the method of separation of variables is not applicable. By changing the boundary conditions slightly as is illustrated in the following example, the method of separation of variables may be employed. The transform method is applicable for either case.

Illustrative example For the body shown in Fig. 8.5, the temperature at both ends is maintained at zero so that

$$u(0,t) = u(L,t) = 0$$

The initial temperature distribution is

$$u(x,0) = f(x)$$

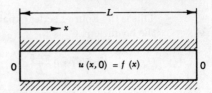

Fig. 8.5. A rectangular solid insulated at top and bottom.

Determine the equation for the resulting temperature distribution $u(x,t)$.

Solution   Evaluating Eq. (8.53) at $x = 0$ gives

$$u(0,t) = Ce^{-\alpha\lambda^2 t}(A) = 0$$

If $C = 0$, the trivial solution $u(x,t) = 0$ results. Thus, A must vanish. Next, evaluating Eq. (8.53) at $x = L$ gives

$$u(L,t) = Ce^{-\alpha\lambda^2 t}(B \sin \lambda L) = 0$$

If $A = B = 0$, again the trivial solution $u(x,t) = 0$ results. Hence, it is necessary that

$$\sin \lambda L = 0 \qquad \text{or} \qquad \lambda L = \pi n$$

As in the case of the wave equation, it suffices to use only positive values of n. Thus, the values of λ that satisfy the preceding equation are

$$\lambda_n = \frac{\pi n}{L} \qquad n = 1, 2, 3, \ldots \tag{8.54}$$

The portion of the solution $u_n(x,t)$ due to the nth value, λ_n, is

$$u_n(x,t) = (BC)_n e^{-\alpha\lambda_n^2 t} \sin \lambda_n x$$

Thus, the total solution is

$$u(x,t) = \sum_{n=1}^{\infty} C_n e^{-\alpha(\pi n/L)^2 t} \sin \frac{\pi n x}{L} \tag{8.55}$$

where $C_n = (BC)_n$. The coefficient C_n is determined from the remaining boundary condition that $u(x,0) = f(x)$. Thus, evaluating the preceding equation at $t = 0$ gives

$$u(x,0) = \sum_{n=1}^{\infty} C_n \sin \frac{\pi n x}{L} = f(x)$$

317

This is the Fourier series expansion for an odd function. Hence, the coefficient C_n is

$$C_n = \frac{2}{L} \int_0^L f(x) \sin \frac{\pi n x}{L} \, dx \tag{8.56}$$

The solution is thus given by Eq. (8.55), in which the coefficient C_n is known from Eq. (8.56).

Two-dimensional flow

The application of the method of separation of variables to the heat equation for two-dimensional flow is now considered. That is,

$$\frac{\partial u}{\partial t} = \alpha \left(\frac{\partial^2 u}{\partial x^2} + \frac{\partial^2 u}{\partial y^2} \right)$$

Because u is a function of x, y, and t, the assumed form of the solution is

$$u = T(t) X(x) Y(y) \tag{8.57}$$

Thus

$$\frac{\partial u}{\partial t} = T'XY$$

$$\frac{\partial^2 u}{\partial x^2} = TX''Y$$

$$\frac{\partial^2 u}{\partial y^2} = TXY''$$

Substitution of these expressions into the original partial differential equation and dividing through by TXY gives

$$\frac{T'}{T} = \alpha \left(\frac{X''}{X} + \frac{Y''}{Y} \right) \tag{8.58}$$

Because T'/T is independent of x and y and the right-hand side is independent of t, the preceding expression must equal a constant k. The differential equation in T is

$$T' - kT = 0$$

The solution is

$$T = Ce^{kt}$$

Substituting $T'/T = kCe^{kt}/Ce^{kt} = k$ into Eq. (8.58) and rearranging gives

$$\frac{X''}{X} - \frac{k}{\alpha} = -\frac{Y''}{Y} = \beta^2 \tag{8.59}$$

Because the first term is independent of y and the second is independent of x, each is equal to a constant β^2. The preceding expression may be represented by two differential equations,

$$X'' + \lambda^2 X = 0$$
$$Y'' + \beta^2 Y = 0$$

where $\lambda^2 = -(k/\alpha + \beta^2)$ or $k = -\alpha(\beta^2 + \lambda^2)$. The solution of these differential equations may be expressed in the form

$$X = A_1 \cos \lambda x + B_1 \sin \lambda x$$
$$Y = A_2 \cos \beta y + B_2 \sin \beta y$$

Thus, the general form of the response is

$$u(x,y,t) = Ce^{kt}(A_1 \cos \lambda x + B_1 \sin \lambda x)(A_2 \cos \beta y + B_2 \sin \beta y) \tag{8.60}$$

For the case of a rectangular plate of width L and height H in which the temperature on all four sides is maintained at zero,

$$u(0,y,t) = u(L,y,t) = u(x,0,t) = u(x,H,t) = 0$$

Utilizing the boundary condition $u(0,y,t) = 0$ in Eq. (8.60) leads to the fact that $A_1 = 0$. Next, the $u(L,y,t)$ boundary condition shows that $\sin \lambda L = 0$, and therefore

$$\lambda_n = \frac{\pi n}{L} \qquad n = 1, 2, 3, \ldots \tag{8.61}$$

Similarly, the boundary condition $u(x,0,t) = 0$ reveals that $A_2 = 0$, and the $u(x,H,t)$ boundary condition shows that $\sin \beta H = 0$. Thus

$$\beta_m = \frac{\pi m}{H} \qquad m = 1, 2, 3, \ldots \tag{8.62}$$

Substitution of these results into Eq. (8.60) gives for the general solution

$$u(x,y,t) = \sum_{n=1}^{\infty} \sum_{m=1}^{\infty} C_{mn} e^{k_{mn}t} \left(\sin \frac{\pi m y}{H} \right) \left(\sin \frac{\pi n x}{L} \right) \tag{8.63}$$

where

$$C_{mn} = (B_1 B_2 C)_{mn}$$

and

$$k_{mn} = -\alpha(\beta_m{}^2 + \lambda_n{}^2) = -\alpha\pi^2\left(\frac{m^2}{L^2} + \frac{n^2}{H^2}\right)$$

Because the solution [Eq. (8.21)] for the problem pertaining to Fig. 8.3 is not in product form, separation of variables is not applicable. By changing the boundary conditions slightly, as is illustrated in the following example, then separation of variables may be used, because the solution is now in a product form. The Laplace transform method is applicable for either case.

Illustrative example Determine the temperature distribution for the rectangular plate of Fig. 8.6, in which the temperature on all four sides is maintained at zero. The initial temperature distribution throughout the body is

$$u(x,y,0) = f(x,y)$$

Solution The last boundary condition is utilized by letting $t = 0$ in Eq. (8.63); that is,

$$u(x,y,0) = \sum_{n=1}^{\infty} \sum_{m=1}^{\infty} C_{mn}\left(\sin\frac{\pi my}{H}\right)\left(\sin\frac{\pi nx}{L}\right) = f(x,y)$$

The coefficient C_{mn} in this double sine series is obtained by an extension of the method described in Sec. 1.1 for evaluating the coefficient of an ordinary Fourier series. That is, each term in the double series is multiplied by $[\sin(\pi py/H)]\,[\sin(\pi qx/L)]$, where p and q are specific integers. Next, each resulting term is integrated with respect to x from 0 to L and then integrated with respect to y from 0 to H. All terms now vanish except

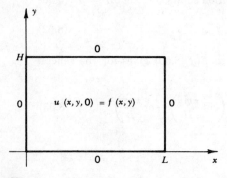

Fig. 8.6. A rectangular plate.

those for which $m = p$ and $n = q$. Thus, the desired result is found to be

$$C_{mn} = \frac{4}{HL} \int_0^H \int_0^L f(x,y) \left(\sin \frac{\pi my}{H} \right) \left(\sin \frac{\pi nx}{L} \right) dx \, dy \qquad (8.64)$$

The solution for this problem is given by Eq. (8.63), in which the coefficient C_{mn} is known from the preceding equation.

The solution of a one-dimensional partial differential equation is often expressed in terms of an infinite summation (or integral), the solution of a two-dimensional equation often contains a double infinite summation (or integral), and so on for three-dimensional equations, etc.

The application of separation of variables to the solution of Laplace's equation is explained in the discussion of elliptic equations in the following section.

8.5 HYPERBOLIC, PARABOLIC, AND ELLIPTIC EQUATIONS

Consider the partial differential equation

$$a \frac{\partial^2 u}{\partial x^2} + b \frac{\partial^2 u}{\partial x \, \partial y} + c \frac{\partial^2 u}{\partial y^2} = f(x,y,u,u_x,u_y) \qquad (8.65)$$

where $u_x = \partial u / \partial x$ and $u_y = \partial u / \partial y$.

The coefficients a, b, and c may be functions of x and y. This partial differential equation is said to be *hyperbolic* when $b^2 > 4ac$, *parabolic* when $b^2 = 4ac$, and *elliptic* when $b^2 < 4ac$. This nomenclature corresponds to that for a conic, which, depending on the value of the discriminant ($b^2 - 4ac$), may be a hyperbola, a parabola, or an ellipse.

The wave equation, heat equation, Laplace's equation, and Poisson's equation may be written as follows:

$$\frac{\partial^2 u}{\partial x^2} - c^2 \frac{\partial^2 u}{\partial y^2} = 0 \qquad \text{Wave equation}$$

$$\frac{\partial^2 u}{\partial x^2} = \frac{1}{\alpha} \frac{\partial u}{\partial y} \qquad \text{Heat equation}$$

$$\frac{\partial^2 u}{\partial x^2} + \frac{\partial^2 u}{\partial y^2} = 0 \qquad \text{Laplace's equation}$$

$$\frac{\partial^2 u}{\partial x^2} + \frac{\partial^2 u}{\partial y^2} = f(x,y) \qquad \text{Poisson's equation}$$

Thus, the wave equation is hyperbolic, the heat equation is parabolic, and both Laplace's and Poisson's equations are elliptic.

The solution of Eq. (8.65) is simplified by transforming it (by a change of variable) to the normal, or canonical, form. The normal form depends upon whether Eq. (8.65) is hyperbolic, parabolic, or elliptic. These normal forms are

$$\frac{\partial^2 u}{\partial v\, \partial w} = g(v,w,u,u_v,u_w) \qquad \text{Hyperbolic}$$

$$\frac{\partial^2 u}{\partial v^2} = g(v,w,u,u_v,u_w) \qquad \text{Parabolic} \qquad (8.66)$$

$$\frac{\partial^2 u}{\partial v^2} + \frac{\partial^2 u}{\partial w^2} = g(v,w,u,u_v,u_w) \qquad \text{Elliptic}$$

The change of variable for the transformation to the normal form is obtained as follows: First determine the solution of the following differential equation, which has the same form as the left-hand side of Eq. (8.65):

$$a\left(\frac{dy}{dx}\right)^2 + b\left(\frac{dy}{dx}\right) + c$$

$$= \left(\frac{dy}{dx} - \frac{-b - \sqrt{b^2 - 4ac}}{2a}\right)\left(\frac{dy}{dx} - \frac{-b + \sqrt{b^2 - 4ac}}{2a}\right) = 0$$

Integration of the preceding expression gives, for the solution,

$$y + \frac{b + \sqrt{b^2 - 4ac}}{2a}\, x = y + r_1 x = \text{const}$$

$$y + \frac{b - \sqrt{b^2 - 4ac}}{2a}\, x = y + r_2 x = \text{const}$$

where $r_1 = (b + \sqrt{b^2 - 4ac})/2a$ and $r_2 = (b - \sqrt{b^2 - 4ac})/2a$. The equations $y + r_1 x = C$ and $y + r_2 x = C$ define families of straight lines. The normal trajectories $r_1 x - y = C$ and $y - r_2 x = C$ are called the *characteristic curves*.

Hyperbolic

When Eq. (8.65) is hyperbolic, it is transformed to the normal form by the use of the relationships

$$v = \frac{r_1 x - y}{\sqrt{(b^2 - 4ac)/a}} \qquad (8.67)$$

$$w = \frac{y - r_2 x}{\sqrt{(b^2 - 4ac)/a}} \qquad (8.68)$$

Without the denominator term $\sqrt{(b^2 - 4ac)/a}$, both sides of the normal form would be multiplied by the constant $(b^2 - 4ac)/a$. Hence, the denominator provides a constant factor which yields the normal form directly.

To illustrate the method for solving a hyperbolic equation, let it be desired to determine the solution of the following equation for the case in which the boundary conditions are $u(x,0) = u(0,y) = 0$:

$$\frac{\partial^2 u}{\partial x^2} + 10 \frac{\partial^2 u}{\partial x \, \partial y} + 9 \frac{\partial^2 u}{\partial y^2} = x \tag{8.69}$$

For $a = 1$, $b = 10$, and $c = 9$, then $r_1 = (b + \sqrt{b^2 - 4ac})/2a = 9$ and $r_2 = (b - \sqrt{b^2 - 4ac})/2a = 1$. Thus, from Eqs. (8.67) and (8.68), the equations for v and w are

$$v = \frac{9x - y}{8} \tag{8.70}$$

$$w = \frac{y - x}{8} \tag{8.71}$$

Solving the preceding expressions for x and y gives

$$x = v + w \tag{8.72}$$
$$y = v + 9w \tag{8.73}$$

To check that this transformation is correct, first apply the chain rule for partial differentiation to evaluate $\partial u / \partial w$. That is,

$$\frac{\partial u}{\partial w} = \frac{\partial u}{\partial x} \frac{\partial x}{\partial w} + \frac{\partial u}{\partial y} \frac{\partial y}{\partial w} = \frac{\partial u}{\partial x} + 9 \frac{\partial u}{\partial y}$$

where from Eq. (8.72) $\partial x / \partial w = 1$ and from Eq. (8.73) $\partial y / \partial w = 9$. The term $\partial^2 u/(\partial v \partial w)$ is now obtained by applying the chain rule to the preceding result and noting from Eqs. (8.72) and (8.73) that $\partial x / \partial v = 1$ and $\partial y / \partial v = 1$. Thus

$$\frac{\partial}{\partial v} \left(\frac{\partial u}{\partial w} \right) = \frac{\partial}{\partial x} \left(\frac{\partial u}{\partial x} + 9 \frac{\partial u}{\partial y} \right) \frac{\partial x}{\partial v} + \frac{\partial}{\partial y} \left(\frac{\partial u}{\partial x} + 9 \frac{\partial u}{\partial y} \right) \frac{\partial y}{\partial v}$$

$$= \left(\frac{\partial^2 u}{\partial x^2} + 9 \frac{\partial^2 u}{\partial x \, \partial y} \right) + \left(\frac{\partial^2 u}{\partial x \, \partial y} + 9 \frac{\partial^2 u}{\partial y^2} \right)$$

$$= \frac{\partial^2 u}{\partial x^2} + 10 \frac{\partial^2 u}{\partial x \, \partial y} + 9 \frac{\partial^2 u}{\partial y^2}$$

The transformation defined by Eqs. (8.70) and (8.71) is thus seen to transform the left-hand side of Eq. (8.69) into the normal form. The transform of x, the right-hand side of Eq. (8.69), is given by Eq. (8.72). Hence, the resulting transformed equation is

$$\frac{\partial^2 u}{\partial v\,\partial w} = v + w \tag{8.74}$$

Although this equation may be Laplace-transformed, it is sufficiently simple so that the solution may be obtained by integrating directly with respect to v and then w. Thus,

$$\frac{\partial u}{\partial w} = \frac{v^2}{2} + vw + g(w)$$

$$u = \frac{v^2 w}{2} + \frac{vw^2}{2} + \int g(w)\,dw + f(v) \tag{8.75}$$

where $g(w)$ and $f(v)$ are the constants of integration that result in integrating with respect to v and w, respectively. For $y = 0$, Eq. (8.73) shows that $w = -v/9$. The boundary condition $u(x,0) = 0$ means that, for $w = -v/9$, Eq. (8.75) vanishes. Similarly, for $x = 0$, Eq. (8.72) shows that $w = -v$. The boundary condition $u(0,y) = 0$ means that, for $w = -v$, Eq. (8.75) again vanishes. Substitution of these boundary conditions into Eq. (8.75) gives

$$-\frac{v^3}{18} + \frac{v^3}{(9)(18)} + \int g\left(\frac{-v}{9}\right) d\left(\frac{-v}{9}\right) + f(v) = 0 \tag{8.76}$$

$$-\frac{v^3}{2} + \frac{v^3}{2} + \int g(-v)\,d(-v) + f(v) = 0 \tag{8.77}$$

Replacing v by $v/9$ in Eq. (8.77) and then eliminating the integral term between Eqs. (8.77) and (8.76) gives

$$-\frac{8v^3}{(9)(18)} - f\left(\frac{v}{9}\right) + f(v) = 0$$

It is now apparent that $f(v)$ has the form Cv^3, and thus $f(v/9)$ has the form $Cv^3/9^3$. Substituting these into the preceding expression and solving for $f(v) = Cv^3$ gives $f(v) = \frac{9}{182}v^3$. The integral term is evaluated by substituting $-w$ for v in Eq. (8.77). That is,

$$\int g(w)\,dw = -f(-w) = \frac{9}{182}w^3$$

Thus, Eq. (8.75) becomes

$$u = \frac{v^2 w}{2} + \frac{vw^2}{2} + \frac{9}{182}(v^3 + w^3) \tag{8.78}$$

The desired result is now obtained by using Eqs. (8.70) and (8.71) to convert this answer back to a function of x and y. Thus

$$u(x,y) = \frac{1}{182}(10x^2 y - xy^2) \tag{8.79}$$

It should be pointed out that Eq. (8.75) could have been converted first to a function of x and y and then the boundary conditions employed to evaluate the unknown constant. The work is roughly the same by either method.

Parabolic

For the case in which $b^2 = 4ac$, then $r_1 = r_2$. The transform relationships given by Eqs. (8.67) and (8.68) are no longer independent. That is, one is the negative of the other. (Note that the common factor $\sqrt{b^2 - 4ac}$ may be removed from these equations.) For $r_1 = r_2 = r = b/2a$, then Eq. (8.65) is transformed to the normal form by the following relationships:

$$v = \frac{x}{\sqrt{a}} \tag{8.80}$$

$$w = y - rx = y - \frac{b}{2a}x \tag{8.81}$$

To illustrate this method for equal roots, let it be desired to determine the solution of the following equation when the initial conditions are $u(x,0) = u(0,y) = 0$:

$$\frac{\partial^2 u}{\partial x^2} + 6\frac{\partial^2 u}{\partial x\,\partial y} + 9\frac{\partial^2 u}{\partial y^2} = x \tag{8.82}$$

For $r = b/2a = 3$, the transform relationships are

$$v = x \tag{8.83}$$
$$w = y - 3x \tag{8.84}$$

Solving the preceding equations for x and y gives

$$x = v$$
$$y = 3v + w$$

These transform relationships are verified by successive application of the chain rule to evaluate $\partial^2 u/\partial v^2$; that is,

$$\frac{\partial u}{\partial v} = \frac{\partial u}{\partial x}\frac{\partial x}{\partial v} + \frac{\partial u}{\partial y}\frac{\partial y}{\partial v} = \frac{\partial u}{\partial x}(1) + \frac{\partial u}{\partial y}(3)$$

$$\frac{\partial^2 u}{\partial v^2} = \frac{\partial}{\partial x}\left(\frac{\partial u}{\partial x} + 3\frac{\partial u}{\partial y}\right)\frac{\partial x}{\partial v} + \frac{\partial}{\partial y}\left(\frac{\partial u}{\partial x} + 3\frac{\partial u}{\partial y}\right)\frac{\partial y}{\partial v}$$

$$= \frac{\partial^2 u}{\partial x^2} + 6\frac{\partial^2 u}{\partial x\,\partial y} + 9\frac{\partial^2 u}{\partial y^2}$$

The transform of the right-hand side of Eq. (8.82) is $x = v$. Thus, the resultant transformed equation is

$$\frac{\partial^2 u}{\partial v^2} = v \qquad (8.85)$$

Integrating twice with respect to v gives

$$\frac{\partial u}{\partial v} = \frac{v^2}{2} + g(w)$$

$$u = \frac{v^3}{6} + vg(w) + f(w)$$

This result is converted back to a function of x and y by application of Eqs. (8.83) and (8.84). Thus

$$u(x,y) = \frac{x^3}{6} + xg(y - 3x) + f(y - 3x) \qquad (8.86)$$

Utilizing the boundary conditions $u(x,0) = u(0,y) = 0$ gives

$$u(x,0) = \frac{x^3}{6} + xg(-3x) + f(-3x) = 0 \qquad (8.87)$$

$$u(0,y) = f(y) = 0 \qquad (8.88)$$

Replacing y by $y - 3x$ in Eq. (8.88) shows that $f(y - 3x) = 0$. Similarly, replacing y by $-3x$ shows that $f(-3x) = 0$. The value of $g(-3x)$ may now be obtained directly from Eq. (8.87). That is, $g(-3x) = -x^2/6 = -(-3x)^2/54$. Replacing $-3x$ by $y - 3x$ gives

$$g(y - 3x) = -\frac{(y - 3x)^2}{54}$$

Substitution of $f(y - 3x) = 0$ and the preceding result into Eq. (8.86) gives for the solution

$$u(x,y) = \frac{x^3}{6} - \frac{x}{54}(y - 3x)^2 = \frac{1}{54}(6x^2y - xy^2) \qquad (8.89)$$

Elliptic

For $b^2 < 4ac$, some of the coefficients in the transform relationships given by Eqs. (8.67) and (8.68) are imaginary. A new set of transform equations is obtained by using the real part of the root $(b/2a)$ as one coefficient and the imaginary part of the root $(\sqrt{4ac - b^2}/2a)$

as the other coefficient. The resulting transform equations are

$$v = \frac{\sqrt{4ac - b^2}}{2a} x$$

$$w = y - \left(\frac{b}{2a}\right) x$$

In using these transform relationships, it is found that both sides of the normal form are multiplied by the constant $\sqrt{(4ac - b^2)/4a}$. Dividing all the coefficients by $\sqrt{(4ac - b^2)/4a}$ yields the transform relationships for converting directly to the normal form. That is,

$$v = \frac{x}{\sqrt{a}} \qquad (8.90)$$

$$w = \frac{y - (b/2a)x}{\sqrt{c - (b^2/4a)}} \qquad (8.91)$$

To illustrate this method, let it be desired to determine the solution of the following elliptic equation when the boundary conditions are $u(0,y) = u(L,y) = u(x,\infty) = 0$ and $u(x,y) = 1$ when $x = y$.

$$\frac{\partial^2 u}{\partial x^2} + 2\frac{\partial^2 u}{\partial x \partial y} + 9\frac{\partial^2 u}{\partial y^2} = 0 \qquad (8.92)$$

The transform relationships for this case are

$$v = x$$

$$w = \frac{y - x}{2\sqrt{2}}$$

The normal form of Eq. (8.92) is

$$\frac{\partial^2 u}{\partial v^2} + \frac{\partial^2 u}{\partial w^2} = 0 \qquad (8.93)$$

This equation may be solved by use of the Fourier sine transform. Because the use of this transform has been discussed earlier in this chapter, the application of the method of separation of variables for solving Laplace's equation [Eq. (8.93)] will now be explained.

In the method of separation of variables, the solution is assumed to be the product of a function of v [that is, $V(v)$] and a function of w [that is, $W(w)$].

$$u(v,w) = V(v)W(w)$$

Differentiation gives

$$\frac{\partial^2 u}{\partial v^2} = V''(v)W(w) \quad \text{and} \quad \frac{\partial^2 u}{\partial w^2} = V(v)W''(w)$$

Thus,

$$\frac{V''}{V} = -\frac{W''}{W} = k$$

Because V''/V is independent of w and similarly W''/W is independent of v, each must equal a constant k. The preceding expression may be written in the form of two differential equations

$$V'' + \lambda^2 V = 0$$
$$W'' - \lambda^2 W = 0$$

where $k = -\lambda^2 < 0$. The corresponding solutions are

$$V = A \cos \lambda v + B \sin \lambda v$$
$$W = Ce^{-\lambda w} + De^{\lambda w}$$

The product VW is the general solution. Converting this general solution back to a function of x and y gives

$$u(x,y) = (A \cos \lambda x + B \sin \lambda x)(Ce^{-\lambda(y-x)/2\sqrt{2}} + De^{\lambda(y-x)/2\sqrt{2}}) \quad (8.94)$$

The boundary condition $u(0,y) = 0$ reveals that $A = 0$. For $u(L,y) = 0$, it follows that $B \sin \lambda L = 0$. Thus, the values of λ are

$$\lambda_n = \frac{\pi n}{L} \quad n = 1, 2, 3, \ldots$$

To satisfy the boundary condition $u(x,\infty) = 0$, it is necessary that $D = 0$. Thus, the general solution is

$$u(x,y) = \sum_{n=1}^{\infty} C_n \, e^{-\lambda_n(y-x)/2\sqrt{2}} \sin \lambda x \quad (8.95)$$

where $C_n = (BC)_n$.

Employing the remaining boundary condition that $u(x,y) = 1$ when $x = y$ shows that

$$\sum_{n=1}^{\infty} C_n \sin \lambda_n x = 1$$

This is recognized as the Fourier series expansion for the case in which $f(x) = 1$ is regarded as an odd function for the interval

$0 < x \leq L$. Hence, the coefficient C_n is

$$C_n = \frac{2}{L} \int_0^L (1) \sin \frac{\pi n x}{L} \tag{8.96}$$

Thus, the result is given by Eq. (8.95), in which the coefficient C_n is known from the preceding equation.

PROBLEMS

8.1   In Fig. P8.1 is shown a string of length L. The initial position of the string is $y(x,0) = \sin (\pi x / L)$. Use the transform method to determine the equation for the resulting motion $y(x,t)$. The string is initially at rest [that is, $y'(x,0) = 0$].

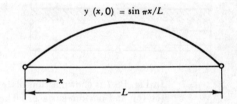

$$y(x,0) = \sin \pi x / L$$

Fig. P8.1

8.2   In Fig. P8.2 is shown a string of length L. The initial position of the string is $y(x,0) = f(x)$. Use the transform method to determine the equation for the resulting motion $y(x,t)$ for the case in which the string is initially at rest [that is, $y'(x,0) = 0$].

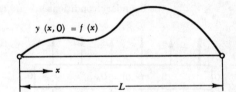

$$y(x,0) = f(x)$$

Fig. P8.2

8.3   A string of length L is fixed at both ends $[y(0,t) = y(L,t) = 0]$. The string is initially stretched along the x axis so that $y(x,0) = 0$. Determine the equation of motion $y(x,t)$ for this string for the case in which the initial velocity is $y'(x,0) = f(x)$.

8.4   Solve the heat, or diffusion, equation for the bar shown in Fig. 8.5. The boundary conditions are $u(0,t) = u(L,t) = 0$ and $u(x,0) = f(x)$.

8.5   Show that when heat is generated (or absorbed) within a solid the heat equation becomes

$$\frac{\partial u}{\partial t} = \alpha \frac{\partial^2 u}{\partial x^2} + \frac{q(x,t)}{c}$$

where $q(x,t)$ is the rate (for example, Btu per minute per unit volume) at which heat is being generated (or absorbed).

8.6   In Fig. P8.6 is shown a rod in which heat is generated at a constant rate of q units per unit volume along its entire length. Solve the heat equation (Prob. 8.5) for the case in which the boundary conditions are $u(0,t) = u(L,t) = u(x,0) = 0$.

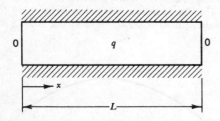

Fig. P8.6

8.7   In Fig. P8.7 is shown a thin plate (fin) which is radiating heat from the top and bottom surface. The fin is sufficiently thin so that the variation of temperature in the vertical (y direction) is negligible. When heat is being radiated, the heat equation becomes

$$\frac{\partial u}{\partial t} = \alpha \frac{\partial^2 u}{\partial x^2} - hu$$

where h is the coefficient of surface heat transfer. Determine the equation for the temperature distribution $u(x,t)$ for the case in which the boundary conditions are $u(0,t) = u(L,t) = 0$ and $u(x,0) = U_0$.

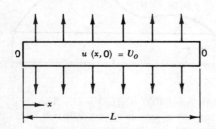

Fig. P8.7

8.8   In Fig. P8.8 is shown a plate of infinite height. Determine the steady-state temperature distribution when the boundary conditions are $u(0,y) = u(L,y) = u(x,\infty) = 0$ and $u(x,0) = 1$.

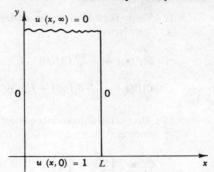

Fig. P8.8

8.9 For the plate shown in Fig. P8.8, determine the steady-state temperature distribution when $u(0,y) = u(L,y) = u(x,\infty) = 0$ and $u(x,0) = f(x)$.

8.10 For the finite plate shown in Fig. P8.10, determine the steady-state temperature distribution when $u(0,y) = u(L,y) = 0$, $u(x,H) = 1$, and $u(x,0) = -1$.

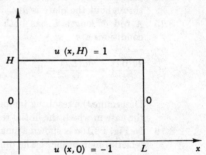

Fig. P8.10

8.11 For the finite plate shown in Fig. P8.11, the bottom is insulated so that $\partial u(x,0)/\partial y = 0$. Determine the steady-state temperature distribution when $u(0,y) = u(L,y) = 0$ and $u(x,H) = 1$.

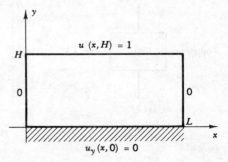

Fig. P8.11

8.12 Verify the following relationships between finite sine and finite cosine transforms:

$$S[f'(x)] = -\frac{\pi n}{L} C[f(x)]$$

$$C[f'(x)] = \frac{\pi n}{L} S[f(x)] - f(0) + (-1)^n f(L)$$

8.13 The partial differential equation for unsteady heat flow in two dimensions is

$$\frac{\partial u}{\partial t} = \alpha \left(\frac{\partial^2 u}{\partial x^2} + \frac{\partial^2 u}{\partial y^2} \right)$$

Determine the temperature distribution in a rectangular plate for which the boundary conditions are $u(0,y,t) = u(L,y,t) = u(x,0,t) = u(x,H,t) = 0$ and $u(x,y,0) = U_0$. (*Hint:* First take the finite sine transform with respect to x, and then take the finite sine transform with respect to y.)

8.14 Same as Prob. 8.13 except that the initial temperature distribution throughout the plate is $u(x,y,0) = f(x,y)$ rather than $u(x,y,0) = U_0$.

8.15 A rod of length L has both ends insulated so that the boundary conditions are

$$\frac{\partial u(0,t)}{\partial x} = \frac{\partial u(L,t)}{\partial x} = 0$$

Determine the resulting temperature distribution $u(x,t)$ in the bar for the case in which the initial temperature distribution is $u(x,0) = f(x)$.

8.16 In Fig. P8.16a is shown a small rectangular section of a thin, flexible (elastic) membrane. The differential equation of motion for a flexi-

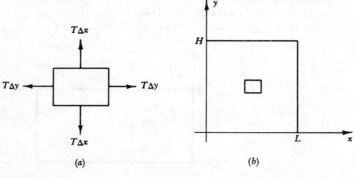

(a)   (b)

Fig. P8.16

332

ble membrane is given by the wave equation,

$$\frac{\partial^2 z}{\partial t^2} = c^2 \left(\frac{\partial^2 z}{\partial x^2} + \frac{\partial^2 z}{\partial y^2} \right)$$

where $c = \sqrt{T/\rho}$, T is the tension per unit length of edge, and ρ is the mass per unit area.

Determine the equation of motion $z(x,y,t)$ for the rectangular membrane shown in Fig. P8.16b for the case in which

$z(0,y,t) = z(L,y,t) = z(x,0,t) = z(x,H,t) = 0$
$z(x,y,0) = f(x,y)$ and $z'(x,y,0) = 0.$

8.17 Use the method of separation of variables to verify the result of Prob. 8.2.

8.18 Use the method of separation of variables to determine the temperature distribution $u(x,t)$ for the bar of Fig. 8.5 when the boundary conditions are $u(0,t) = u(l,t) = 0$ and $u(x,0) = U_0$.

8.19 Use the method of separation of variables to verify the result of Prob. 8.9.

8.20 The differential equation for the transverse motion of a beam is

$$\rho \frac{\partial^2 y}{\partial t^2} + \frac{EI}{A} \frac{\partial^4 y}{\partial x^4} = f(x)$$

where ρ is the mass per unit length, A is the cross-sectional area, E is the modulus of elasticity, I is the moment of inertia of the cross section about the transverse axis through the center of gravity, and $f(x)$ is the loading.   For a cantilever beam as shown in Fig. P8.20, the boundary conditions are $y(0,t) = y_x(0,t) = y_{xx}(L,t) = y_{xxx}(L,t) = 0$. Determine the equation of motion for the case in which the beam is initially at rest, $y_t(x,0) = 0$, and the initial position is $y(x,0) = y(x)$ (there is no external loading.)

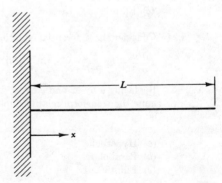

Fig. P8.20

8.21 The differential equation for transverse motion of the simply supported beam shown in Fig. P8.21 is given in Prob. 8.20. The boundary conditions for a simply supported beam are $y(0,t) = y(L,t) = 0$ and $y_{xx}(0,t) = y_{xx}(L,t) = 0$. Determine the equation for the free vibration (no external loading) of this simply supported beam when the beam is initially at rest and the initial position is $y(x,0) = y(x)$.

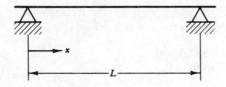

Fig. P8.21

8.22 For the free vibration (no external loading) of a cantilever beam, in the method of separation of variables the general solution of the differential equation given in Prob. 8.20 is of the form

$$y(x,t) = T(t)X(x)$$

(a) Show that the functions $X(x)$ and $T(t)$ satisfy the equation

$$\frac{X''(x)}{X(x)} = -\frac{T''(x)}{c^2 T(x)} = \lambda^4 = \text{const}$$

where $c = \sqrt{EI/\rho A}$.
(b) Show that the functions $X(x)$ and $T(t)$ are

$$T(t) = A \cos c\lambda^2 t + B \sin c\lambda^2 t$$
$$X(x) = (A_1 \cos \lambda x + B_1 \sin \lambda x) + (A_2 \cosh \lambda x + B_2 \sinh \lambda x)$$

8.23 To verify that the transform relationships given in Sec. 8.5 are correct, use these relationships to obtain the normal form when Eq. (8.65) is:

(a) Hyperbolic
(b) Parabolic
(c) Elliptic

8.24 Consider the differential equation

$$a \frac{\partial^2 u}{\partial x^2} + bx \frac{\partial^2 u}{\partial x \partial y} + cx^2 \frac{\partial^2 u}{\partial y^2} = 0$$

Replacing b by bx and c by cx^2 in the equations for r_1 and r_2, determine the transform relationships for the case in which this equation is:

(a) Hyperbolic
(b) Parabolic
(c) Elliptic

8.25 The equation of motion for a viscously damped string is

$$\frac{\partial^2 y}{\partial t^2} - c^2 \frac{\partial^2 y}{\partial x^2} = -2k \frac{\partial y}{\partial t}$$

Determine the normal form of this equation.

8.26 Solve the following differential equation:

$$\frac{\partial^2 u}{\partial x^2} + 5 \frac{\partial^2 u}{\partial x \partial y} + 4 \frac{\partial^2 u}{\partial y^2} = y$$

The initial conditions are $u(x,0) = u(0,y) = 0$.

8.27 Solve the following differential equation:

$$\frac{\partial^2 u}{\partial x^2} + 4 \frac{\partial^2 u}{\partial x \partial y} + 4 \frac{\partial^2 u}{\partial y^2} = y$$

The initial conditions are $u(x,0) = u(0,y) = 0$.

8.28 Solve the following differential equation:

$$\frac{\partial^2 u}{\partial x^2} + 2 \frac{\partial^2 u}{\partial x \partial y} + 4 \frac{\partial^2 u}{\partial y^2} = 0$$

The initial conditions are $u(x,0) = u(L,y) = u(x, \infty) = 0$, and $u(x,y) = f(x)$ when $y = x$.

REFERENCES

1 Pipes, L. A.: "Applied Mathematics for Engineers and Physicists," 2d ed., McGraw-Hill Book Company, New York, 1958.

2 Sokolnikoff, I. S., and R. M. Redheffer: "Mathematics of Physics and Modern Engineering," McGraw-Hill Book Company, New York, 1958.

3 Sneddon, I. N.: "Fourier Transforms," McGraw-Hill Book Company, New York, 1951.

4 Tranter, C. J.: "Integral Transforms in Mathematical Physics," 2d ed., John Wiley & Sons, Inc., New York, 1956.

5 Aseltine, J. A.: "Transform Method in Linear System Analysis," McGraw-Hill Book Company, New York, 1958.

6 Titchmarsh, E. C.: "Introduction to the Theory of Fourier Integrals," Oxford University Press, New York, 1948.

7 Campbell, G. A., and R. M. Foster: Fourier Integrals for Practical Applications (Table of Fourier Transformations), *Bell Tel. System Tech. Pub.*, 1942.

8 Hildebrand, F. B.: "Methods of Applied Mathematics," Prentice-Hall, Inc., Englewood Cliffs, N.J., 1952.

chapter 9

Vector-field theory

The region, or space, in which a vector function $\mathbf{F}$ is defined is called a *vector field*. For the flow of fluid in a given region, the vector quantity is the fluid velocity $\mathbf{v}$. For the flow of heat, the vector quantity is the rate of heat flow, $\mathbf{q}$. For an electric field, the vector is the force that would be exerted on a unit charge. For a magnetic field, the vector is the force that would act on a unit magnetic pole. The theory of vector fields has extensive applications to engineering situations.

In Sec. 9.2, it is shown that, if a vector field is solenoidal and irrotational, there exists a scalar quantity which satisfies Laplace's equation. For fluid flow, this scalar quantity is pressure; for heat flow, the scalar is temperature; for electric fields, it is voltage; and for magnetic fields, it is charge. Each of these scalar quantities may be regarded as a potential, and thus the solution of Laplace's equation for such fields is called *potential theory*.

9.1 FLUID FLOW

In Fig. 9.1 is shown a rectangular parallelepiped (block). The mass of fluid ΔQ_x entering through the back face during time Δt is

$$\Delta Q_x = \rho v_x \, dy \, dz \, \Delta t$$

where ρ is the density of the fluid and v_x is the component of velocity in the x direction. The mass of fluid leaving through the front face is

$$\Delta Q_{x+\Delta x} = \left[\rho v_x + \frac{\partial (\rho v_x)}{\partial x} \, dx \right] dy \, dz \, \Delta t$$

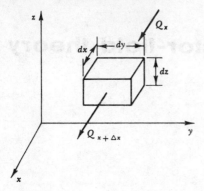

Vector field theory

Fig. 9.1. Flow through a rectangular block.

The net increase of mass of fluid due to flow through the back and front faces is obtained by subtracting that which leaves from that which enters.

$$\Delta(Q_x - Q_{x+\Delta x}) = -\frac{\partial(\rho v_x)}{\partial x}\,dx\,dy\,dz\,\Delta t \tag{9.1}$$

Performing a similar analysis on the remaining sides gives the total net increase of mass within the block.

$$\Delta Q = -\left[\frac{\partial(\rho v_x)}{\partial x} + \frac{\partial(\rho v_y)}{\partial y} + \frac{\partial(\rho v_z)}{\partial z}\right] dx\,dy\,dz\,\Delta t \tag{9.2}$$

The operator ∇ (called *del*) is

$$\nabla = \mathbf{i}\frac{\partial}{\partial x} + \mathbf{j}\frac{\partial}{\partial y} + \mathbf{k}\frac{\partial}{\partial z} \tag{9.3}$$

where $\mathbf{i}$ is a unit vector along the x axis, $\mathbf{j}$ is a unit vector along the y axis, and $\mathbf{k}$ is a unit vector along the z axis. The dot product of $\nabla \cdot \rho\mathbf{v}$ is

$$\nabla \cdot \rho\mathbf{v} = \left(\mathbf{i}\frac{\partial}{\partial x} + \mathbf{j}\frac{\partial}{\partial y} + \mathbf{k}\frac{\partial}{\partial z}\right) \cdot [\mathbf{i}(\rho v_x) + \mathbf{j}(\rho v_y) + \mathbf{k}(\rho v_z)]$$

Because $\mathbf{i} \cdot \mathbf{i} = \mathbf{j} \cdot \mathbf{j} = \mathbf{k} \cdot \mathbf{k} = 1$ and $\mathbf{i} \cdot \mathbf{j} = \mathbf{i} \cdot \mathbf{k} = \mathbf{j} \cdot \mathbf{k} = 0$, then

$$\nabla \cdot \rho\mathbf{v} = \frac{\partial(\rho v_x)}{\partial x} + \frac{\partial(\rho v_y)}{\partial y} + \frac{\partial(\rho v_z)}{\partial z}$$

Thus Eq. (9.2) becomes

$$\Delta Q = -\nabla \cdot \rho\mathbf{v}\,dx\,dy\,dz\,\Delta t$$

Dividing by Δt and taking the limit as Δt approaches zero gives the net rate of increase of mass.

$$\frac{dQ}{dt} = -\nabla \cdot \rho \mathbf{v} \, dx \, dy \, dz \tag{9.4}$$

The total mass of fluid contained within the block is

$$Q = \rho \, dx \, dy \, dz$$

Differentiation gives

$$\frac{dQ}{dt} = \frac{\partial \rho}{\partial t} \, dx \, dy \, dz \tag{9.5}$$

From Eqs. (9.4) and (9.5) it follows that

$$\nabla \cdot \rho \mathbf{v} = -\frac{\partial \rho}{\partial t} \tag{9.6}$$

This is called the *continuity equation*.

For a compressible fluid with steady flow (i.e., flow which is independent of time) then $\partial \rho / \partial t = 0$. Thus the continuity equation becomes

$$\nabla \cdot \rho \mathbf{v} = 0 \tag{9.7}$$

For an incompressible fluid, the density ρ is a constant. Thus, regardless of whether or not the flow is steady, the continuity equation for an incompressible fluid is

$$\nabla \cdot \mathbf{v} = 0 \tag{9.8}$$

The net mass rate of increase of flow into an incremental block $(dx \, dy \, dz)$ is given by Eq. (9.4). Integration over a region of volume V gives the net mass rate of increase of flow into the region. That is,

$$\frac{dQ}{dt} = \iiint\limits_{V} (-\nabla \cdot \rho \mathbf{v}) \, dV$$

where $dV = dx \, dy \, dz$. The integrand $-\nabla \cdot \rho \mathbf{v}$ is an indication of the increase or convergence of flow into the region. Hence, the term $\nabla \cdot \rho \mathbf{v}$, which is a measure of the rate at which flow is leaving, is called the *divergence*.

A better understanding of the physical meaning of divergence is obtained by solving Eq. (9.4) for the divergence of $\rho \mathbf{v}$, which is $\nabla \cdot \rho \mathbf{v}$. That is,

$$\text{div } \rho \mathbf{v} = \nabla \cdot \rho \mathbf{v} = \frac{-(dQ/dt)}{dV} = \lim_{V \to 0} \frac{-dQ/dt}{V} \tag{9.9}$$

Hence, the divergence is the limit as the volume approaches zero of the rate of flow per unit volume. The points in a region from which fluid is generated are called *sources*, and the points which absorb fluid are called *sinks*. Thus, the divergence is a measure of the strength of a source or a sink.

9.2 VECTOR ANALYSIS

Fundamental concepts of vector analysis such as divergence, gradient, curl, etc., have numerous applications in the study of engineering phenomena. These concepts provide a concise formulation of the physical laws which describe fluid flow, heat flow, electric fields, magnetic fields, gravitational and force fields, etc. A most significant aspect of the method of vector analysis is that it enables one to obtain tremendous insight into the actual behavior of such phenomena.

Divergence

In the preceding section, the concept of divergence was developed. In general, the divergence of a vector quantity $\mathbf{F}$ is defined as the dot product of the vector operator ∇ and the vector $\mathbf{F}$. That is,

$$\text{div } \mathbf{F} = \nabla \cdot \mathbf{F} = \left(\mathbf{i} \frac{\partial}{\partial x} + \mathbf{j} \frac{\partial}{\partial y} + \mathbf{k} \frac{\partial}{\partial z} \right) \cdot (\mathbf{i}F_x + \mathbf{j}F_y + \mathbf{k}F_z)$$

$$= \frac{\partial F_x}{\partial x} + \frac{\partial F_y}{\partial y} + \frac{\partial F_z}{\partial z} \tag{9.10}$$

For the case of fluid flow, the vector $\mathbf{F}$ is $\rho\mathbf{v}$, so that the divergence is $\nabla \cdot \rho\mathbf{v}$. For heat flow, the vector $\mathbf{F}$ is the rate of heat flow $\mathbf{q}$, and the divergence $\nabla \cdot \mathbf{q}$ is a measure of the rate at which heat is leaving a given region. For electric and magnetic fields, the vector quantities are the electric intensity vector $\mathbf{E}$ and the magnetic intensity vector $\mathbf{H}$. The corresponding divergences for these fields are $\nabla \cdot \mathbf{E}$ and $\nabla \cdot \mathbf{H}$.

Gauss' theorem

This is a basic theorem which relates surface and volume integrals. It states that the integral over a volume of the divergence of a vector function ($\nabla \cdot \mathbf{F}$) is equal to the integral over the surface of the normal component of **F**. That is,

$$\iiint_V \nabla \cdot \mathbf{F} \, dV = \iint_S \mathbf{F} \cdot \mathbf{n} \, dS \tag{9.11}$$

where **n** is a unit vector which is normal to the surface and thus **F** · **n** is the component of **F** normal to the surface. Gauss' theorem is also called the *divergence theorem*.

To verify Gauss' theorem, first write Eq. (9.11) in the form

$$\iiint_V \left(\frac{\partial F_x}{\partial x} + \frac{\partial F_y}{\partial y} + \frac{\partial F_z}{\partial z} \right) dx \, dy \, dz$$

$$= \iint_S (\mathbf{i} F_x + \mathbf{j} F_y + \mathbf{k} F_z) \cdot (\mathbf{i} \cos \alpha + \mathbf{j} \cos \beta + \mathbf{k} \cos \gamma) \, dS$$

$$= \iint_S (F_x \cos \alpha + F_y \cos \beta + F_z \cos \gamma) \, dS \tag{9.12}$$

where α is the angle between the normal to the surface **n** and the x axis, β is the angle between **n** and the y axis, and γ is the angle between **n** and the z axis. Thus, $\mathbf{n} = \mathbf{i} \cos \alpha + \mathbf{j} \cos \beta + \mathbf{k} \cos \gamma$ expresses **n** in terms of its components. The third term on the left-hand side of Eq. (9.12) may be expressed in the form

$$\iiint_V \frac{\partial F_z}{\partial z} \, dx \, dy \, dz = \iint F_z \, dx \, dy \tag{9.13}$$

The shaded section in Fig. 9.2 is a section of the surface of the volume under consideration. The projection of this section on the xy plane is designated $dx \, dy$. If the section were rotated through the angle γ, then the normal **n** would lie on top of **k**. Thus, γ is the angle of rotation of the section with respect to the $dx \, dy$ projec-

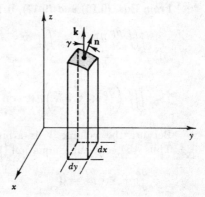

Fig. 9.2. Projection of a surface on the xy plane.

tion, or $dx\,dy = \cos\gamma\,dS$. Hence, Eq. (9.13) becomes

$$\iiint_V \frac{\partial F_z}{\partial z}\,dx\,dy\,dz = \iint_S F_z \cos\gamma\,dS \tag{9.14}$$

Similarly, it may be shown that the first and second terms on each side of Eq. (9.12) are equal. Thus, Gauss' theorem is verified.

With the aid of Gauss' theorem, the continuity equation [Eq. (9.6)] may now be derived in a more direct manner. The mass of fluid contained in a given volume V is

$$Q = \iiint_V \rho\,dV$$

For a given volume the rate at which the mass increases is

$$\frac{dQ}{dt} = \iiint_V \frac{\partial\rho}{\partial t}\,dV \tag{9.15}$$

The component of velocity normal to the surface is $\mathbf{v}\cdot\mathbf{n}$. Thus, the quantity $\rho\mathbf{v}\cdot\mathbf{n}\,dS$ is the mass rate of flow leaving the incremental surface area dS. Integration over the entire surface gives the mass rate of flow which is leaving the volume under consideration. Changing the sign gives the mass rate of flow entering. Thus,

$$\frac{dQ}{dt} = -\iint_S \rho\mathbf{v}\cdot\mathbf{n}\,dS \tag{9.16}$$

Application of Gauss' theorem to the preceding expression in which the vector $\rho\mathbf{v}$ corresponds to $\mathbf{F}$ gives

$$\frac{dQ}{dt} = -\iint_S \rho\mathbf{v}\cdot\mathbf{n}\,dS = -\iiint_V \nabla\cdot\rho\mathbf{v}\,dV \tag{9.17}$$

From Eqs. (9.15) and (9.17), it follows that

$$\iiint_V \frac{\partial\rho}{\partial t}\,dV = -\iiint_V \nabla\cdot\rho\mathbf{v}\,dV$$

or

$$\iiint_V \left(\frac{\partial\rho}{\partial t} + \nabla\cdot\rho\mathbf{v}\right)dV = 0 \tag{9.18}$$

Because the volume V is arbitrary, the integrand must vanish. Thus, the continuity equation [Eq. (9.6)] results. That is,

$$\frac{\partial\rho}{\partial t} + \nabla\cdot\rho\mathbf{v} = 0 \tag{9.19}$$

Gradient

In potential theory, the potential, such as pressure, voltage, or temperature, is a scalar quantity. The gradient of a scalar quantity ϕ is

$$\text{grad } \phi = \nabla \phi \tag{9.20}$$

Expanding the operator ∇ shows that

$$\begin{aligned}
\text{grad } \phi &= \left(\mathbf{i}\, \frac{\partial}{\partial x} + \mathbf{j}\, \frac{\partial}{\partial y} + \mathbf{k}\, \frac{\partial}{\partial z} \right) \phi \\
&= \mathbf{i}\, \frac{\partial \phi}{\partial x} + \mathbf{j}\, \frac{\partial \phi}{\partial y} + \mathbf{k}\, \frac{\partial \phi}{\partial z}
\end{aligned} \tag{9.21}$$

From Eq. (9.21), the gradient is seen to be a measure of the rate of change of a scalar quantity. If $d\mathbf{r} = \mathbf{i}\, dx + \mathbf{j}\, dy + \mathbf{k}\, dz$ and $|d\mathbf{r}| = ds$, then $d\mathbf{r}/|d\mathbf{r}| = d\mathbf{r}/ds$ represents a unit vector in the direction of $d\mathbf{r}$. It will now be shown that the derivative or rate of change of ϕ in the direction of $d\mathbf{r}$ is $\nabla \phi \cdot d\mathbf{r}/ds$.

$$\begin{aligned}
\nabla \phi \cdot \frac{d\mathbf{r}}{ds} &= \left(\mathbf{i}\, \frac{\partial \phi}{\partial x} + \mathbf{j}\, \frac{\partial \phi}{\partial y} + \mathbf{k}\, \frac{\partial \phi}{\partial z} \right) \cdot \left(\mathbf{i}\, \frac{dx}{ds} + \mathbf{j}\, \frac{dy}{ds} + \mathbf{k}\, \frac{dz}{ds} \right) \\
&= \frac{\partial \phi}{\partial x} \frac{dx}{ds} + \frac{\partial \phi}{\partial y} \frac{dy}{ds} + \frac{\partial \phi}{\partial z} \frac{dz}{ds} \\
&= \frac{d\phi}{ds}
\end{aligned} \tag{9.22}$$

The dot product $\nabla \phi \cdot d\mathbf{r}/ds$ is the projection of $\nabla \phi$ in the direction of $d\mathbf{r}$. Because the maximum projection of $\nabla \phi$ is the vector itself, the vector $\nabla \phi$ (grad ϕ) has the direction of the greatest rate of change of ϕ.

In Fig. 9.3a is shown the path C_{ab}, which goes from point a to b. The line is subdivided into vector segments $d\mathbf{r}_1, d\mathbf{r}_2, \ldots$, etc. The value of $\mathbf{F}$ at the midpoint of the $d\mathbf{r}_i$ segment is $\mathbf{F}_i$. The line integral of $\mathbf{F}$ along the curve C_{ab} is

$$\begin{aligned}
\int_{C_{ab}} \mathbf{F} \cdot d\mathbf{r} &= \int_{C_{ab}} (\mathbf{i} F_x + \mathbf{j} F_y + \mathbf{k} F_z) \cdot (\mathbf{i}\, dx + \mathbf{j}\, dy + \mathbf{k}\, dz) \\
&= \int_{C_{ab}} F_x\, dx + F_y\, dy + F_z\, dz \\
&= \lim_{\substack{\Delta r_i \to 0 \\ n \to \infty}} \sum_{i=0}^{n} \mathbf{F}_i \cdot \Delta \mathbf{r}_i
\end{aligned} \tag{9.23}$$

When $\mathbf{F}$ is the force exerted on a particle moving along a path, the line integral is the work done by the force. For the case in

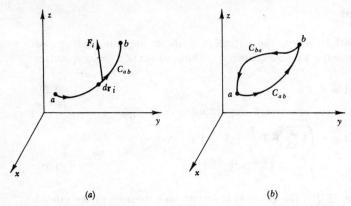

(a)   (b)

Fig. 9.3. Paths of integration.

which $\mathbf{F}$ is the gradient of a scalar function, that is, $\mathbf{F} = \nabla\phi$, then

$$\int_{C_{ab}} \nabla\phi \cdot d\mathbf{r} = \int_{C_{ab}} d\phi = \phi_b - \phi_a \tag{9.24}$$

where C_{ab} is any path from a to b.

The value of this line integral depends only on the end points and is independent of the path. As shown in Fig. 9.3b, the path C_{ba} goes from point b to point a. Because the line integral along a path C_{ba} is the negative of that along a path C_{ab},

$$\int_{C_{ab}} \nabla\phi \cdot d\mathbf{r} + \int_{C_{ba}} \nabla\phi \cdot d\mathbf{r} = \int_{C} \nabla\phi \cdot d\mathbf{r} = 0 \tag{9.25}$$

Thus, the line integral of the gradient of a scalar function of position ϕ around any closed curve C is zero (i.e., vanishes).

Curl

Another concept which has extensive physical significance in engineering problems is the curl of a vector quantity. The curl of a vector is the cross product of the operator ∇ and the vector quantity $\mathbf{F}$; that is,

$$\text{curl } \mathbf{F} = \nabla \times \mathbf{F} \tag{9.26}$$

Because $\mathbf{i} \times \mathbf{j} = -\mathbf{j} \times \mathbf{i} = \mathbf{k}$, $\mathbf{j} \times \mathbf{k} = -\mathbf{k} \times \mathbf{j} = \mathbf{i}$, and $\mathbf{k} \times \mathbf{i} = -\mathbf{i} \times \mathbf{k} = \mathbf{j}$, then the cross product of the vector ∇ and the vector

F is

$$\text{curl } \mathbf{F} = \nabla \times \mathbf{F} = \left(\mathbf{i}\frac{\partial}{\partial x} + \mathbf{j}\frac{\partial}{\partial y} + \mathbf{k}\frac{\partial}{\partial z} \right) \times (\mathbf{i}F_x + \mathbf{j}F_y + \mathbf{k}F_z)$$

$$= \mathbf{i}\left(\frac{\partial F_z}{\partial y} - \frac{\partial F_y}{\partial z} \right) + \mathbf{j}\left(\frac{\partial F_x}{\partial z} - \frac{\partial F_z}{\partial x} \right) + \mathbf{k}\left(\frac{\partial F_y}{\partial x} - \frac{\partial F_x}{\partial y} \right)$$

$$(9.27)$$

This result shows that the curl of a vector **F** may be expressed as a determinant; that is,

$$\text{curl } \mathbf{F} = \nabla \times \mathbf{F} = \begin{vmatrix} \mathbf{i} & \mathbf{j} & \mathbf{k} \\ \dfrac{\partial}{\partial x} & \dfrac{\partial}{\partial y} & \dfrac{\partial}{\partial z} \\ F_x & F_y & F_z \end{vmatrix}$$

It will now be shown that the curl of the gradient of a scalar ϕ must vanish.

$$\nabla \times \nabla \phi = \nabla \times \left(\mathbf{i}\frac{\partial \phi}{\partial x} + \mathbf{j}\frac{\partial \phi}{\partial y} + \mathbf{k}\frac{\partial \phi}{\partial z} \right)$$

$$= \begin{vmatrix} \mathbf{i} & \mathbf{j} & \mathbf{k} \\ \dfrac{\partial}{\partial x} & \dfrac{\partial}{\partial y} & \dfrac{\partial}{\partial z} \\ \dfrac{\partial \phi}{\partial x} & \dfrac{\partial \phi}{\partial y} & \dfrac{\partial \phi}{\partial z} \end{vmatrix}$$

$$= \mathbf{i}\left(\frac{\partial^2 \phi}{\partial y\,\partial z} - \frac{\partial^2 \phi}{\partial y\,\partial z} \right) + \mathbf{j}\left(\frac{\partial^2 \phi}{\partial x\,\partial z} - \frac{\partial^2 \phi}{\partial x\,\partial z} \right)$$

$$+ \mathbf{k}\left(\frac{\partial^2 \phi}{\partial x\,\partial y} - \frac{\partial^2 \phi}{\partial x\,\partial y} \right) = 0 \quad (9.28)$$

An irrotational vector field is one in which **F** is such that curl **F** = 0 everywhere in the field. In accordance with Eq. (9.28), the curl of the gradient of ϕ must always be zero (that is, $\nabla \times \nabla \phi = 0$). Thus, if **F** is irrotational (that is, $\nabla \times \mathbf{F} = 0$), there must exist a scalar quantity ϕ such that $\mathbf{F} = \nabla \phi$.

When a function is such that div $\mathbf{F} = \nabla \cdot \mathbf{F} = 0$, the field is said to be *solenoidal*. When a field is both solenoidal (that is, $\nabla \cdot \mathbf{F} = 0$) and irrotational (that is, $\mathbf{F} = \nabla \phi$), then

$$\nabla \cdot \mathbf{F} = \nabla \cdot \nabla \phi = \left(\mathbf{i}\frac{\partial}{\partial x} + \mathbf{j}\frac{\partial}{\partial y} + \mathbf{k}\frac{\partial}{\partial z} \right)$$

$$\left(\mathbf{i}\frac{\partial \phi}{\partial x} + \mathbf{j}\frac{\partial \phi}{\partial y} + \mathbf{k}\frac{\partial \phi}{\partial z} \right) = 0$$

or

$$\frac{\partial^2 \phi}{\partial x^2} + \frac{\partial^2 \phi}{\partial y^2} + \frac{\partial^2 \phi}{\partial z^2} = \nabla^2 \phi = 0 \qquad (9.29)$$

where

$$\nabla^2 = \frac{\partial^2}{\partial x^2} + \frac{\partial^2}{\partial y^2} + \frac{\partial^2}{\partial z^2}$$

Thus, fields which are solenoidal and irrotational are described by Laplace's equation. Because $\mathbf{F} = \nabla \phi$, solving Laplace's equation for the potential function ϕ is equivalent to determining F.

Stokes' theorem

A better understanding of an irrotational field is obtained with the help of Stokes' theorem. In Fig. 9.4 is shown a surface S whose periphery, or edge, is a simple closed curve C. Stokes' theorem states that the integral of $(\nabla \times \mathbf{F}) \cdot \mathbf{n}$ taken over the surface S is equal to the line integral of $\mathbf{F} \cdot d\mathbf{r}$ taken over the periphery.

$$\iint\limits_{S} (\nabla \times \mathbf{F}) \cdot \mathbf{n} \, dS = \int_{C} \mathbf{F} \cdot d\mathbf{r} \qquad (9.30)$$

where $d\mathbf{r}$ is a vector segment which is tangent to C.

Comparison of Eqs. (9.23) and (9.25) shows that $\int_{C} \mathbf{F} \cdot d\mathbf{r}$ vanishes when $\mathbf{F} = \nabla \phi$. Thus, Stokes' theorem shows that for irrotational flow ($\nabla \times \mathbf{F} = 0$), then it is necessary that $\mathbf{F} = \nabla \phi$.

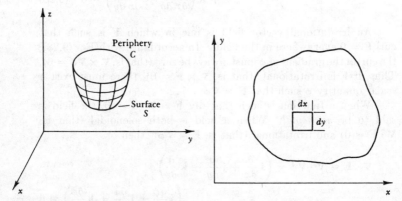

Fig. 9.4. A surface S and its periphery C.　　Fig. 9.5. A plane surface.

For the case of a closed surface, there is no periphery C. Hence,

$$\iint_S (\nabla \times \mathbf{F}) \cdot \mathbf{n} \, dS = 0 \qquad (9.31)$$

This last result shows that the surface integral of $(\nabla \times \mathbf{F}) \cdot \mathbf{n}$ vanishes over any closed surface, regardless of whether the flow is rotational or irrotational. However, Eq. (9.30) shows that the surface integral vanishes over an open surface (C is a finite closed boundary) for irrotational flow only.

Stokes' theorem is proved as follows: In Fig. 9.5 is shown a surface S whose boundary is C. The integral over this surface of $(\nabla \times \mathbf{F}) \cdot \mathbf{n} \, dS$ is

$$\iint_S (\nabla \times \mathbf{F}) \cdot \mathbf{n} \, dS = \iint_S \left[\mathbf{i} \left(\frac{\partial F_z}{\partial y} - \frac{\partial F_y}{\partial z} \right) + \mathbf{j} \left(\frac{\partial F_x}{\partial z} - \frac{\partial F_z}{\partial x} \right) \right.$$
$$\left. + \mathbf{k} \left(\frac{\partial F_y}{\partial x} - \frac{\partial F_x}{\partial y} \right) \right] \cdot \mathbf{n} \, dS$$

The normal $\mathbf{n}$ to the surface shown in Fig. 9.5 is $\mathbf{k}$. Because $\mathbf{i} \cdot \mathbf{k} = \mathbf{j} \cdot \mathbf{k} = 0$ and $\mathbf{k} \cdot \mathbf{k} = 1$, the preceding equation becomes

$$\iint_S (\nabla \times \mathbf{F}) \cdot \mathbf{n} \, dS = \iint_S \left(\frac{\partial F_y}{\partial x} - \frac{\partial F_x}{\partial y} \right) dS$$

Application of Green's lemma [Eq. (5.41)], in which $M = F_x$, $N = F_y$, and $dS = dx \, dy$, gives

$$\iint_S \left(\frac{\partial F_y}{\partial x} - \frac{\partial F_x}{\partial y} \right) dS = \int_C F_x \, dx + F_y \, dy = \int_C \mathbf{F} \cdot d\mathbf{r}$$

Thus, Stokes' theorem [Eq. (9.30)] has been verified for the case of a plane surface. This result is applicable regardless of the orientation of this surface in space. The spatial surface shown in Fig. 9.6 has been approximated by triangular elements. Because a triangle is a plane surface, Stokes' theorem is applicable to each triangle. In the limit, the surface formed by an infinite number of triangles approaches the spatial surface S. Integration is performed in opposite directions along the sides of adjacent triangles. The sum of the integrals along the sides is the line integral along the periphery C. The direction of integration along C is such that, if one stands on the side of the surface from which the normal extends and then walks along the periphery C, the region S is seen to the left. If the normal n in Fig. 9.6 were interior to the surface, rather than exterior, the direction of integration along C would have to be reversed.

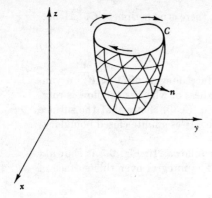

Fig. 9.6.   A normal to a surface.

The concept of irrotational flow may be visualized by applying Stokes' theorem to a very small circular surface.   For fluid flow, the general vector **F** is replaced by the velocity vector **v**; thus

$$(\nabla \times \mathbf{v}) \cdot \mathbf{n} = \lim_{A \to 0} \frac{\int_C \mathbf{v} \cdot d\mathbf{r}}{A} \tag{9.32}$$

where A is the area of a small circular surface.   The term $\mathbf{v} \cdot d\mathbf{r}$ is the tangential component of **v** about the circumference.   Such a fluid particle merely circulates around the circumference C shown dotted in Fig. 9.7a.   The term $\int_C \mathbf{v} \cdot d\mathbf{r}$ is called the *circulation* of **v** along C.   The large arrow in Fig. 9.7a shows the mean path of flow. The small circular paths indicate the paths of individual particles. For irrotational flow (circulation is zero), the individual particles move along the mean paths as shown in Fig. 9.7b.

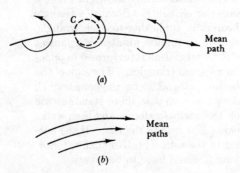

Fig. 9.7.   Mean paths.

9.3 POTENTIAL THEORY

The application of potential theory to the study of two-dimensional flow will be considered first. For two-dimensional flow, the motion is the same in all planes parallel to the xy plane. Thus, the flow is independent of the z coordinate. For solenoidal flow, it follows that

$$\text{div } \mathbf{v} = \boldsymbol{\nabla} \cdot \mathbf{v} = \left(\mathbf{i} \frac{\partial}{\partial x} + \mathbf{j} \frac{\partial}{\partial y} \right) \cdot (\mathbf{i} v_x + \mathbf{j} v_j) = 0$$

Thus,

$$\frac{\partial v_x}{\partial x} + \frac{\partial v_y}{\partial y} = 0 \tag{9.33}$$

For irrotational flow ($\mathbf{v} = \boldsymbol{\nabla}\phi$), it follows that

$$\int_C d\phi = \int_C \boldsymbol{\nabla}\phi \cdot d\mathbf{r} = \int_C \mathbf{v} \cdot d\mathbf{r} = \int_C (\mathbf{i} v_x + \mathbf{j} v_y) \cdot (\mathbf{i} dx + \mathbf{j} dy)$$

$$= \int_C (v_x \, dx + v_y \, dy) = 0$$

where $d\mathbf{r} = \mathbf{i} \, dx + \mathbf{j} dy$. Because the value of this integral around a closed path C is zero, then the integrand

$$d\phi = v_x \, dx + v_y \, dy \tag{9.34}$$

is an exact differential. Replacing M by v_x and N by v_y in Eq. (5.58) shows that

$$v_x = \frac{\partial \phi}{\partial x} \quad \text{and} \quad v_y = \frac{\partial \phi}{\partial y} \tag{9.35}$$

The solution of the exact differential equation given by Eq. (9.34) is $\phi(x,y) = \text{const}$. The lines of $\phi(x,y) = \text{const}$ are called the *equipotential lines*. Because $\mathbf{v} = \boldsymbol{\nabla}\phi$ is the gradient (derivative) of ϕ, lines of constant ϕ are perpendicular to the flow.

The substitution of v_x and v_y from Eq. (9.35) into Eq. (9.33) gives

$$\frac{\partial^2 \phi}{\partial x^2} + \frac{\partial^2 \phi}{\partial y^2} = 0 \tag{9.36}$$

Thus, as was shown in the preceding section, when the flow is solenoidal and irrotational, the function ϕ satisfies Laplace's equation.

In Sec. 5.2, it is shown that both the real and imaginary parts of an analytic function satisfy Laplace's equation. By regarding $\phi(x,y)$ as the real part of an analytic function $F(z)$, the corresponding imaginary part $\psi(x,y)$ also satisfies Laplace's equation. That is,

$$\frac{\partial^2 \psi}{\partial x^2} + \frac{\partial^2 \psi}{\partial y^2} = 0 \tag{9.37}$$

The analytic function $F(z)$ whose real part is ϕ and whose imaginary part is ψ is called the *complex potential*.

$$F(z) = \phi + j\psi \tag{9.38}$$

A streamline is a line which is tangent to the velocity of the fluid at every point. It will now be shown that, if $\phi(x,y)$ is the real part of an analytic function, the corresponding streamlines are given by the imaginary part $\psi(x,y)$ of the analytic function. The slope of the equipotential lines $\phi(x,y) = C$ is obtained as follows:

$$d\phi = \frac{\partial \phi}{\partial x} dx + \frac{\partial \phi}{\partial y} dy = 0$$

Solving for dy/dx gives for the slope of the equipotential lines

$$\frac{dy}{dx} = -\frac{\partial \phi/\partial x}{\partial \phi/\partial y}$$

From the Cauchy-Riemann equations [Eqs. (5.16) and (5.17)], in which the real part is now ϕ and the imaginary part is now ψ, it follows that

$$\frac{\partial \phi}{\partial x} = \frac{\partial \psi}{\partial y} \tag{9.39}$$

$$\frac{\partial \phi}{\partial y} = -\frac{\partial \psi}{\partial x} \tag{9.40}$$

Thus, in terms of ψ the slope of the equipotential lines is

$$\frac{dy}{dx} = -\frac{\partial \phi/\partial x}{\partial \phi/\partial y} = \frac{\partial \psi/\partial y}{\partial \psi/\partial x}$$

The normal is the negative of the reciprocal of the tangent. Hence, the normal to the equipotential lines is

$$-\frac{\partial \psi/\partial x}{\partial \psi/\partial y}$$

This proof is completed by showing that the slope of the streamlines $\psi(x,y) = C$ is the normal to the equipotential lines. The derivative $d\psi$ is

$$d\psi = \frac{\partial \psi}{\partial x} dx + \frac{\partial \psi}{\partial y} dy = 0$$

Thus, the slope of the streamlines is

$$\frac{dy}{dx} = -\frac{\partial\psi/\partial x}{\partial\psi/\partial y}$$

The velocity **v** at any point is most conveniently determined by the relationship

$$\mathbf{v} = v_x + jv_y = \overline{F'(z)} \tag{9.41}$$

where $\overline{F'(z)}$ is the complex conjugate of the derivative $F'(z)$. Equation (9.41) is verified by recalling that the derivative $F'(z)$ of an analytic function is independent of the path along which Δz approaches zero. By taking a horizontal path (y is constant, and thus $\Delta y = 0$), then $\Delta z = \Delta x$. Thus, using Eq. (9.38) gives

$$F'(z) = \lim_{\Delta z \to 0} \frac{dF}{dz} = \lim_{\Delta x \to 0} \frac{dF}{dx} = \frac{\partial\phi}{\partial x} + j\frac{\partial\psi}{\partial x}$$

Replacing $\partial\psi/\partial x$ by $-\partial\phi/\partial y$ in accordance with Eq. (9.39) and then utilizing Eq. (9.35) gives

$$F'(z) = \frac{\partial\phi}{\partial x} - j\frac{\partial\phi}{\partial y} = v_x - jv_y$$

The complex conjugate is the desired relationship given by Eq. (9.41). The preceding concepts are now illustrated by a few examples.

Flow inside a corner

Consider the complex potential

$$F(z) = z^2 = x^2 - y^2 + j2xy \tag{9.42}$$

The equipotential lines are

$$\phi(x,y) = x^2 - y^2 = C \tag{9.43}$$

The streamlines are

$$\psi(x,y) = 2xy = C \tag{9.44}$$

The streamlines are obtained by plotting the stream function $2xy = C$ for various values of C. As shown in Fig. 9.8, these streamlines are a family of hyperbolas (solid lines) which represent fluid flowing inside a corner. The equipotential lines (dotted lines) in Fig. 9.8 are obtained by plotting the potential function $x^2 - y^2 = C$ for various values of C. It is to be noted that the equipotential lines intersect the streamlines at right angles.

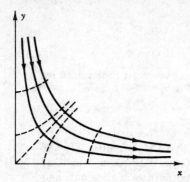

Fig. 9.8.   Flow inside a corner.

The velocity **v** at any point is

$$\mathbf{v} = \overline{F'(z)} = 2\bar{z} = 2\overline{(x + jy)} = 2x - j2y$$

The velocity components are

$$v_x = \frac{\partial \phi}{\partial x} = 2x$$

$$v_y = \frac{\partial \phi}{\partial y} = -2y$$

The magnitude of the velocity is

$$|\mathbf{v}| = 2\sqrt{x^2 + y^2}$$

The direction of the velocity is along the streamline, which is

$$\tan^{-1}\frac{v_y}{v_x} = \tan^{-1} - \frac{y}{x}$$

This result may also be obtained from the stream function ψ. That is,

$$d\psi = \frac{\partial \psi}{\partial x}\, dx + \frac{\partial \psi}{\partial y}\, dy = 2y\, dx + 2x\, dy = 0$$

Hence

$$\frac{dy}{dx} = -\frac{y}{x}$$

In effect, in the preceding, Laplace's equation has been solved for the motion of a fluid inside a corner.

Flow around a cylinder

Consider the complex potential

$$F(z) = z + \frac{1}{z} \tag{9.45}$$

By letting $z = re^{j\theta}$, then

$$F(z) = re^{j\theta} + \frac{1}{r} e^{-j\theta}$$

$$= \left(r + \frac{1}{r}\right) \cos \theta + j \left(r - \frac{1}{r}\right) \sin \theta$$

The equipotential lines are

$$\phi = \left(r + \frac{1}{r}\right) \cos \theta = C \tag{9.46}$$

The streamlines are

$$\psi = \left(r - \frac{1}{r}\right) \sin \theta = C \tag{9.47}$$

The corresponding streamlines are shown in Fig. 9.9. For $C = 0$, the streamline equation reduces to $r^2 = 1$, which describes a unit circle, for regardless of the angle θ the magnitude of r is 1. To plot the streamline for a given value of C, assume a value of r, and then determine the corresponding value of θ which satisfies the streamline equation.

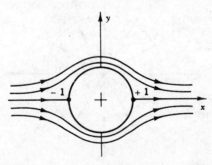

Fig. 9.9. Flow around a cylinder.

Sources and sinks

Consider the complex potential

$$F(z) = k \ln z \tag{9.48}$$

By letting $z = re^{j\theta}$,

$$F(z) = k(\ln r + j\theta)$$

The equipotential lines ϕ and the streamlines ψ are

$$
\begin{aligned}
\phi &= k \ln r = C \\
\psi &= k\theta = C
\end{aligned}
\tag{9.49}
$$

Because the streamlines are independent of r, these are radial lines about the origin, as shown in Fig. 9.10. The equipotential lines are independent of θ, and thus these are circles as shown dotted in Fig. 9.10. The velocity **v** is

$$\mathbf{v} = \overline{F'(z)} = \frac{k}{\bar{z}} = \left(\frac{k}{r}\right) e^{j\theta}$$

The magnitude of the velocity is k/r. For $k > 0$, the direction is outward along the radial line $e^{j\theta}$. This flow corresponds to a point source at the origin. The constant k is a measure of the strength of the source. For negative k the direction is reversed, so that the origin represents a sink.

The rate at which flow passes a circle of radius r is the velocity at this radius (k/r) times the circumference $(2\pi r)$. This is equal to

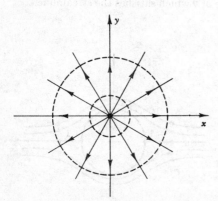

Fig. 9.10. A source.

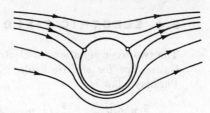

Fig. 9.11. Flow around a cylinder with circulation.

the rate of flow Q coming from the source. Thus

$$Q = 2\pi r \frac{k}{r} = 2\pi k$$

or

$$k = \frac{Q}{2\pi} \tag{9.50}$$

The source strength k is thus seen to be the rate of flow Q divided by 2π.

If k is replaced by the imaginary constant $(-jk)$, equipotential lines and streamlines are interchanged. For this case, the flow circles about the origin, which is now called the *vortex*. The addition of the circulation function $(-jk \ln z)$ to a complex potential function yields the resulting function for flow with circulation. For example, for flow around a cylinder with circulation, the complex potential is

$$F(z) = \left(z + \frac{1}{z}\right) - jk \ln z \tag{9.51}$$

In Fig. 9.11 is shown a typical set of streamlines for flow around a cylinder with circulation.

Heat flow

The preceding analyses (i.e., Figs. 9.8 to 9.11) may be regarded as representing the steady flow of heat, as well as fluid flow. For steady flow, no heat is being accumulated in any region; hence

$$\text{div } \mathbf{q} = \boldsymbol{\nabla} \cdot \mathbf{q} = 0$$

Thus, rate of heat flow $\mathbf{q}$ corresponds to the velocity $\mathbf{v}$ of fluid flow. The partial differential equation for steady heat flow is

$$\boldsymbol{\nabla} \cdot \boldsymbol{\nabla} u = \boldsymbol{\nabla}^2 u = \frac{\partial^2 u}{\partial x^2} + \frac{\partial^2 u}{\partial y^2} = 0$$

Comparison of the two preceding equations shows that $\mathbf{q} = \boldsymbol{\nabla} u$. Thus, the temperature u corresponds to the potential function ϕ.

355

9.4 ELECTRIC FIELDS

The two basic equations of electrodynamics are Faraday's law and Gauss' law for electric fields. The mathematical formulation of Faraday's law is

$$\int_C \mathbf{E} \cdot d\mathbf{R} = -\frac{\partial}{\partial t} \iint_S \mathbf{B} \cdot \mathbf{n} \, dS \tag{9.52}$$

Faraday's law states that the integral of the tangential component of the electric intensity vector $\mathbf{E}$ around a closed curve C is the negative of the rate of change of magnetic flux ($\mathbf{B}$ is magnetic flux density) that passes through any surface that spans C.

Application of Stokes' theorem to Faraday's law gives

$$\iint_S (\nabla \times \mathbf{E}) \cdot \mathbf{n} \, dS = -\frac{\partial}{\partial t} \iint_S \mathbf{B} \cdot \mathbf{n} \, dS = -\iint_S \frac{\partial \mathbf{B}}{\partial t} \cdot \mathbf{n} \, dS$$

Because S is any arbitrary surface, the preceding equation is satisfied if

$$\nabla \times \mathbf{E} = -\frac{\partial \mathbf{B}}{\partial t} \tag{9.53}$$

When the magnetic flux density $\mathbf{B}$ does not vary with time,

$$\nabla \times \mathbf{E} = 0 \tag{9.54}$$

The mathematical formulation of Gauss' law for electric fields is

$$\iint_S \mathbf{D} \cdot \mathbf{n} \, dS = \iiint_V \rho \, dV = q \tag{9.55}$$

Gauss' law states that the integral of the normal component of the electric flux density $\mathbf{D}$ over any closed surface S is equal to the total electric charge q (ρ is the charge density) within the volume V enclosed by S. Application of the divergence theorem to Gauss' law gives

$$\iiint_V \nabla \cdot \mathbf{D} \, dV = \iiint_V \rho \, dV$$

Because the volume V is arbitrary, the preceding equation is satisfied if

$$\nabla \cdot \mathbf{D} = \rho \tag{9.56}$$

The electric flux density **D** is proportional to the electric intensity **E**; that is,

$$\mathbf{D} = k\mathbf{E}$$

where k is the permittivity of the medium. Thus,

$$\mathbf{\nabla} \cdot \mathbf{E} = \frac{\rho}{k} \tag{9.57}$$

Because the field is irrotational ($\mathbf{\nabla} \times \mathbf{E} = 0$), there exists an electric potential ϕ such that $\mathbf{E} = \mathbf{\nabla}\phi$. As the sign is arbitrary, it is customary to let $\mathbf{E} = -\mathbf{\nabla}\phi$. Hence, the preceding expression becomes

$$\mathbf{\nabla} \cdot (-\mathbf{\nabla}\phi) = \frac{\rho}{k}$$

or

$$\mathbf{\nabla}^2\phi = -\frac{\rho}{k} \tag{9.58}$$

This is Poisson's equation. For both the cases of good conductors and good dielectrics the space-charge density ρ is extremely small ($\rho \approx 0$). Thus, Laplace's equation results.

The equipotential lines ϕ are lines of constant electrostatic potential (units of electrostatic potential are volts). Because force is the gradient of the electrostatic potential ϕ, the streamlines ψ are the lines of force for the electrostatic field.

Illustrative example Two parallel conducting plates are shown in Fig. 9.12. The electrostatic potential on the vertical plate at $x = a$ is 10 volts, and the electrostatic potential on the vertical plate at $x = b$ is 40 volts.

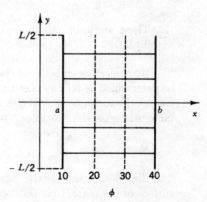

Fig. 9.12. Electrostatic potential between conducting plates.

The complex potential function is

$$F(z) = Az + B = (Ax + B) + jAy$$

Determine the equation for the lines of constant electrostatic potential and the equation for the lines of force.

Solution   The electrostatic-potential lines are

$$\phi = Ax + B$$

Utilizing the boundary conditions that $\phi = 10$ at $x = a$ and $\phi = 40$ at $x = b$ gives

$$10 = Aa + B$$
$$40 = Ab + B$$

Solving for A and B gives

$$A = \frac{30}{b - a}$$
$$B = \frac{10b - 40a}{b - a}$$

Thus, the electrostatic-potential lines are

$$\phi = \frac{30x + 10b - 40a}{b - a} \qquad (9.59)$$

As shown by the vertical dotted lines in Fig. 9.12, the equipotential lines are equally spaced between the plates.   The lines of force are

$$\psi = \frac{30y}{b - a} \qquad (9.60)$$

These are uniformly spaced horizontal lines, as shown in Fig. 9.12.

By neglecting end effects, the capacitance of the plates may now be determined as follows: Let the vertical length of the plates be L, and let the width be W.   If q in Eq. (9.55) is the total charge on the plate at $x = a$, the charge per unit area is

$$\mathbf{D} = \mathbf{i}\,\frac{q}{LW}$$

The direction of the electric flux density $\mathbf{D}$ is perpendicular to the surface of the plate, that is, along $\mathbf{i}$.

The electric intensity $\mathbf{E}$ is

$$\mathbf{E} = -\nabla\phi = -\left(\mathbf{i}\frac{\partial\phi}{\partial x} + \mathbf{j}\frac{\partial\phi}{\partial y}\right) = -\mathbf{i}A$$

Because $\mathbf{D} = k\mathbf{E}$, then

$$\mathbf{D} = -\mathbf{i}kA$$

Equating the two preceding expressions for $\mathbf{D}$ gives

$$A = -\frac{q}{kLW}$$

The potential on the vertical plate at $x = a$ may be written in the form

$$\phi_a = -\frac{qa}{kLW} + B$$

Similarly, the potential at $x = b$ is

$$\phi_b = -\frac{qb}{kLW} + B$$

For a capacitor, Eq. (4.56) shows that $C = \int i\,dt/e = q/e$, where $q = \int i\,dt$ is the charge on each plate and e is the potential difference between the plates. The potential difference for the plates of Fig. 9.12 is $\phi_a - \phi_b$; hence the capacitance is

$$C = \frac{q}{\phi_a - \phi_b} = \frac{kLW}{b-a} \tag{9.61}$$

9.5 MAGNETIC FIELDS

An analysis similar to that just described for electric fields may also be applied to magnetic fields. The two basic equations for magnetic fields are Ampère's law and Gauss' law for magnetic fields. Ampère's law (for current i) may be expressed in the form

$$\int_C \mathbf{H} \cdot d\mathbf{R} = \iint_S \mathbf{J} \cdot \mathbf{n}\,dS = i \tag{9.62}$$

This law states that the integral of the tangential component of the magnetic intensity vector $\mathbf{H}$ around a closed path is equal to the current ($\mathbf{J}$ is current density) that flows through any surface S that spans C. Application of Stokes' theorem to Ampère's law gives

$$\iint_S (\nabla \times \mathbf{H}) \cdot \mathbf{n}\,d\mathbf{R} = \iint_S \mathbf{J} \cdot \mathbf{n}\,dS$$

Because S is an arbitrary surface, then

$$\nabla \times \mathbf{H} = \mathbf{J} = \sigma\mathbf{E} + k\left(\frac{\partial \mathbf{E}}{\partial t}\right) \qquad (9.63)$$

The current density $\mathbf{J}$ is composed of two parts. The first part, $\sigma\mathbf{E}$, is the current density due to the flow of electric charges (σ a constant is the electric conductivity). The second part, $\partial\mathbf{D}/\partial t = k(\partial\mathbf{E}/\partial t)$, is the displacement current density.

When the electric and magnetic fields are independent of time and $\mathbf{J} = 0$, the electric vectors and magnetic vectors are independent of each other. For this case, Eq. (9.63) becomes

$$\nabla \times \mathbf{H} = 0 \qquad (9.64)$$

Gauss' law for magnetic fields may be expressed in the form

$$\iint\limits_{S} \mathbf{B} \cdot \mathbf{n}\, dS = 0 \qquad (9.65)$$

This law states that the integral of the normal component of the magnetic flux density $\mathbf{B}$ over a closed surface S is zero. Application of the divergence theorem gives

$$\iiint\limits_{V} \nabla \cdot \mathbf{B}\, dV = 0$$

Hence, it follows that

$$\nabla \cdot \mathbf{B} = 0 \qquad (9.66)$$

The magnetic flux density $\mathbf{B}$ is proportional to the magnetic intensity $\mathbf{H}$; that is,

$$\mathbf{B} = \mu\mathbf{H} \qquad (9.67)$$

where μ is the permeability of the medium. Thus

$$\nabla \cdot \mathbf{H} = 0 \qquad (9.68)$$

Because the field is irrotational ($\nabla \times \mathbf{H} = 0$), a magnetic potential ϕ exists such that $\mathbf{H} = \nabla\phi$. Thus magnetostatic fields satisfy Laplace's equation.

Electromagnetic fields

When the electric field vectors and the magnetic field vectors are not independent of each other, the right-hand sides of Eqs. (9.53) and (9.63) do not vanish. Taking the curl of both sides of Eq.

(9.53) gives

$$\nabla \times (\nabla \times \mathbf{E}) = -\frac{\partial}{\partial t}(\nabla \times \mathbf{B}) = -\mu\frac{\partial}{\partial t}(\nabla \times \mathbf{H})$$

where $\mathbf{B} = \mu\mathbf{H}$ in accordance with Eq. (9.67). By expanding the left-hand side and using Eq. (9.63) to replace $\nabla \times \mathbf{H}$, then

$$\nabla(\nabla \cdot \mathbf{E}) - \nabla^2\mathbf{E} = -\mu\frac{\partial}{\partial t}(\nabla \times \mathbf{H}) = -\mu\frac{\partial}{\partial t}\left(\sigma\mathbf{E} + k\frac{\partial \mathbf{E}}{\partial t}\right)$$

$$(9.69)$$

For both the case of a good dielectric (e.g., free space) and a good conductor, the space-charge density is extremely small ($\rho \approx 0$). Thus, Eq. (9.57) becomes

$$\nabla \cdot \mathbf{E} = 0$$

Substitution of this result into Eq. (9.69) yields Maxwell's equation for the electric-intensity vector $\mathbf{E}$; that is,

$$\nabla^2\mathbf{E} = \mu k\frac{\partial^2 \mathbf{E}}{\partial t^2} + \mu\sigma\frac{\partial \mathbf{E}}{\partial t} \tag{9.70}$$

For a good dielectric, not only is $\rho \approx 0$, but also the conductivity $\sigma \approx 0$. Thus, Maxwell's equation reduces to the wave equation

$$\nabla^2\mathbf{E} = \mu k\frac{\partial^2 \mathbf{E}}{\partial t^2} = \frac{1}{c^2}\frac{\partial^2 \mathbf{E}}{\partial t^2} \tag{9.71}$$

where $c = 1/\sqrt{\mu k}$.

For the case of a good conductor, not only is $\rho \approx 0$, but the displacement current is also negligible. Thus Maxwell's equation reduces to the diffusion equation

$$\nabla^2\mathbf{E} = \mu\sigma\frac{\partial \mathbf{E}}{\partial t} \tag{9.72}$$

By proceeding in a similar manner [i.e., taking the curl of both sides of Eq. (9.63) and noting that $\nabla \times \mathbf{E} = -\partial\mathbf{B}/\partial t = -\mu\partial\mathbf{H}/\partial t$], then Maxwell's equation for the magnetic-intensity vector $\mathbf{H}$ is obtained. That is,

$$\nabla^2\mathbf{H} = \mu k\frac{\partial^2 \mathbf{H}}{\partial t^2} + \mu\sigma\frac{\partial \mathbf{H}}{\partial t} \tag{9.73}$$

For a good dielectric, $\sigma \approx 0$; thus

$$\nabla^2\mathbf{H} = \frac{1}{c^2}\frac{\partial^2 \mathbf{H}}{\partial t^2} \tag{9.74}$$

where $c = 1/\sqrt{\mu k}$.

Similarly, for a good conductor, $k \approx 0$; thus Maxwell's equation reduces to the diffusion equation. Equations (9.71) and (9.74) form the basis for the theory of electromagnetic waves in which c is the velocity of light in free space.

It is interesting to note that when James Clerk Maxwell formulated his equations describing the mutual interaction between electric and magnetic phenomena, just over a century ago, he recognized the similarity of these phenomena and the vibrations of a string. Thus, he realized that electromagnetic waves must exist. He later showed that visible light is composed of electromagnetic waves of certain frequencies and predicted that waves of other frequencies would also be found. About twenty years later, Hertz produced such waves—radio waves—in his laboratory. After Marconi's pioneering efforts on the technological problems, radio, radar, and television were developed.

PROBLEMS

9.1 Verify each of the following vector identities:

(a) $\nabla \cdot (\nabla \times \mathbf{F}) = 0$

(b) $\nabla \cdot (\mathbf{A} \times \mathbf{B}) = \mathbf{B} \cdot (\nabla \times \mathbf{A}) - \mathbf{A} \cdot (\nabla \times \mathbf{B})$

(c) $\nabla \times (\mathbf{A} \times \mathbf{B}) = (\mathbf{B} \cdot \nabla)\mathbf{A} - (\mathbf{A} \cdot \nabla)\mathbf{B} + \mathbf{A}(\nabla \cdot \mathbf{B}) - \mathbf{B}(\nabla \cdot \mathbf{A})$

9.2 The directional derivative of a function is the rate of change of the function in a certain direction. Because $\mathbf{r}/|\mathbf{r}|$ is a unit vector in the direction of $\mathbf{r}$, the directional derivative of ϕ in the direction of $\mathbf{r}$ is $\nabla\phi \cdot (\mathbf{r}/|\mathbf{r}|)$. For the function $\phi(x,y,z) = xy + xz + yz$, determine the directional derivative at the point $(1,2,-1)$ in the direction $\mathbf{r} = 2\mathbf{i} - 4\mathbf{j} + 4\mathbf{k}$.

9.3 The gradient $\nabla\phi$ is the rate of change of ϕ in the direction normal to the surface $\phi(x,y,z) = C$. Thus, the unit normal vector is $\nabla\phi/|\nabla\phi|$. Determine the unit vector normal to the surface $x^2yz = C$ at the point $(1,2,-2)$.

9.4 The equation for a plane is $ax + by + cz = C$. Determine the unit vector which is normal to the plane $2x - 4y + 4z = C$.

9.5 The position $\mathbf{r}$ of a point traveling in space is given as the following function of time:

$$\mathbf{r} = t\mathbf{i} + e^t\mathbf{j} - \sin t\mathbf{k}$$

Thus, the parametric equations for the positions x, y, and z as functions of time are $x = t$, $y = e^t$, and $z = -\sin t$. The derivative of the position yields the velocity $\mathbf{v}$, which is a vector tangent to the path. Determine the unit vector which is tangent to the path when $t = 0$.

9.6 Evaluate $\int_C \mathbf{F} \cdot d\mathbf{r}$ from $t = 0$ to $t = 1$, where $\mathbf{r}$ is the path given in Prob. 9.5 and $\mathbf{F}$ is the function

$$\mathbf{F} = (2y - 3)\mathbf{i} + xz\mathbf{j} + (yz - x)\mathbf{k}$$

9.7 Determine the gradient $(\nabla\phi)$ for each of the following scalar functions:

(a) $\phi = xyz$
(b) $\phi = x^2 + yz$
(c) $\phi = x \sin y + e^z$

9.8 Determine the divergence $(\nabla \cdot \mathbf{F})$ for each of the following vector functions:

(a) $\mathbf{F} = x\mathbf{i} - y\mathbf{j} + 2z\mathbf{k}$
(b) $\mathbf{F} = x^2\mathbf{i} + y^2\mathbf{j} - 3z^2\mathbf{k}$
(c) $\mathbf{F} = xy\mathbf{i} + yz\mathbf{j} + xz\mathbf{k}$

Which fields, if any, are solenoidal?

9.9 Determine the curl $(\nabla \times \mathbf{F})$ for each of the functions given in Prob. 9.8. Which fields, if any, are irrotational?

9.10 The complex potential which represents flow circulating around a flat plate as shown in Fig. P9.10 is $F(z) = \cos^{-1}(z/a)$.

(a) Show that

$$\frac{x^2}{a^2 \cosh^2 v} + \frac{y^2}{a^2 \sinh^2 v} = 1$$

and

$$\frac{x^2}{a^2 \cos^2 u} - \frac{y^2}{a^2 \sin^2 u} = 1$$

(b) Show that the stream functions (i.e., lines of constant v) are confocal ellipses with foci at $\pm a$.

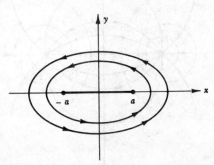

Fig. P9.10

(c) Show that the equipotential lines (i.e., lines of constant u) are confocal hyperbolas with foci at $\pm a$.

9.11 Show that the complex potential $F(z) = \cosh^{-1}(z/a)$ may be interpreted as representing flow through an aperture, as shown in Fig. P9.11.

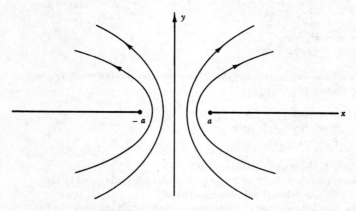

Fig. P9.11

9.12 Determine the equations for the streamlines and equipotential lines for the flow situation in which there is a source at $z = -1$ and a sink at $z = +1$. The source and sink are of equal but opposite strengths k. The streamlines are shown solid in Fig. P9.12, and the equipotential lines are shown dotted.

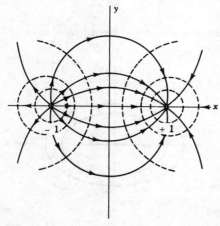

Fig. 9.12

9.13 Determine the equations for the lines of force and the equipotential lines for the electrostatic field in which there is a point source of strength k at $z = -1$ and a sink of strength $-k$ at $z = +1$. The lines of force are shown solid in Fig. P9.12, and the equipotential lines are shown dotted.

9.14 The closed-loop transfer function for a unity feedback control system is $G/(1 + G)$. The term G may be regarded as a vector, $z = x + jy$, so that the transfer function becomes $z/(1 + z)$. Loci of constant magnitude are called M *contours*,

$$M = \left| \frac{G}{1 + G} \right| = \left| \frac{z}{1 + z} \right|$$

Loci of constant phase angle are called N *contours*,

$$N = \measuredangle \frac{G}{1 + G} = \measuredangle \frac{z}{1 + z}$$

Determine the equation for the M contours. (This equation is the same as that for the equipotential lines of Probs. 9.12 and 9.13, and thus the M contours are the dotted lines of Fig. P9.12.) Determine the equation for the N contours. (This equation is the same as that for the streamlines of Prob. 9.12 and the force lines of Prob. 9.13, and thus the N contours are the solid lines of Fig. P9.12.)

9.15 The complex potential for a conducting cylinder with center at a is $F(z) = k \ln (z - a)$. One conducting cylinder of radius $r = 1$ and center $x = 2$ has a potential on its surface, $u_1 = 1$. A second conducting cylinder of radius $r = 1$ and center $x = -2$ has a surface potential $u_2 = -1$. Determine the equation for the lines of:

(a) Equal potential
(b) Equal force

REFERENCES

1 Brand, L.: "Vector Analysis," John Wiley & Sons, Inc., New York, 1961.
2 Gibbs, J. W.: "Vector Analysis," Yale University Press, New Haven, Conn., 1913.
3 Jeans, J. H.: "Mathematical Theory of Electricity and Magnetism," Cambridge University Press, New York, 1925.
4 Kellogg, O. D.: "Foundations of Potential Theory," Dover Publications, Inc., New York, 1963.
5 Weatherburn, C. E.: "Elementary Vector Analysis with Applications to Geometry and Physics," G. Bell & Sons, Ltd., London, 1921.
6 Weatherburn, C. E.: "Advanced Vector Analysis with Applications to Mathematical Physics," G. Bell & Sons, Ltd., London, 1921.
7 Wylie, Jr., C. R.: "Advanced Engineering Mathematics," 2d ed., McGraw-Hill Book Company, New York, 1960.

chapter 10

Mapping

For a function of a real variable $y = f(x)$, the plot of y is called the *graph* of the function x. A single graph with y the ordinate and x the abscissa suffices to show the correspondence between y and x.

For a function of a complex variable $w = f(z)$, the situation is somewhat different. For each value of z there is a corresponding value of w. As $z = x + jy$ follows a certain path in the z plane, the function $w = u + jv$ traverses a separate path in the w plane.

10.1 GEOMETRIC TRANSFORMATIONS

The function $w = f(z)$ defines a transformation, or mapping, of the z plane into the w plane. For example, the function $w = e^{\pi z/H}$ transforms the rectangle of length L and height H shown in Fig. 10.1a into the half ring shown in Fig. 10.1b. For z on C_1, then $z = x$. Thus, $w = e^{\pi x/H}$ goes from $w = e^0 = 1$ to $w = e^{\pi L/H} = r_0$ as x goes from 0 to L. For z on C_2, then $z = L + jy$. Thus, $w = e^{\pi L/H}e^{j\pi y/H}$ describes a semicircle of radius $r_0 = e^{\pi L/H}$ in which the angle varies from $e^{j0} = 0°$ to $e^{j\pi} = 180°$ as y goes from 0 to H. The sides C_3 and C_4 of Fig. 10.1a are similarly mapped as shown in Fig. 10.1b.

A transformation is thus seen to be a means of converting a given geometric configuration into some other geometric pattern. When the solution of a problem with a given geometric configuration is known, then transforming this solution gives the corresponding solution for the case of the transformed geometry.

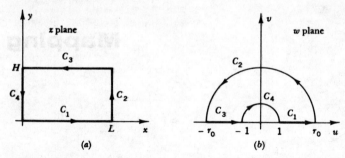

Fig. 10.1. Rectangles in the z plane are mapped into half rings in the w plane by the transformation $w = e^{\pi z/H}$.

Illustrative example Determine the temperature distribution throughout the half ring of Fig. 10.1b for the case in which the temperature on C_1, C_2, and C_4 is maintained at zero and that on C_3 is U_0.

Solution The temperature distribution for the rectangle of Fig. 10.1a in which the temperature on C_1, C_2, and C_4 is maintained at zero and that on C_3 is U_0 is given by Eq. (8.11). That is,

$$u(x,y) = \frac{2U_0}{\pi} \sum_{n=1}^{\infty} \frac{\sinh{(\pi n y/L)}}{\sinh{(\pi n H/L)}} \frac{1 - (-1)^n}{n} \sin{\frac{\pi n x}{L}} \qquad (8.11)$$

Figure 10.1b shows that w is most conveniently expressed in polar coordinates (that is, $w = re^{j\theta}$). Thus, the transformation is

$$w = re^{j\theta} = e^{\pi z/H} = e^{\pi x/H}e^{j\pi y/H} \qquad (10.1)$$

The radius is $r = e^{\pi x/H}$; hence

$$\ln r = \frac{\pi x}{H}$$

For $x = L$, then $r = r_0 = e^{\pi L/H}$; hence

$$\ln r_0 = \frac{\pi L}{H}$$

Similarly, the angle θ is

$$\theta = \frac{\pi y}{H}$$

Equation (8.11) is transformed from a function of x and y to a function of r and θ by means of the relationships

$$\frac{\pi y}{L} = \frac{\theta H}{L} = \frac{\pi \theta}{\ln r_0}$$

$$\frac{\pi x}{L} = \frac{(\ln r) H}{L} = \frac{\pi \ln r}{\ln r_0}$$

$$\frac{\pi H}{L} = \frac{\pi^2}{\ln r_0}$$

Thus, the desired transformed result is

$$u(r,\theta) = \frac{2U_0}{\pi} \sum_{n=1}^{\infty} \frac{\sinh (\pi n\theta/\ln r_0)}{\sinh (\pi^2 n/\ln r)} [1 - (-1)^n] \sin \frac{\pi n \ln r}{\ln r_0}$$

$$(10.2)$$

In a similar manner, the transformation $w = e^{2\pi z/L}$ transforms the parallel plates of Fig. 9.12 to the coaxial cylinders of Fig. 10.2. For the vertical plate at $x = a$, then $z = a + jy$. Thus, $w = e^{2\pi a/L}e^{j2\pi y/L}$ describes a circle of radius $r_1 = e^{2\pi a/L}$. The angle $2\pi y/L$ goes from $-\pi$ to $+\pi$ as y goes from $-L/2$ to $L/2$. By expressing w in polar coordinates ($w = re^{j\theta}$), the transformation is

$$w = re^{j\theta} = e^{2\pi x/L}e^{j2\pi y/L}$$

The radius is $r = e^{2\pi x/L}$; hence

$$\ln r = \frac{2\pi x}{L}$$

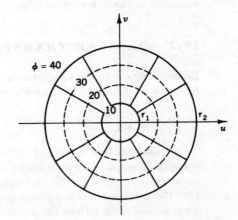

Fig. 10.2. Parallel plates of Fig. 9.12 are mapped into coaxial cylinders by the transformation $w = e^{2\pi z/L}$.

369

For $x = a$, then $r = r_1 = e^{2\pi a/L}$; and for $x = b$, then $r = r_2 = e^{2\pi b/L}$. Thus

$$\ln r_1 = \frac{2\pi a}{L}$$

$$\ln r_2 = \frac{2\pi b}{L}$$

The angle θ is

$$\theta = \frac{2\pi y}{L}$$

Thus, transforming Eq. (9.59) gives for the electrostatic potential

$$\phi = \frac{30 \ln r + 10 \ln r_2 - 40 \ln r_1}{\ln (r_2/r_1)} \qquad (10.3)$$

Because ϕ is independent of θ, these are circles as shown in Fig. 10.2. The transformed lines of force [Eq. (9.60)] become

$$\psi = \frac{30\theta}{\ln (r_2/r_1)} \qquad (10.4)$$

These are radial lines, as shown in Fig. 10.2.

Transforming Eq. (9.61) gives for the capacitance C of the coaxial cylinders

$$C = \frac{2\pi k W}{\ln (r_2/r_1)} \qquad (10.5)$$

General characteristics and properties of transformations are discussed in the following sections.

10.2 BILINEAR TRANSFORMATIONS

Bilinear transformations comprise a very important class of transformations. Consider the transformation

$$w = \frac{az + b}{cz + d} \qquad ad - bc \neq 0 \qquad (10.6)$$

This transformation is called a *bilinear transformation*, a *Möbius transformation*, a *linear fractional transformation*, or more briefly a *linear transformation*. It is necessary that $ad - bc \neq 0$, for otherwise the entire z plane is mapped into the point b/d in the w plane. That is, factoring b from the numerator of Eq. (10.6) and d from the denominator, and noting that $a/b = c/d$ when the equality holds gives

$$w = \frac{b[(a/b)z + 1]}{d[(c/d)z + 1]} = \frac{b}{d}$$

To investigate bilinear transformations, first consider the case in which $w = az$. The term a may be a complex constant; that is, $a = a_x + ja_y = |a|e^{j\theta_a}$, where a_x is the real part, a_y the imaginary part, and θ_a the angle. Thus, the transformation

$$w = az = |a|ze^{j\theta_a} \tag{10.7}$$

rotates each z vector through the angle θ_a and magnifies the length by a factor $|a|$, as is illustrated in Fig. 10.3a.

Next consider the transformation

$$w = z + b = z + b_x + jb_y \tag{10.8}$$

As is illustrated in Fig. 10.3b, this transformation shifts the horizontal component of z by an amount b_x, and it shifts the vertical component of z by an amount b_y. Because each point in the z plane is shifted by the same amount, as shown in Fig. 10.3c any path of values C_1 in

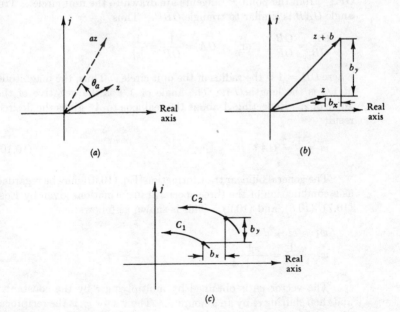

(a)

(b)

(c)

Fig. 10.3. Bilinear transformations.

371

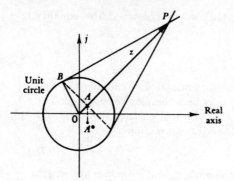

Fig. 10.4. Geometric interpretation of the transformation $w = 1/z$.

the z plane is merely shifted horizontally by b_x and vertically by b_y to obtain the corresponding plot C_2 in the w plane. The meaning of the transformation

$$w = \frac{1}{z} \tag{10.9}$$

is obtained geometrically from Fig. 10.4. The vector z is the line OP. From the point P tangents are drawn to the unit circle. Triangle OAB is similar to triangle OBP. Thus

$$\frac{OA}{OB} = \frac{OB}{OP} \quad \text{or} \quad OA = \frac{1}{OP} = \frac{1}{|z|}$$

where $OB = 1$ is the radius of the unit circle. Thus, the magnitude of $1/z$ is the length OA. The angle of $1/z$ is the negative of the angle of z. Reflecting A about the real axis to A^* gives the desired result

$$w = \frac{1}{z} = OA^* \tag{10.10}$$

The general bilinear transformation [Eq. (10.6)] may be regarded as a combination of the three special transformations given by Eqs. (10.7), (10.8), and (10.9). This is shown as follows:

$$w_1 = cz + d$$
$$w_2 = \frac{1}{w_1} = \frac{1}{cz + d}$$

The vector w_1 is obtained by multiplying z by the constant c and then shifting cz by an amount d. The vector w_2 is the reciprocal of w_1. The desired transformation for w is now obtained by multiplying w_2 by $(bc - ad)/c$ and then shifting by a/c; that is,

$$w = \frac{a}{c} + \frac{bc - ad}{c} w_2 = \frac{a(cz + d) + (bc - ad)}{c(cz + d)}$$
$$= \frac{az + b}{cz + d}$$

An important characteristic of bilinear transformations is that any circle or straight line in the z plane maps into a circle or straight line in the w plane. To prove this, first write the equation for a circle in the z plane,

$$A(x^2 + y^2) + B_1 x + B_2 y + C = 0 \qquad (10.11)$$

where A, B_1, B_2, and C are real constants. When A is zero, this becomes the equation for a straight line. Thus, Eq. (10.11) suffices for the consideration of straight lines as well as circles.

By letting $\bar{z} = x - jy$ denote the complex conjugate of z, then

$$z\bar{z} = x^2 + y^2 \qquad x = \frac{z + \bar{z}}{2} \qquad y = \frac{z - \bar{z}}{2j}$$

Thus, Eq. (10.11) may be written in the form

$$Az\bar{z} + B_1 \frac{z + \bar{z}}{2} + B_2 \frac{z - \bar{z}}{2j} + C = 0$$

or

$$Az\bar{z} + Bz + \bar{B}\bar{z} + C = 0 \qquad (10.12)$$

where $B = (B_1 - jB_2)/2$ and $\bar{B} = (B_1 + jB_2)/2$. Solving Eq. (10.6) for z gives

$$z = \frac{-dw + b}{cw - a}$$

and thus

$$\bar{z} = \frac{-\bar{d}\bar{w} + \bar{b}}{\bar{c}\bar{w} - \bar{a}}$$

The substitution of z and $\bar{z}$ into Eq. (10.12) gives

$$\alpha w\bar{w} + \beta w + \bar{\beta}\bar{w} + \gamma = 0 \qquad (10.13)$$

where

$$\alpha = |d|^2 A - 2 \operatorname{Re}(\bar{c}\, dB) + |c|^2 C$$
$$\beta = -\bar{b}\, dA + \bar{a}\, dB + \bar{b}c\bar{B} - \bar{a}cC$$
$$\gamma = |b^2|A - 2 \operatorname{Re}(\bar{a}\, bB) + |a|^2 C$$

where Re means "real part of." Because Eq. (10.12) is the equation for a circle in the z plane, it follows that Eq. (10.13) is the equation for a circle in the w plane. It is to be noted that if A is zero in Eq. (10.12), then α is not necessarily zero in Eq. (10.13). Thus, straight lines in the z plane (that is, $A = 0$) may be mapped into circles in the w plane, and vice versa.

Illustrative example The vertical axis of the z plane is to be mapped onto the w plane in accordance with the transformation $w = 1/(\tau z + 1)$, in which τ is a real constant. Determine the equation for the resultant circle or straight line in the w plane.

Solution The equation for the vertical axis is $x = 0$. Thus, Eq. (10.11) shows that $A = B_2 = C = 0$ and $B_1 = 1$. From the given transformation, $a = 0, b = d = 1$. Because τ is real, $c = \bar{c} = \tau$. Using these values to determine the coefficients of Eq. (10.13) and noting that $B = (B_1 - jB_2)/2 = \frac{1}{2}$ gives $\alpha = -\tau$, $\beta = \tau/2$, and $\gamma = 0$. Substitution of these results into Eq. (10.13) gives

$$-\tau w \bar{w} + \tau \frac{(w + \bar{w})}{2} = 0$$

Because $w = u + jv$ and $\bar{w} = u - jv$, then $w\bar{w} = u^2 + v^2$ and $w + \bar{w} = 2u$. Thus,

$$-\tau(u^2 + v^2) + \tau u = 0$$

Dividing through by $-\tau$, and then completing the square, gives

$$(u - \tfrac{1}{2})^2 + v^2 = (\tfrac{1}{2})^2$$

This is recognized as the equation for a circle with radius of $\frac{1}{2}$ and center at $u = \frac{1}{2}$ and $v = 0$.

The transformation shown in Fig. 10.5 maps the right half plane to the exterior of the unit circle. This is a bilinear transformation because a straight line (the vertical axis) is being mapped into a unit circle. Thus,

$$w = \frac{az + b}{cz + d} \quad \text{or} \quad az + b = w(cz + d)$$

As illustrated in Fig. 10.5, the point C, which is the origin of the z plane, is to correspond to the point $w = 1$. For $z = 0$ and $w = 1$, the preceding equation shows that $b = d$. Similarly, for

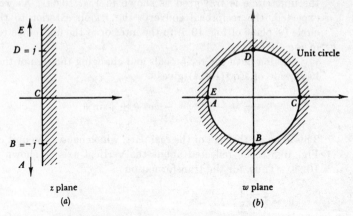

Fig. 10.5. The right half of the z plane is mapped to the exterior of the w plane by the transformation $w = (z + 1)/(1 - z)$.

the point B, then $z = -j$, and $w = -j$; hence

$$-aj + b = -j(-cj + d) = -c - jd$$

For the point D, then $z = j$, and $w = j$; hence

$$aj + b = j(cj + d) = -c + jd$$

Adding these last two equations gives $b = -c$. Subtraction shows that $a = d$. Because $b = d = a$ and $c = -b = -a$, the transformation becomes

$$w = \frac{az + a}{-az + a} = \frac{z + 1}{1 - z} \tag{10.14}$$

Note that, as z approaches A or as z approaches E, then w approaches -1 from the directions shown in the w plane of Fig. 10.5.

Because w is a unit circle ($w = e^{j\theta} = \cos\theta + j\sin\theta$), then taking the reciprocal of the right-hand side of Eq. (10.14) gives

$$w = \frac{1}{e^{j\theta}} = e^{-j\theta} = \cos(-\theta) + j\sin(-\theta) = \cos\theta - j\sin\theta$$

This changes the sign of the imaginary part, which causes the unit circle of Fig. 10.5 to be reflected about the horizontal axis as shown in Fig. 10.6a. Thus, for the reciprocal transformation

$$w = \frac{1 - z}{z + 1} \tag{10.15}$$

375

the unit circle is traversed as shown in Fig. 10.6a. As would be expected, the reciprocal converts the region exterior to the unit circle (w plane of Fig. 10.5) to the interior of the unit circle (w plane of Fig. 10.6a).

Taking both the reciprocals and changing the sign of the right-hand side of Eq. (10.14) gives

$$w = -\frac{1}{e^{j\theta}} = -e^{-j\theta} = -\cos\theta + j\sin\theta$$

This changes the sign of the real part, which causes the unit circle of Fig. 10.5 to be reflected about the vertical axis as shown in Fig. 10.6b. Thus, for the transformation

$$w = \frac{z-1}{z+1} \tag{10.16}$$

the unit circle is traversed as shown in Fig. 10.6b. In traversing each boundary from A to B to C to D to E in Figs. 10.5 and 10.6, it is to be noted that the crosshatched region is always seen to the right. In general, as a boundary is traversed in a given direction, a given region is always viewed on the same side of the boundary.

Let it now be desired to determine the transformation which maps the upper half plane into the interior of the unit circle, as shown in Fig. 10.7. Multiplication by $-j = e^{-j90°}$ rotates a geometric shape $(-90°)$. Thus, the transformation for going from the z plane

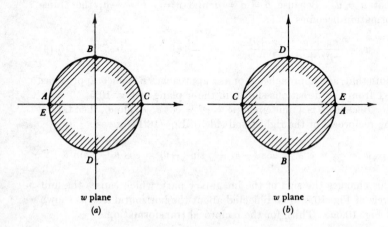

Fig. 10.6. The z plane of Fig. 10.5 is mapped as shown in (a) by the transformation $w = (1-z)/(z+1)$ and as shown in (b) by the transformation $w = (z-1)/(z+1)$.

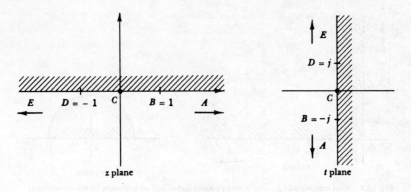

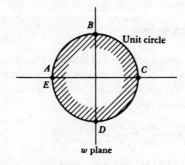

Fig. 10.7.   Transforming from the z plane to the t plane by $t = -jz$ and then from the t plane to the w plane by $w = (1 - t)/(t + 1)$.

to the t plane is

$$t = -jz$$

The transformation for going from the t plane to the w plane is obtained by replacing z by t in Eq. (10.15); that is,

$$w = \frac{1 - t}{t + 1}$$

From these two results, the transformation for going from the z plane to the w plane of Fig. 10.7 is found to be

$$w = \frac{1 + jz}{1 - jz} \tag{10.17}$$

As shown in Fig. 10.8, this transformation is such that the first quadrant of the z plane is transformed into the upper half of the unit circle.

377

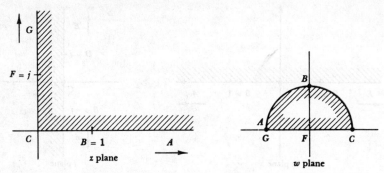

Fig. 10.8. First quadrant of z plane is mapped into upper half of unit circle by the transformation $w = (1 + jz)/(1 - jz)$.

Three distinct points

Another very useful property of bilinear transformations is that three distinct points z_1, z_2, z_3 can always be mapped into three prescribed points w_1, w_2, w_3. The required transformation is determined by the following relationship:

$$\frac{w - w_1}{w - w_3} \frac{w_2 - w_3}{w_2 - w_1} = \frac{z - z_1}{z - z_3} \frac{z_2 - z_3}{z_2 - z_1} \tag{10.18}$$

To verify Eq. (10.18), first note that

$$w - w_1 = \frac{az + b}{cz + d} - \frac{az_1 + b}{cz_1 + d} = \frac{(ad - bc)(z - z_1)}{(cz + d)(cz_1 + d)}$$

Similarly,

$$w - w_3 = \frac{(ad - bc)(z - z_3)}{(cz + d)(cz_3 + d)}$$

Dividing $w - w_1$ by $w - w_3$ gives

$$\frac{w - w_1}{w - w_3} = \frac{(z - z_1)(cz_3 + d)}{(z - z_3)(cz_1 + d)}$$

By proceeding in the same manner, it may be shown that

$$\frac{w_2 - w_3}{w_2 - w_1} = \frac{(z_2 - z_3)(cz_1 + d)}{(z_2 - z_1)(cz_3 + d)}$$

Multiplication of the two preceding equations yields the desired relationship given by Eq. (10.18).

Illustrative example   Determine the transformation that maps the points 0, 1, 2, of the z plane into the points $-1, j, 1$ of the w plane.

Solution   Application of Eq. (10.18) gives

$$\frac{w - (-1)}{w - 1} \frac{j - 1}{j - (-1)} = \frac{z - 0}{z - 2} \frac{1 - 2}{1 - 0}$$

Solving for w,

$$w = \frac{z - (1 - j)}{jz + (1 - j)}$$

Comparison with Eq. (10.6) shows that $a = 1$, $b = -(1 - j)$, $c = j$, and $d = 1 - j$.

The characteristics of some other transformations are illustrated in Fig. 10.9. As shown in Fig. 10.9a, the transformation $w = 1/z$ maps the exterior of the unit circle of the z plane into the interior of the unit circle in the w plane. The transformation $w = e^z$ maps rectangles into circular segments, as is shown in Fig. 10.9b. The inverse transformation is the transformation that maps the w plane back to the z plane. For $w = e^z$, solving for z gives the inverse transform $z = \ln w$. By starting with $w = \ln z$, circular segments are mapped into rectangles, as is shown in Fig. 10.9c. The transformation $w = \sin z$ transforms a vertical strip which lies above the real axis into the entire half plane, as is shown in Fig. 10.9d. For the transformation $w = \sin z$, a horizontal line (constant y) maps into an ellipse in the w plane. This is proved as follows:

$$\begin{aligned}
w = u + jv &= \sin (x + jy) \\
&= \sin x \cos jy + \cos x \sin jy \\
&= \sin x \cosh y + j \cos x \sinh y
\end{aligned}$$

Because u is the real part of w and v is the imaginary part, then

$$u = \sin x \cosh y$$
$$v = \cos x \sinh y$$

Thus,

$$\left(\frac{u}{\cosh y}\right)^2 + \left(\frac{v}{\sinh y}\right)^2 = \sin^2 x + \cos^2 x = 1$$

The portion of the z plane from E to F to A in Fig. 10.9e maps into the elliptical portion of the w plane.

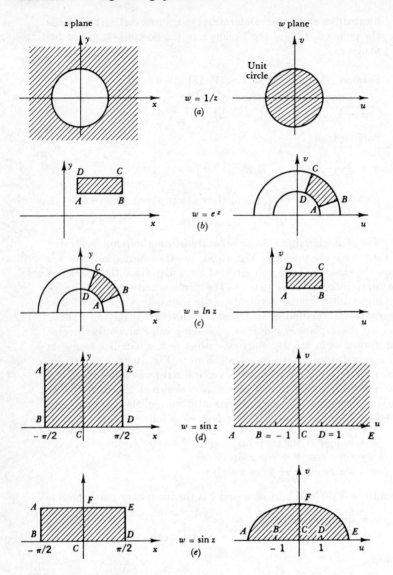

Fig. 10.9. Transformations.

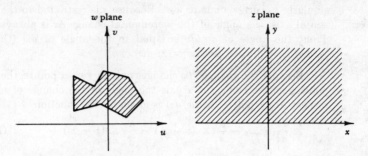

Fig. 10.10. Transforming a polygon in the w plane into the entire half plane.

10.3 SCHWARZ-CHRISTOFFEL TRANSFORMATION

A region enclosed by a polygon (w plane of Fig. 10.10) may be transformed to the entire upper half plane (z plane of Fig. 10.10) by means of the Schwarz-Christoffel transformation. After transformation to the upper half plane, the region may be further mapped into any region into which a half plane can be transformed.

The points w_1 and w_2 in the w plane of Fig. 10.11 are vertex points of a polygon. The point w is a point on the polygon, and the incremental change Δw lies on the polygon. As a point w travels along the polygon (in the direction indicated) and goes past the vertex w_1, the angle of Δw changes by α_1. The angle α_1 is the angle from the extension to the new side. Similarly, as w proceeds past the vertex w_2, the angle of Δw changes by α_2. In accordance with sign convention, a counterclockwise angle (for example, α_1) is positive, and a clockwise angle (for example, α_2) is negative.

The point x_1 in the z plane of Fig. 10.11 corresponds to w_1, and

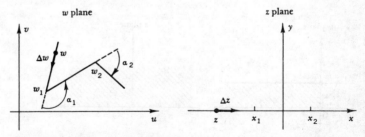

Fig. 10.11. Transforming sides of a polygon in the w plane into the horizontal axis of the z plane.

381

similarly x_2 corresponds to w_2. Because z is restricted to the horizontal axis, the angle of the incremental change Δz is always zero. Thus, the angle of dw/dz is equal to the angle of dw (that is, $\angle\, dw/dz = \lim_{\Delta z \to 0} \angle\, \Delta w/\Delta z = \angle\, dw$).

Because the angle of dw changes only at vertex points, then as a point z moves along the x axis the angle of the change of w with respect to z (that is, $\angle\, dw/dz$) is given by the function

$$\angle\, \frac{dw}{dz} = \alpha_0 - \frac{\alpha_1}{\pi}\, \angle\, (z - x_1) - \frac{\alpha_2}{\pi}\, \angle\, (z - x_2) - \cdots \quad (10.19)$$

where α_0, a constant angle, is the slope of the side of the polygon before the w_1 vertex. To substantiate this equation, first note that, as z moves from the left side of point x_1 to the right, the angle of $(z - x_1)$ goes from π to 0 [that is, $\angle\, (z - x_1)$ changes by $-\pi$]. The corresponding change in the function given by Eq. (10.19) is $-(\alpha_1/\pi)(-\pi) = \alpha_1$. Similarly, as the point z moves past the point x_2, the change in the angle of $(z - x_2)$ is $-\pi$. The corresponding change in Eq. (10.19) is $-(\alpha_2/\pi)(-\pi) = \alpha_2$. Hence, Eq. (10.19) is the general equation for the angle of dw/dz. The equation which represents the change of w with respect to z is

$$\frac{dw}{dz} = K_0(z - x_1)^{-(\alpha_1/\pi)}(z - x_2)^{-(\alpha_2/\pi)} \cdots (z - x_n)^{-(\alpha_n/\pi)}$$

$$(10.20)$$

where K_0 is a complex constant whose angle is α_0. Note that the angle of the $z - x_1$ term is $-(\alpha_1/\pi)\, \angle\, (z - x_1)$. The Schwarz-Christoffel transformation which converts a region in the w plane to the upper half of the z plane is thus obtained by integrating Eq. (10.20).

Illustrative example Determine the transformation which maps the strip shown in the w plane of Fig. 10.12 into the upper

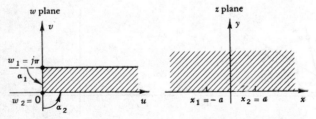

Fig. 10.12. A strip in the w plane is mapped into the upper half plane by the transformation $w = \cosh^{-1} z$.

half plane represented by the z plane of Fig. 10.12. The vertex $w_1 = j\pi$ is to be transformed to the point $x_1 = -a$. The vertex $w_2 = 0$ is to be transformed to the point $x_2 = a$. The value of a is to be selected such that the constant of integration which arises in integrating Eq. (10.20) will vanish. Figure 10.12 shows that $\alpha_1 = \alpha_2 = \pi/2$.

Solution Substitution of these values into Eq. (10.20) gives

$$dw = K_0(z + a)^{-\frac{1}{2}}(z - a)^{-\frac{1}{2}}\, dz$$

Thus

$$w = K_0 \int \frac{dz}{\sqrt{z^2 - a^2}} = K_0 \ln (z + \sqrt{z^2 - a^2}) + K_1$$

For $w = w_2 = 0$, $z = x_2 = a$, then

$$0 = K_0 \ln a + K_1$$

The constant K_1 vanishes when $a = 1$.

The constant K_0 is evaluated from the boundary condition that, for $w = w_1 = j\pi$, then $z = x_1 = -a = -1 = e^{j\pi}$. That is,

$$j\pi = K_0 \ln e^{j\pi} = K_0(\ln 1 + j\pi) = K_0 j\pi \qquad \text{or} \qquad K_0 = 1$$

Thus, the desired transformation is

$$w = \ln (z + \sqrt{z^2 - 1}) = \cosh^{-1} z$$

Inverting gives

$$z = \cosh w \tag{10.21}$$

The application of the Schwarz-Christoffel transformation to problems involving potential theory is explained in the following.

Fluid flow in the upper half plane

In Fig. 10.13a is represented a channel having a flow source at the origin ($w = 0$). Let it be desired to determine the complex potential function for the flow through this channel.

The first step is to transform the channel shown in the w plane (Fig. 10.13b) to the upper half plane, as is illustrated by the z plane (Fig. 10.13c). The dotted lines going from w_1 to w_2 to w_3 to w_4 and back to w_1 convert the channel to a closed polygon. This polygon

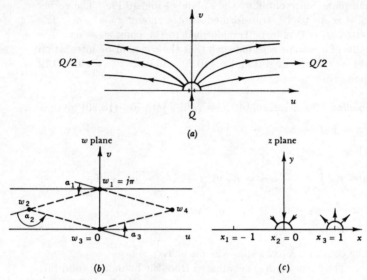

Fig. 10.13. Flow through a channel. (*a*) Actual representation; (*b*) representation in *w* plane; (*c*) representation in upper half *z* plane.

approaches the channel as w_2 goes to $-\infty + j\pi/2$ and w_4 goes to $\infty + j\pi/2$. Let the corresponding image points be $w_1 = j\pi$ and $x_1 = -1$; $w_2 = -\infty + j\pi/2$ and $x_2 = 0$; $w_4 = \infty + j\pi/2$ and $x_4 = \infty$. The final pair of image points are $w_3 = 0$ and x_3, which is to be selected such that the constant of integration that results from integrating Eq. (10.20) vanishes. The w plane of Fig. 10.13*b* shows that in the limiting position $\alpha_1 = 0$, $\alpha_2 = \pi$, and $\alpha_3 = 0$. Thus, Eq. (10.20) becomes

$$\frac{dw}{dz} = K_0(z - x_1)^0(z - x_2)^{-1}(z - x_3)^0 = \frac{K_0}{z - x_2} = \frac{K_0}{z}$$

Because the point x_4 is at $+\infty$, then z cannot pass this point on the real axis. As there is no change in the angle of w due to this point, this term does not appear in Eq. (10.19). Consequently, there is no $z - x_4$ term in the preceding equation for dw/dz. Integration of this expression for dw/dz gives

$$w = K_0 \ln z + K_1$$

For $w = w_3 = 0$ and $K_1 = 0$, then $\ln z = 0$. Thus, the constant of integration K_1 vanishes when $z = x_3 = 1$. The value of K_0 is ascertained from the boundary condition that, for $w = w_1 = j\pi$, then

$z = x_1 = -1 = e^{j\pi}$. Thus,

$$j\pi = K_0(\ln r + j\theta) = K_0(\ln 1 + j\pi) = jK_0\pi$$

Hence, $K_0 = 1$. The desired transformation is thus found to be

$$w = \ln z$$

or

$$z = e^w \tag{10.22}$$

Let Q be the flow rate at the source $w_3 = 0$ of the channel. The z plane shows that only half the flow from the corresponding source $x_3 = 1$ goes into the upper half plane. Thus, the rate of flow from this source must be $2Q$. Replacing Q by $2Q$ in Eq. (9.50) gives $k = Q/\pi$. Thus, the complex potential for the source at $x_3 = 1$ is

$$k \ln (z - 1) = \frac{Q}{\pi} \ln (z - 1)$$

The substitution of $z - 1$ for z in Eq. (9.48) transforms the source from the origin to the $+1$ point. The point w_2 corresponds to a sink of half the flow rate at $x_2 = 0$. The complex potential for this sink at the origin of the z plane is

$$-\frac{Q}{2\pi} \ln z$$

The sink at w_4 does not appear in the z plane, for the corresponding point is at $x_4 = +\infty$. Adding the two preceding equations yields the complex potential function for the z plane.

$$F(z) = \frac{Q}{\pi} \left[\ln (z - 1) - \frac{1}{2} \ln z \right] = \frac{Q}{\pi} \ln (z^{\frac{1}{2}} - z^{-\frac{1}{2}}) \tag{10.23}$$

Using Eq. (10.22) to transform this complex potential to the w plane gives

$$F(w) = \frac{Q}{\pi} \ln (e^{w/2} - e^{-w/2}) = \frac{Q}{\pi} \left(\ln \frac{e^{w/2} - e^{-w/2}}{2} + \ln 2 \right)$$

$$= \frac{Q}{\pi} \left(\ln \sinh \frac{w}{2} + \ln 2 \right) \tag{10.24}$$

The velocity vector $\overline{F'(w)}$ is

$$\overline{F'(w)} = \frac{Q}{2\pi} \frac{\cosh (\bar{w}/2)}{\sinh (\bar{w}/2)} = \frac{Q}{2\pi} \coth \frac{\bar{w}}{2} \tag{10.25}$$

The streamlines ψ are the imaginary part of Eq. (10.24).

$$\psi = \frac{Q}{\pi} \measuredangle \sinh \frac{w}{2} = C \tag{10.26}$$

The sinh $(w/2)$ may be expanded by the following identity:

$$\sinh \frac{u + jv}{2} = \sinh \frac{u}{2} \cos \frac{v}{2} + j \cosh \frac{u}{2} \sin \frac{v}{2}$$

The angle of sinh $(w/2)$ may be written in the form

$$\measuredangle \sinh \frac{w}{2} = \tan^{-1} \frac{\cosh (u/2) \sin (v/2)}{\sinh (u/2) \cos (v/2)} = \tan^{-1} \frac{\tan (v/2)}{\tanh (u/2)}$$

Substituting this result in Eq. (10.26), and then multiplying through by π/Q and taking the tangent gives

$$\frac{\tan (v/2)}{\tanh (u/2)} = \tan \frac{\pi C}{Q} = K \tag{10.27}$$

Thus, the streamlines are values of u and v that satisfy the preceding equation in which K is any real constant.

Heat flow in the upper half plane

Steady-state heat flow provides another example of problems whose solution may be simplified by transforming the geometric configuration to the upper half plane.   For the half plane shown in Fig. 10.14 the temperature on the boundary to the right of the point x_0 is T_0, and the temperature to the left of x_0 is T_1.   Laplace's equation for this case is

$$\frac{\partial^2 T}{\partial x^2} + \frac{\partial^2 T}{\partial y^2} = 0 \tag{10.28}$$

Consider the complex potential function

$$F(z) = jA + B \ln (z - x_0) = jA + B \ln re^{j\theta} \tag{10.29}$$

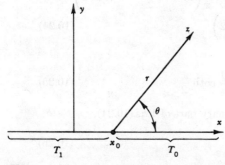

Fig. 10.14.   Temperature distribution in upper half plane.

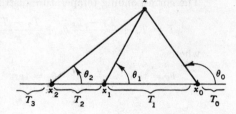

Fig. 10.15. Temperature distribution in upper half plane, with many sections of constant temperature along boundary.

where $z - x_0 = re^{j\theta}$ is the vector from x_0 to any point z in the body, as is illustrated in Fig. 10.14. The function $F(z)$ is analytic except at the point x_0. Thus, the real and imaginary parts of $F(z)$ are solutions of Laplace's equation except at x_0. These real and imaginary parts are

$$\phi = B \ln r$$
$$\psi = A + B\theta \qquad (10.30)$$

On the boundary to the right of x_0 the angle θ is zero, and to the left of x_0 the angle θ is π. Substitution of the boundary conditions $\psi = T_0$ for $\theta = 0$ and $\psi = T_1$ for $\theta = \pi$ into the preceding equation for ψ and solving for A and B gives $A = T_0$ and $B = (T_1 - T_0)/\pi$. Thus,

$$\psi = T = T_0 + \frac{T_1 - T_0}{\pi} \theta \qquad (10.31)$$

Because ψ satisfies Laplace's equation and also the boundary conditions, then ψ is the desired temperature distribution (that is, $\psi = T$). Figure 10.14 shows that $\theta = \tan^{-1} [y/(x - x_0)]$. Thus, the preceding solution may be expressed in the form

$$T = T_0 + \frac{T_1 - T_0}{\pi} \tan^{-1} \frac{y}{x - x_0} \qquad (10.32)$$

For heat flow the ψ lines are the equipotential lines, and the ϕ lines are the streamlines, which are tangent to the direction of heat flow.

In Fig. 10.15 is shown a half plane which has numerous sections of constant temperature along the boundary. The complex potential function for this general case is

$$F(z) = jT_0 + \frac{T_1 - T_0}{\pi} \ln (z - x_0) + \frac{T_2 - T_1}{\pi} \ln (z - x_1)$$
$$+ \cdots \qquad (10.33)$$

387

The corresponding temperature distribution is

$$T = T_0 + \frac{T_1 - T_0}{\pi} \theta_0 + \frac{T_2 - T_1}{\pi} \theta_1 + \cdots \qquad (10.34)$$

where

$$\theta_0 = \tan^{-1} \frac{y}{x - x_0} \qquad \theta_1 = \tan^{-1} \frac{y}{x - x_1} \qquad \cdots$$

10.4 CONFORMAL MAPPING

The curve C_1 in Fig. 10.16a indicates the path of values of z_1 for the function $w = f(z_1)$. The curve C_2 is a second path of values z_2 for the same function, $w = f(z_2)$. These curves intersect at the point z_0, in which case $f(z_1) = f(z_2) = f(z_0)$. The angle between the tangents to C_1 and C_2 at the point of intersection is γ. The plots of $f(z_1)$ and $f(z_2)$ are shown in Fig. 10.16b. The angle between the tangents to $f(z_1)$ and $f(z_2)$ is also γ. Such a mapping is said to be angle-preserving, or conformal. To prove that the angle is preserved, note that the derivative, as z approaches z_0 along the path C_1, is

$$f'(z_0) = \lim_{z_1 \to z_0} \frac{\Delta f(z_1)}{z_1 - z_0} = \lim_{z_1 \to z_0} \frac{|\Delta f(z_1)|}{|z_1 - z_0|} \frac{e^{j\beta_1}}{e^{j\alpha_1}} \qquad (10.35)$$

where β_1 is the angle of the tangent to $f(z_1)$ at $z_1 = z_0$ and α_1 is the angle of the tangent to C_1 at z_0. The derivative, as z approaches z_0 along the path C_2, is

$$f'(z_0) = \lim_{z_2 \to z_0} \frac{\Delta f(z_2)}{z_2 - z_0} = \lim_{z_2 \to z_0} \frac{|\Delta f(z_2)|}{|z_2 - z_0|} \frac{e^{j\beta_2}}{e^{j\alpha_2}} \qquad (10.36)$$

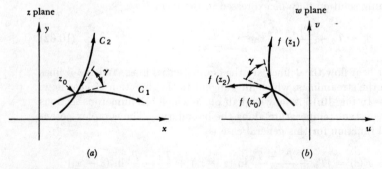

Fig. 10.16. Preservation of angles when $f'(z) \neq 0$. (a) Paths C_1 and C_2 in the z plane; (b) transformed paths C_1 and C_2 in the w plane.

where β_2 is the angle of the tangent to $f(z_2)$ at $z_2 = z_0$ and α_2 is the angle of the tangent to C_2 at z_0.

From Eq. (10.35) the angle of $f'(z_0)$ is $\beta_1 - \alpha_1$, and from Eq. (10.36) the angle is $\beta_2 - \alpha_2$. Because the value of the derivative of an analytic function must be independent of the path along which z approaches z_0, then

$$\beta_2 - \alpha_2 = \beta_1 - \alpha_1$$

or

$$\beta_2 - \beta_1 = \alpha_2 - \alpha_1 = \gamma$$

Thus, the mapping of analytic functions is seen to be conformal, i.e., angle-preserving, except at points where the derivative $f'(z_0)$ is zero.

Solving Eq. (10.35) for the magnitude of $f'(z_0)$ gives

$$|f'(z_0)| = \frac{|\Delta f(z_1)|}{|\Delta z_1|}$$

or

$$|\Delta f(z_1)| = |\Delta z_1|\,|f'(z_0)|$$

A similar relationship is obtained from Eq. (10.36). Thus, for any incremental segment Δz drawn from a point z_0, it follows that

$$|\Delta f(z)| = |\Delta z|\,|f'(z_0)| \tag{10.37}$$

Thus, in the mapping of analytic functions incremental segments Δz are magnified by a factor $|f'(z_0)|$. The angle of rotation of the segment is the angle of $f'(z_0)$. All the transformations discussed in this chapter are conformal.

PROBLEMS

10.1 A unit circle about the origin of the z plane ($z\bar{z} = x^2 + y^2 = 1$) is mapped onto the w plane in accordance with each of the following bilinear transformations:

(a) $w = \dfrac{1}{z}$ (b) $w = 1 + z$ (c) $w = \dfrac{1}{1 + z}$

Determine the resultant circle (i.e., specify center and radius) or straight line (i.e., specify slope and v intercept) in the w plane for each case.

10.2 Each of the following circles or straight lines in the z plane is mapped onto the w plane in accordance with the bilinear transformation

$w = 1/z$:

(a) $x = 1$ (b) $y = x$
(c) $|z - 1| = 1$ (d) $|z - j| = 1$

Determine the resultant circle (i.e., specify center and radius) or straight line (i.e., specify slope and v intercept) in the w plane for each case.

10.3   Determine the equation for the transformation that maps the vertical axis into the unit circle, as shown in Fig. P10.3.

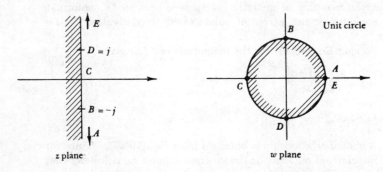

Fig. P10.3

10.4   Determine the equation for the transformation that maps the horizontal axis into the unit circle, as shown in Fig. P10.4.

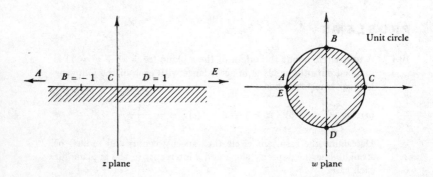

Fig. P10.4

390

10.5 Determine the transformation for mapping the sector shown in the
z plane of Fig. P10.5 into the unit circle shown in the w plane.

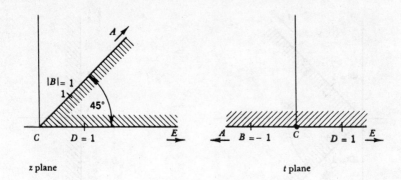

z plane   t plane

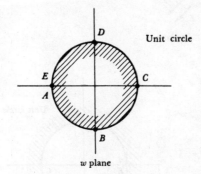

w plane

Fig. P10.5

10.6   Determine the transformation for mapping the sector shown in the z plane of Fig. P10.6 into the unit circle shown in the w plane.

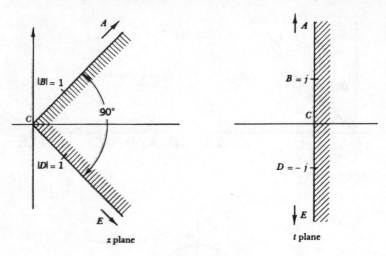

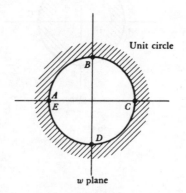

w plane

Fig. P10.6

10.7   Find the bilinear transformation that maps:

(a) The points 0, 1, 2 into the points 5, 10, 20, respectively
(b) The points 0, 1, ∞ into the points -1, j, 1, respectively
(c) The points 0, j, $j\infty$ into the points $-j$, -1, j, respectively

10.8   Determine the transformation which maps the w plane into the z plane in Fig. P10.8.

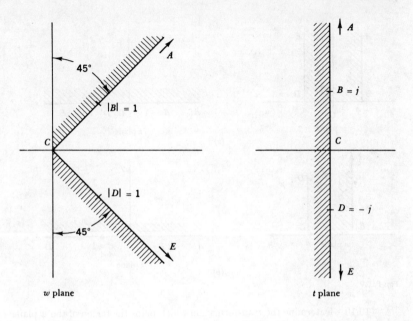

w plane   t plane

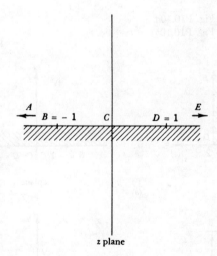

z plane

Fig. P10.8

10.9   Determine the transformation which maps the region of the w plane into the z plane of:

(a) Fig. P10.9a
(b) Fig. P10.9b

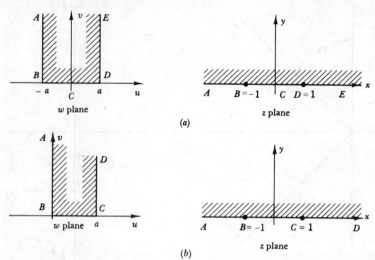

(a)

(b)

Fig. P10.9

10.10 Determine the transformation which maps the region of the w plane into the z plane of:

(a) Fig. P10.10a

(b) Fig. P10.10b

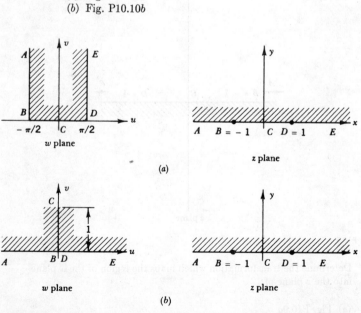

(a)

(b)

Fig. P10.10

10.11 Determine the equation for the streamlines and the complex potential function for the w plane of:

(a) Fig. P10.10a
(b) Fig. P10.10b

10.12 Determine the steady-state temperature distribution for the solids represented in:

(a) Fig. P10.10a
(b) Fig. P10.10b

The surface temperature from A to B and from D to E is maintained at $0°$, and the surface temperature from B to C to D is maintained at U_0. (*Hint:* Solve the problem in the z plane, and then transform back to the w plane.)

10.13 Determine the transformation that maps the w plane into the z plane of:

(a) Fig. P10.13a
(b) Fig. P10.13b

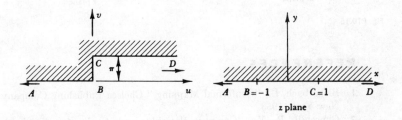

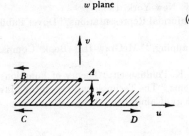

w plane

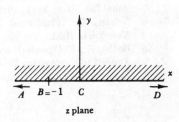

(b)

Fig. P10.13

10.14 Determine the equation for the streamlines and the complex potential function for the w plane of:

    (a) Fig. P10.13a
    (b) Fig. P10.13b

10.15 Determine the temperature distribution for the solid shown in the w plane of Fig. P10.15. Use the transformation $z = \frac{1}{2}(w + 1/w)$ to transform the w plane to the z plane. Note that the semicircular portion $w = e^{i\theta}$ is transformed $[z = (e^{i\theta} + e^{-i\theta})/2 = \cos\theta]$ to the region from -1 to $+1$ of the real axis of the z plane.

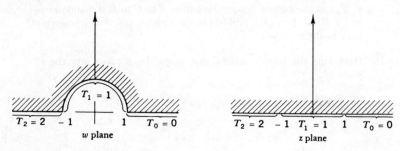

Fig. P10.15

REFERENCES

1 Bieberbach, L.: "Conformal Mapping," Chelsea Publishing Company, New York, 1953.
2 Churchill, R. V.: "Complex Variables and Applications," 2d ed., McGraw-Hill Book Company, New York, 1960.
3 Kober, H.: "Dictionary of Conformal Representations," Dover Publications, Inc., New York, 1962.
4 Nehari, Z.: "Conformal Mapping," McGraw-Hill Book Company, New York, 1952.
5 Rothe, R., F. Ollendorf, and K. Pohlhausen: "Theory of Functions as Applied to Engineering Problems," The Technology Press of the Massachusetts Institute of Technology, Cambridge, Mass., 1938.

Differential equations with variable coefficients

Bessel's equation and Legendre's equation are two well-known examples of differential equations with variable coefficients. Bessel's equation arises in the solution of partial differential equations which are expressed in cylindrical coordinates. Similarly, Legendre's equation arises in the solution of partial differential equations which are expressed in spherical coordinates. Many interesting engineering phenomena are described by differential equations with variable coefficients.

11.1 VARIABLE COEFFICIENTS

Consider the following differential equation:

$$\frac{d^2y}{dt^2} + t\frac{dy}{dt} - 2y = t \qquad y(0) = 1, y'(0) = -1 \qquad (11.1)$$

The coefficient t of the dy/dt term is a variable, or nonconstant, coefficient. Equation (2.29) shows that

$$\mathcal{L}[t^n y(t)] = (-1)^n \frac{d^n}{ds^n} [\mathcal{L}(y(t)] = (-1)^n \frac{d^n}{ds^n} [Y(s)] \qquad (11.2)$$

The Laplace transform of each term in Eq. (11.1) is

$$\mathcal{L}\left(\frac{d^2y}{dt^2}\right) = s^2 Y(s) - sy(0) - y'(0)$$

$$\mathcal{L}\left(t\frac{dy}{dt}\right) = (-1)\frac{d}{ds}[sY(s) - y(0)] = -sY'(s) - Y(s)$$

$$\mathcal{L}(-2y) = -2Y(s)$$

$$\mathcal{L}(t) = \frac{1}{s^2}$$

For $y(0) = 1$ and $y'(0) = -1$, the transform of Eq. (11.1) is

$$Y'(s) + \left(\frac{3}{s} - s\right) Y(s) = -\frac{s^3 - s^2 + 1}{s^3} \qquad (11.3)$$

This is recognized as a linear first-order differential equation (see Prob. 5.14). The general form of a linear first-order differential equation is

$$\frac{dY}{ds} + P(s)Y = Q(s)$$

The corresponding solution is

$$Y(s) = \frac{\int Q(s)\, e^{\int P(s)\, ds}\, ds + C}{e^{\int P(s)\, ds}}$$

where $e^{\int P(s)\, ds}$ is called the *integrating factor*. From Eq. (11.3), $P(s) = 3/s - s$ and $Q(s) = -(s^3 - s^2 + 1)/s^3$. The integrating factor is

$$e^{\int P(s)\, ds} = e^{3\int ds/s} e^{-\int s\, ds} = e^{3\ln s} e^{-s^2/2} = s^3 e^{-s^2/2}$$

Thus, the solution of Eq. (11.3) is

$$Y(s) = \frac{-\int (s^3 - s^2 + 1)e^{-s^2/2}\, ds + C}{s^3 e^{-s^2/2}} = \frac{(2 - s + s^2)e^{-s^2/2} + C}{s^3 e^{-s^2/2}}$$

$$= \left(\frac{2}{s^3} - \frac{1}{s^2} + \frac{1}{s}\right) + \frac{Ce^{s^2/2}}{s^3} \qquad (11.4)$$

The constant C may now be evaluated by means of the initial-value theorem. Integration of Eq. (11.1) from $0-$ to $0+$ shows that $y(0+) = y(0-) = 1$. Thus, application of the initial-value theorem gives

$$y(0+) = \lim_{s \to \infty} sY(s) = \left[\frac{2}{s^2} - \frac{1}{s} + 1 + \frac{Ce^{s^2/2}}{s^2}\right]_{s=\infty}$$

$$= 1 + \frac{Ce^{s^2/2}}{s^2}\bigg|_{s=\infty} \qquad (11.5)$$

Because $e^{s^2/2}/s^2$ becomes infinite as s goes to infinity, then C must vanish in order that $y(0+) = 1$. Thus, Eq. (11.4) becomes

$$Y(s) = \frac{2}{s^3} - \frac{1}{s^2} + \frac{1}{s}$$

Inverting yields for the desired result

$$y(t) = t^2 - t + 1 \qquad (11.6)$$

Additional examples of differential equations with variable coefficients are provided in the following sections.

11.2 BESSEL'S DIFFERENTIAL EQUATION

In problems involving Bessel's differential equation, usually a space coordinate x is the variable rather than time t. Thus, in the following development x is employed as the variable. The general form of Bessel's differential equation is

$$x^2 \frac{d^2y}{dx^2} + x \frac{dy}{dx} + (x^2 - \nu^2)y = 0$$

where $\nu \geq 0$ may be a positive integer or a noninteger. This equation is referred to as *Bessel's differential equation of order ν*. For the case in which $\nu = n$ is an integer, the preceding equation becomes Bessel's differential equation of order n. When $\nu = n = 0$, Bessel's differential equation of order 0 results.

Bessel's differential equation of order 0

This differential equation is obtained by letting $\nu = 0$ in the preceding equation and then dividing through by x. That is,

$$x \frac{d^2y}{dx^2} + \frac{dy}{dx} + xy = 0 \tag{11.7}$$

To solve this differential equation, first transform each term.

$$\mathcal{L}\left(x \frac{d^2y}{dx^2}\right) = -\frac{d}{ds}\left[\mathcal{L}\left(\frac{d^2y}{dx^2}\right)\right] = -\frac{d}{ds}[s^2Y(s) - sy(0) - y'(0)]$$
$$= -[s^2Y'(s) + 2sY(s) - y(0)]$$
$$\mathcal{L}\left(\frac{dy}{dx}\right) = sY(s) - y(0)$$
$$\mathcal{L}(xy) = -\frac{d}{ds}[\mathcal{L}(y)] = -Y'(s)$$

Therefore, the transform of Eq. (11.7) is

$$(s^2 + 1)Y'(s) + sY(s) = 0$$

or

$$\frac{dY(s)}{Y(s)} = \frac{-s\,ds}{s^2 + 1}$$

Integration gives

$$\ln Y(s) = -\tfrac{1}{2}\ln(s^2 + 1) + \ln C$$
$$= \ln \frac{C}{\sqrt{s^2 + 1}}$$

399

Hence,

$$Y(s) = \frac{C}{\sqrt{s^2 + 1}} = \frac{C}{s}\left(1 + \frac{1}{s^2}\right)^{-\frac{1}{2}} \tag{11.8}$$

The preceding expression may be expanded by the binomial series. The general form for the binomial series is

$$(a + x)^n = a^n + na^{n-1}x + \frac{n(n-1)}{2!}a^{n-2}x^2$$
$$+ \frac{n(n-1)(n-2)}{3!}a^{n-3}x^3 + \cdots \qquad x^2 < a^2$$

Thus, for $a = 1$, $n = -\frac{1}{2}$, and $x = 1/s^2$, the binomial series expansion for $(1 + 1/s^2)^{-\frac{1}{2}}$ is

$$\left(1 + \frac{1}{s^2}\right)^{-\frac{1}{2}} = 1 - \frac{1}{2}\frac{1}{s^2} + \frac{(3)(1)}{2!2^2}\frac{1}{s^4} - \frac{(5)(3)(1)}{3!2^3}\frac{1}{s^6} + \cdots$$

Substitution of this result into Eq. (11.8) gives

$$Y(s) = C\left[\frac{1}{s} - \frac{1}{2}\frac{1}{s^3} + \frac{(3)(1)}{2!2^2}\frac{1}{s^5} - \frac{(5)(3)(1)}{3!2^3}\frac{1}{s^7} + \cdots\right]$$

Inverting yields for $y(x)$,

$$y(x) = C\left[1 - \frac{1}{2!2}x^2 + \frac{(3)(1)}{4!2!2^2}x^4 - \frac{(5)(3)(1)}{6!3!2^3}x^6 + \cdots\right] \tag{11.9}$$

Inspection of the coefficient of the last term shows that

$$\frac{-(5)(3)(1)}{(6)(5)(4)3!3!2^3} = \frac{-1}{(3!)^2}\frac{1}{2^6}$$

A similar inspection of the coefficient of each term reveals the general form

$$y(x) = C\sum_{k=0}^{\infty}\frac{(-1)^k}{(k!)^2}\left(\frac{x}{2}\right)^{2k} = CJ_0(x) \tag{11.10}$$

Because $J_0(x)$ satisfies Bessel's equation of order zero, the function $J_0(x)$ is called Bessel's function of order 0.

It is to be noted that

$$\mathcal{L}^{-1}\left(\frac{1}{\sqrt{s^2 + 1}}\right) = J_0(x) \tag{11.11}$$

Bessel's differential equation of order n

This differential equation is

$$x^2 \frac{d^2y}{dx^2} + x \frac{dy}{dx} + (x^2 - n^2)y = 0 \qquad n = 0, 1, 2, \ldots \quad (11.12)$$

For $n = 0$, this reduces to Bessel's differential equation of order 0 [i.e., Eq. (11.7)]. Bessel's equation of order n is simplified by making the following change of variable:

$$y(x) = x^{-n}z(x) \tag{11.13}$$

Differentiating gives

$$\frac{dy}{dx} = x^{-n} \frac{dz}{dx} - nx^{-n-1}z$$

$$\frac{d^2y}{dx^2} = x^{-n} \frac{d^2z}{dx^2} - 2nx^{-n-1} \frac{dz}{dx} + n(n+1)x^{-n-2}z$$

Thus, Eq. (11.12) becomes

$$x^{-n+2} \frac{d^2z}{dx^2} + (1 - 2n)x^{-n+1} \frac{dz}{dx}$$
$$+ [n(n+1) - n + x^2 - n^2]x^{-n}z = 0$$

Dividing through by x^{-n+1} gives

$$x \frac{d^2z}{dx^2} + (1 - 2n) \frac{dz}{dx} + xz = 0 \tag{11.14}$$

This is Bessel's differential equation of order n in terms of the variable z. The transform of each term is

$$\mathcal{L}\left(x \frac{d^2z}{dx^2} \right) = - \frac{d}{ds}[s^2Z(s) - sz(0) - z'(0)]$$
$$= -[s^2Z'(s) + 2sZ(s) - z(0)]$$

$$\mathcal{L}\left[(1 - 2n) \frac{dz}{dx} \right] = (1 - 2n)[sZ(s) - z(0)]$$

$$\mathcal{L}(xz) = - \frac{d}{ds}[Z(s)] = -Z'(s)$$

Collecting terms yields for the transformed equation

$$(s^2 + 1)Z'(s) = -(2n + 1)sZ(s) + 2nz(0)$$

Because $z(x) = x^ny(x)$, it follows that $z(0) = 0^ny(0) = 0$. Thus, the term $2nz(0)$ vanishes.

$$\frac{Z'(s)}{Z(s)} = \frac{-(2n + 1)s}{s^2 + 1}$$

Integrating both sides of the preceding expression with respect to s gives

$$\ln Z(s) = \frac{-(2n+1)}{2} \ln (s^2 + 1) + \ln C$$

$$= \frac{-(2n+1)}{2} \ln s^2 \frac{s^2 + 1}{s^2} + \ln C$$

$$= \ln \frac{C}{s^{2n+1}} \left(1 + \frac{1}{s^2}\right)^{-(2n+1)/2}$$

Taking the antilog yields

$$Z(s) = \frac{C}{s^{2n+1}} \left(1 + \frac{1}{s^2}\right)^{-(2n+1)/2} \tag{11.15}$$

The last term may be expanded by the binomial series

$$\left(1 + \frac{1}{s^2}\right)^{-(2n+1)/2} = 1 - \frac{2n+1}{2} \frac{1}{s^2} + \frac{(2n+3)(2n+1)}{2!2^2} \frac{1}{s^4}$$
$$- \frac{(2n+5)(2n+3)(2n+1)}{3!2^3} \frac{1}{s^6} + \cdots$$

The substitution of this result into Eq. (11.15) yields

$$Z(s) = C\left[\frac{1}{s^{2n+1}} - \frac{2n+1}{2} \frac{1}{s^{2n+3}} + \frac{(2n+3)(2n+1)}{2!2^2} \frac{1}{s^{2n+5}}\right.$$
$$\left. - \frac{(2n+5)(2n+3)(2n+1)}{3!2^3} \frac{1}{s^{2n+7}} + \cdots\right]$$

Inverting gives

$$z(x) = C\left[\frac{x^{2n}}{(2n)!} - \frac{(2n+1)x^{2n+2}}{2(2n+2)!} + \frac{(2n+3)(2n+1)x^{2n+4}}{2!2^2(2n+4)!}\right.$$
$$\left. - \frac{(2n+5)(2n+3)(2n+1)x^{2n+6}}{3!2^3(2n+6)!} + \cdots\right]$$

Inspection of the last coefficient shows that

$$\frac{-(2n+5)(2n+3)(2n+1)}{3!2^3(2n+6)!} = \frac{-1}{3!2^3(2n+6)(2n+4)(2n+2)(2n)!}$$
$$= \frac{-1}{3!2^3(n+3)(n+2)(n+1)(2n)!2^3}$$
$$= \frac{-1(n)!}{3!(n+3)!(2n)!2^6}$$

Thus, the general form is

$$z(x) = C \sum_{k=0}^{\infty} \frac{(-1)^k n!}{k!(n+k)!(2n)!} \frac{x^{2n+2k}}{2^{2k}}$$

Because $y(x) = x^{-n}z(x)$, then

$$y(x) = C \frac{n!}{(2n)!} \sum_{k=0}^{\infty} \frac{(-1)^k}{k!(n+k)!} \frac{x^{n+2k}}{2^{2k}}$$

$$= C \frac{2^n n!}{(2n)!} \sum_{k=0}^{\infty} \frac{(-1)^k}{k!(n+k)!} \left(\frac{x}{2}\right)^{n+2k}$$

In performing the preceding summation, n is regarded as a fixed integer. Thus, we may replace $C[2^n n!/(2n)!]$ by some other constant C. That is,

$$y(x) = C \sum_{k=0}^{\infty} \frac{(-1)^k}{k!(n+k)!} \left(\frac{x}{2}\right)^{n+2k} = CJ_n(x) \tag{11.16}$$

where

$$J_n(x) = \sum_{k=0}^{\infty} \frac{(-1)^k}{k!(n+k)!} \left(\frac{x}{2}\right)^{n+2k} \qquad n = 0, 1, 2, \ldots \tag{11.17}$$

The function $J_n(x)$ satisfies Bessel's equation of order n, and thus it is called Bessel's function of order n. For $n = 0$, then $J_n(x)$ reduces to $J_0(x)$.

For even values of n, Eq. (11.17) shows that only even powers of x appear in the series for $J_n(x)$. Thus, $J_n(x)$ is an even function when n is even. That is

$$J_n(x) = J_n(-x) \qquad n \text{ even}$$

Similarly, it follows that $J_n(x)$ is an odd function when n is odd. That is

$$J_n(x) = -J_n(-x) \qquad n \text{ odd}$$

Before proceeding to Bessel's differential equation of order ν, it will be helpful to discuss some general properties of Bessel functions. These functions in turn require a knowledge of the gamma function.

The gamma function

The gamma function may be regarded as a generalization of the factorial function. The factorial function has meaning for positive integers only, but the gamma function is applicable for nonintegers as well as integers.

403

The gamma function $\Gamma(\alpha)$ is defined by the integral

$$\Gamma(\alpha) = \int_0^\infty e^{-x} x^{\alpha-1}\, dx \qquad (\alpha > 0) \tag{11.18}$$

The substitution of $\alpha + 1$ for α gives $\Gamma(\alpha + 1)$; that is,

$$\Gamma(\alpha + 1) = \int_0^\infty e^{-x} x^{\alpha}\, dx$$

By letting $u = x^\alpha$ and $dv = e^{-x}\, dx$, then integration by parts yields

$$\Gamma(\alpha + 1) = -e^{-x} x^\alpha \Big|_0^\infty + \alpha \int_0^\infty e^{-x} x^{\alpha-1}\, dx$$

The first term on the right vanishes when evaluated at the upper and lower limits.

Because the integral in the second term is the gamma function $\Gamma(\alpha)$, then

$$\Gamma(\alpha + 1) = \alpha\Gamma(\alpha) \tag{11.19}$$

The value of the gamma function for $\alpha = 1$ is

$$\Gamma(1) = \int_0^\infty e^{-x}\, dx = 1$$

Successive application of Eq. (11.19) yields

$$\Gamma(2) = \Gamma(1 + 1) = 1\Gamma(1) = 1$$
$$\Gamma(3) = \Gamma(2 + 1) = 2\Gamma(2) = 2 \cdot 1 = 2!$$
$$\Gamma(4) = \Gamma(3 + 1) = 3\Gamma(3) = 3 \cdot 2! = 3!$$

Hence, in general

$$\Gamma(k + 1) = k\Gamma(k) = k! \qquad k = 0, 1, 2, 3, \ldots$$

The preceding result shows that, for positive integer values of α (that is, $\alpha = k + 1 = 1, 2, 3, \ldots$), the gamma function reduces to the factorial function. Table 11.1 gives corresponding values of

Table 11.1   Gamma function

α	$\Gamma(\alpha)$	α	$\Gamma(\alpha)$
1.0	1.0000	1.5	0.8862
1.1	0.9514	1.6	0.8935
1.2	0.9182	1.7	0.9086
1.3	0.8975	1.8	0.9314
1.4	0.8873	1.9	0.9618
1.5	0.8862	2.0	1.0000

$\Gamma(\alpha)$ versus α for $1 \leq \alpha \leq 2$. Values of the gamma function outside the range of this table may be computed by noting that

$$\Gamma(\alpha) = \frac{\Gamma(\alpha + 1)}{\alpha} = \frac{\Gamma(\alpha + 2)}{\alpha(\alpha + 1)} = \frac{\Gamma(\alpha + 3)}{\alpha(\alpha + 1)(\alpha + 2)}$$

$$= \cdots = \frac{\Gamma(\alpha + k + 1)}{\alpha(\alpha + 1)(\alpha + 2) \cdots (\alpha + k)}$$

Thus,

$$\Gamma(\alpha + k + 1) = \alpha(\alpha + 1)(\alpha + 2) \cdots (\alpha + k)\Gamma(\alpha)$$
$$k = 0, 1, 2, \ldots \qquad (11.20)$$

This result is used to determine the gamma function for values greater than 2.

Example Let it be desired to determine $\Gamma(3.6)$. Because $\alpha + k + 1 = 3.6$ and $1 \leq \alpha \leq 2$, then $\alpha = 1.6$ and $k = 1$. From Table 11.1, it follows that $\Gamma(1.6) = 0.8935$; hence

$$\Gamma(3.6) = (1.6)(2.6)(0.8935) = 3.717$$

Solving Eq. (11.20) for $\Gamma(\alpha)$ yields an expression for evaluating the gamma function for values less than 1. That is,

$$\Gamma(\alpha) = \frac{\Gamma(\alpha + k + 1)}{\alpha(\alpha + 1)(\alpha + 2) \cdots (\alpha + k)} \qquad (11.21)$$

where k is an integer such that $1 \leq \alpha + k + 1 \leq 2$.

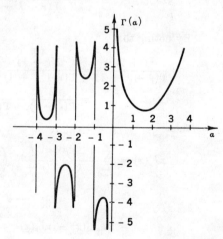

Fig. 11.1. The gamma function.

Example Let it be desired to evaluate $\Gamma(-1.6)$. Because $\alpha = -1.6$, then

$$1 \leq \alpha + k + 1 = -0.6 + k \leq 2$$

Thus, $k = 2$ and $\alpha + k + 1 = 1.4$. Application of Eq. (11.21) gives

$$\Gamma(-1.6) = \frac{\Gamma(1.4)}{(-1.6)(-0.6)(0.4)} = 2.31$$

A plot of $\Gamma(\alpha)$ versus α is shown in Fig. 11.1.

Properties of Bessel functions

Some very useful properties of Bessel functions are derived in this section. The Bessel function of order ν is

$$J_\nu(x) = \sum_{k=0}^\infty \frac{(-1)^k}{k!} \frac{1}{\Gamma(\nu + k + 1)} \left(\frac{x}{2}\right)^{\nu+2k} \qquad \nu \geq 0 \qquad (11.22)$$

where $\nu \geq 0$ may be a positive noninteger or an integer. When ν is the integer n, then $\Gamma(n + k + 1) = (n + k)!$ and Eq. (11.22) becomes identical to Eq. (11.17).

Differentiation of Eq. (11.22) with respect to x, and then multiplication by x, gives

$$xJ_\nu'(x) = \sum_{k=0}^\infty \frac{(-1)^k}{k!} \frac{\nu + 2k}{\Gamma(\nu + k + 1)} \left(\frac{x}{2}\right)^{\nu+2k}$$

By noting that

$$\frac{\nu + 2k}{\Gamma(\nu + k + 1)} = \frac{2(\nu + k)}{\Gamma(\nu + k + 1)} - \frac{\nu}{\Gamma(\nu + k + 1)}$$

$$= \frac{2}{\Gamma(\nu + k)} - \frac{\nu}{\Gamma(\nu + k + 1)}$$

then

$$xJ_\nu'(x) = \sum_{k=0}^\infty \frac{(-1)^k}{k!} \frac{2}{\Gamma(\nu + k)} \left(\frac{x}{2}\right)^{\nu+2k-1} \frac{x}{2}$$

$$- \nu \sum_{k=0}^\infty \frac{(-1)^k}{k!} \frac{1}{\Gamma(\nu + k + 1)} \left(\frac{x}{2}\right)^{\nu+2k}$$

$$= xJ_{\nu-1}(x) - \nu J_\nu(x) \qquad (11.23)$$

The preceding result can be expressed in an alternate form by first noting that

$$\frac{(-1)^k}{k!}\frac{\nu + 2k}{\Gamma(\nu + k + 1)} = \frac{(-1)^k}{k!}\frac{\nu}{\Gamma(\nu + k + 1)}$$
$$+ \frac{(-1)^k}{(k-1)!}\frac{2}{\Gamma(\nu + k + 1)}$$

Thus

$$xJ'_\nu(x) = \nu \sum_{k=0}^{\infty} \frac{(-1)^k}{k!}\frac{1}{\Gamma(\nu + k + 1)}\left(\frac{x}{2}\right)^{\nu+2k}$$
$$- x \sum_{k=1}^{\infty} \frac{(-1)^{k-1}}{(k-1)!}\frac{1}{\Gamma(\nu + k + 1)}\left(\frac{x}{2}\right)^{\nu+2k-1}$$

The lower summation in the last term is $k = 1$ rather than $k = 0$, because, for $k = 0$, then $(k - 1)! = \Gamma(k) = \Gamma(0) = \infty$. Hence, the resulting term vanishes for $k = 0$. The substitution of $k + 1$ for k in this latter summation gives

$$xJ'_\nu(x) = \nu J_\nu(x) - x \sum_{k=0}^{\infty} \frac{(-1)^k}{k!}\frac{1}{\Gamma(\nu + k + 2)}\left(\frac{x}{2}\right)^{\nu+2k+1}$$
$$= \nu J_\nu(x) - xJ_{\nu+1}(x) \tag{11.24}$$

Subtracting Eq. (11.23) from Eq. (11.24) yields the recurrence formula,

$$J_{\nu+1}(x) = \frac{2\nu}{x}J_\nu(x) - J_{\nu-1}(x) \tag{11.25}$$

This recurrence formula enables one to compute $J_{\nu+1}(x)$ when $J_\nu(x)$ and $J_{\nu-1}(x)$ are known. Successive values $J_{\nu+2}(x)$, $J_{\nu+3}(x)$, etc., may then be determined. Additional recurrence formulas may be obtained as follows:

Adding Eqs. (11.23) and (11.24) gives

$$J'_\nu(x) = \tfrac{1}{2}[J_{\nu-1}(x) - J_{\nu+1}(x)] \tag{11.26}$$

Replacing ν by n in Eq. (11.23), and then multiplying by x^{n-1}, gives

$$x^n J'_n(x) + nx^{n-1}J_n(x) = x^n J_{n-1}(x)$$

or

$$\frac{d}{dx}[x^n J_n(x)] = x^n J_{n-1}(x) \tag{11.27}$$

In a similar manner, it may be shown that

$$\frac{d}{dx}[x^{-n}J_n(x)] = -x^{-n}J_{n+1}(x) \tag{11.28}$$

There are numerous other recurrence relationships.

Bessel's equation of order ν

Bessel's differential equation of order ν is

$$x^2\frac{d^2y}{dx^2} + x\frac{dy}{dx} + (x^2 - \nu^2)y = 0 \tag{11.29}$$

Let us now investigate to determine whether or not $J_\nu(x)$ satisfies Bessel's equation of order ν. If $y = CJ_\nu(x)$ satisfies Eq. (11.29), then

$$x^2J_\nu''(x) + xJ_\nu'(x) + (x^2 - \nu^2)J_\nu(x) = 0 \tag{11.30}$$

Differentiation of Eq. (11.26) gives

$$J_\nu'' = \tfrac{1}{2}(J_{\nu-1}' - J_{\nu+1}')$$
$$= \tfrac{1}{2}[\tfrac{1}{2}(J_{\nu-2} - J_\nu) - \tfrac{1}{2}(J_\nu - J_{\nu+2})]$$

Application of Eq. (11.25) to eliminate the $J_{\nu-2}$ and $J_{\nu+2}$ terms gives

$$J_\nu'' = \frac{1}{2}\left\{\frac{1}{2}\left[\frac{2(\nu-1)}{x}J_{\nu-1} - 2J_\nu\right] + \frac{1}{2}\left[\frac{2(\nu+1)}{x}J_{\nu+1} - 2J_\nu\right]\right\}$$
$$= \frac{1}{2}\left[\frac{\nu}{x}(J_{\nu-1} + J_{\nu+1}) - 2J_\nu - \frac{1}{x}(J_{\nu-1} - J_{\nu+1})\right]$$
$$= \frac{1}{2}\left(\frac{2\nu^2}{x^2}J_\nu - 2J_\nu - \frac{2}{x}J_\nu'\right)$$

Multiplying through by x^2 verifies Eq. (11.30). Thus, $CJ_\nu(x)$ satisfies Bessel's differential equation of order ν.

Negative values of ν are represented by $-\nu$. The substitution of $-\nu$ for ν in the preceding recurrence relationships yields the results for the case in which $-\nu < 0$. In particular, Bessel's function of order $-\nu$ is

$$J_{-\nu}(x) = \sum_{k=0}^{\infty}\frac{(-1)^k}{k!\Gamma(-\nu + k + 1)}\left(\frac{x}{2}\right)^{-\nu+2k} \tag{11.31}$$

As $-\nu$ approaches a negative integer $-n$, then Eq. (11.31) becomes

$$J_{-n}(x) = \sum_{k=n}^{\infty}\frac{(-1)^k}{k!(-n + k)!}\left(\frac{x}{2}\right)^{-n+2k}$$

The summation begins at $k = n$ because the factorial function $(-n + k)!$ is infinite for negative integers (see Fig. 11.1). The substitution of $k = n + m$ in the preceding equation gives

$$J_{-n}(x) = (-1)^n \sum_{m=0}^{\infty} \frac{(-1)^m}{m!(n + m)!} \left(\frac{x}{2}\right)^{n+2m} \tag{11.32}$$

Comparison of Eqs. (11.17) and (11.32) shows that

$$J_{-n}(x) = (-1)^n J_n(x)$$

Hence, for integer values the Bessel functions $J_n(x)$ and $J_{-n}(x)$ are linearly dependent.

Comparison of Eqs. (11.22) and (11.31) shows that such is not the case for noninteger values. In particular, $J_{-\nu}(x)$ is seen to contain some terms with negative powers of x so that $J_{-\nu}(x)$ must become infinite as x approaches zero. The function $J_\nu(x)$ approaches zero or some finite value as x approaches zero. Hence, $J_{-\nu}(x)$ and $J_\nu(x)$ are linearly independent for noninteger values.

Because Bessel's differential equation is a second-order differential equation, it has two independent solutions. For noninteger values of ν, these solutions are the linearly independent functions $J_\nu(x)$ and $J_{-\nu}(x)$. Thus, the complete solution is

$$y(x) = C_1 J_\nu(x) + C_2 J_{-\nu}(x) \qquad \nu \text{ not an integer} \tag{11.33}$$

When ν is an integer, it is necessary to use a second function which is independent of $J_\nu(x)$. The function $Y_\nu(x)$ is defined as the following linear combination:

$$Y_\nu(x) = \frac{\cos \nu\pi J_\nu(x) - J_{-\nu}(x)}{\sin \nu\pi} \tag{11.34}$$

For any value of ν the cosine and sine terms in the preceding expression are constant. For the case in which ν is an integer n, then

$$\lim_{\nu \to n} Y_\nu(x) = Y_n(x) = \frac{(-1)^n J_n(x) - J_{-n}(x)}{0} = \frac{0}{0}$$

The limit of the preceding function $Y_n(x)$ does exist and furthermore is independent of $J_n(x)$. Thus, $Y_n(x)$ provides the second independent solution for the case in which $\nu = n$ is an integer. That is,

$$y(x) = C_1 J_n(x) + C_2 Y_n(x) \tag{11.35}$$

The function $J_\nu(x)$ is called a *Bessel function of the first kind*, and the function $Y_\nu(x)$ is called a *Bessel function of the second kind*. In both cases the order is ν.

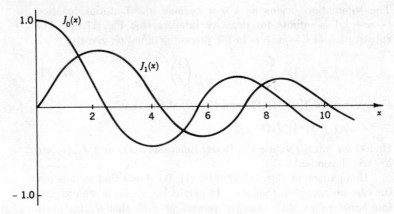

Fig. 11.2. Bessel functions of the first kind $J_0(x)$ and $J_1(x)$.

The functions $J_0(x)$ and $J_1(x)$ are shown in Fig. 11.2. Similarly, the functions $Y_0(x)$ and $Y_1(x)$ are shown in Fig. 11.3.

Illustrative example Determine the solution of the following differential equation:

$$x \frac{d^2 y}{dx^2} + \frac{dy}{dx} + xy = 0 \qquad y(0) = 2$$

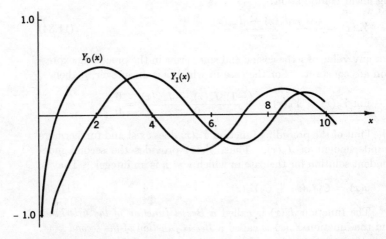

Fig. 11.3. Bessel functions of the second kind $Y_0(x)$ and $Y_1(x)$.

Solution For this differential equation $n = 0$. Thus, the general solution is given by Eq. (11.35), in which n is zero. To satisfy the initial condition $y(0) = 2$, it is necessary that C_2 vanish because $Y_0(0) = -\infty$. Thus

$$y(x) = C_1 J_0(x) = 2J_0(x)$$

The fact that $y(0) = 2$ and $J_0(0) = 1$ shows that C_1 must be 2.

11.3 EQUATIONS REDUCIBLE TO BESSEL'S EQUATION

The behavior of engineering phenomena is often described by differential equations which are equivalent, or alternate, forms of Bessel's differential equation. One such very common form is

$$x^2 \frac{d^2y}{dx^2} + x \frac{dy}{dx} + (\lambda^2 x^2 - \nu^2)y = 0 \tag{11.36}$$

This differential equation may be transformed back to Bessel's differential equation of order ν by making the change of variable $z = \lambda x$. Because $dz/dx = \lambda$, then

$$\frac{dy}{dx} = \frac{dy}{dz}\frac{dz}{dx} = \lambda \frac{dy}{dz}$$

Similarly,

$$\frac{d^2y}{dx^2} = \lambda^2 \frac{d^2y}{dz^2}$$

Thus, Eq. (11.36) becomes

$$z^2 \frac{d^2y}{dz^2} + z \frac{dy}{dz} + (z^2 - \nu^2)y = 0$$

This is the familiar form of Bessel's differential equation. Hence, the solution is

$$\begin{aligned} y &= C_1 J_\nu(z) + C_2 J_{-\nu}(z) \\ &= C_1 J_\nu(\lambda x) + C_2 J_{-\nu}(\lambda x) \qquad \nu \text{ not an integer} \end{aligned} \tag{11.37}$$

When $\nu = n$ is an integer, the solution is

$$y = C_1 J_n(\lambda x) + C_2 Y_n(\lambda x) \tag{11.38}$$

An important differential equation which frequently arises in engineering practice is that known as the *modified Bessel equation of order* ν. That is,

$$x^2 \frac{d^2y}{dx^2} + x \frac{dy}{dx} - (x^2 + \nu^2)y = 0 \tag{11.39}$$

This equation is the same as Eq. (11.36), in which $\lambda = j$ so that $\lambda^2 = -1$. Thus, the solution is obtained by replacing λ by j in Eq. (11.37). Hence,

$$y = C_1 J_\nu(jx) + C_2 J_{-\nu}(jx) \qquad \nu \text{ not an integer}$$

The substitution of jx for x in Eq. (11.22) shows that $J_\nu(jx)$ is

$$J_\nu(jx) = \sum_{k=0}^{\infty} \frac{(-1)^k}{k!\Gamma(\nu + k + 1)} \left(\frac{jx}{2}\right)^{\nu+2k}$$

$$= j^\nu \sum_{k=0}^{\infty} \frac{1}{k!\Gamma(\nu + k + 1)} \left(\frac{x}{2}\right)^{\nu+2k}$$

Multiplying $J_\nu(jx)$ by the constant $j^{-\nu}$ gives

$$I_\nu(x) = j^{-\nu} J_\nu(jx) = \sum_{k=0}^{\infty} \frac{1}{k!\Gamma(\nu + k + 1)} \left(\frac{x}{2}\right)^{\nu+2k}$$

The function $I_\nu(x)$ is a real function and is similar to $J_\nu(x)$ except that the terms in $I_\nu(x)$ are all positive and do not alternate in sign. Replacing ν by $-\nu$ throughout gives

$$I_{-\nu}(x) = j^\nu J_{-\nu}(jx) = \sum_{k=0}^{\infty} \frac{1}{k!\Gamma(k + 1 - \nu)} \left(\frac{x}{2}\right)^{2k-\nu}$$

Because $I_\nu(x)$ is merely a constant times $J_\nu(jx)$, and similarly $I_{-\nu}(x)$ is a constant times $J_{-\nu}(jx)$, then the solution of Eq. (11.39) may be written in the form

$$y(x) = C_1 I_\nu(x) + C_2 I_{-\nu}(x) \qquad \nu \text{ not an integer} \tag{11.40}$$

The functions $I_\nu(x)$ and $I_{-\nu}(x)$ are linearly independent for noninteger values of ν. However, when ν is an integer, then

$$J_n(jx) = (-1)^n J_{-n}(jx)$$
$$J_n(jx) = (j^2)^n J_{-n}(jx)$$
$$j^{-n} J_n(jx) = j^n J_{-n}(jx)$$
$$I_n(x) = I_{-n}(x)$$

For integer values $\nu = n$, it is necessary to employ the function $K_\nu(x)$, which is defined by the linear combination

$$K_\nu(x) = \frac{\pi}{2} \frac{I_{-\nu}(x) - I_\nu(x)}{\sin \nu\pi}$$

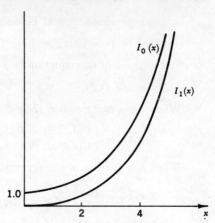

Fig. 11.4.　Modified Bessel functions of the first kind $I_0(x)$ and $I_1(x)$.

The limit of this function exists for integer as well as for noninteger values. Thus, $K_n(x)$ provides the second independent solution for the case in which $\nu = n$ is an integer. That is,

$$y(x) = C_1 I_n(x) + C_2 K_n(x) \tag{11.41}$$

The functions $I_0(x)$ and $I_1(x)$ are plotted in Fig. 11.4. Similarly, the functions $K_0(x)$ and $K_1(x)$ are shown in Fig. 11.5. The function $I_\nu(x)$ is called the *modified Bessel function of the first kind*, and the function $K_\nu(x)$ is called the *modified Bessel function of the second kind*. The order of each is ν.

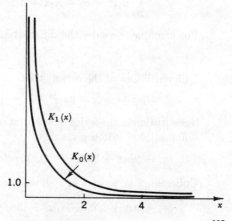

Fig. 11.5.　Modified Bessel functions of the second kind $K_0(x)$ and $K_1(x)$.

A generalized form of Bessel's differential equation is

$$x^2 y'' + (1 - 2a)xy' + [(b\lambda x^b)^2 + (a^2 - b^2\nu^2)]y = 0 \qquad (11.42)$$

The solution of this equation is

$$y = x^a Z_\nu(\lambda x^b) \qquad (11.43)$$

When λ is a real number, then $Z_\nu(\lambda x^b)$ is

$$Z_\nu(\lambda x^b) = \begin{cases} C_1 J_\nu(\lambda x^b) + C_2 J_{-\nu}(\lambda x^b) & \nu \text{ not an integer} \\ C_1 J_n(\lambda x^b) + C_2 Y_n(\lambda x^b) & \nu = n \text{ is an integer} \end{cases}$$

When $\lambda = j\xi$ is an imaginary number, then $Z_\nu(\lambda x^b)$ is

$$Z_\nu(\lambda x^b) = \begin{cases} C_1 I_\nu(\xi x^b) + C_2 I_{-\nu}(\xi x^b) & \nu \text{ not an integer} \\ C_1 I_n(\xi x^b) + C_2 K_n(\xi x^b) & \nu = n \text{ is an integer} \end{cases}$$

In using Laplace transformations to solve differential equations which have series solutions, the transform $Y(s)$ is usually obtained in a closed form [e.g., Eq. (11.8)]. The series form of the solution is obtained by expanding $Y(s)$ in a series and then inverting each term. In the following section, it is shown how such differential equations are solved by initially assuming the solution to be a series.

11.4 SERIES SOLUTION OF DIFFERENTIAL EQUATIONS

In this method, it is assumed that the solution is in the form of a power series. That is,

$$y = \sum_{k=0}^{\infty} a_k x^k = a_0 + a_1 x + a_2 x^2 + a_3 x^3 + \cdots \qquad (11.44)$$

For example, consider the differential equation

$$y' + y = 0$$

Differentiation of the series gives

$$y' = a_1 + 2a_2 x + 3a_3 x^2 + \cdots$$

Substitution of the series representations for y and y' into the original differential equation gives

$$(a_1 + 2a_2 x + 3a_3 x^2 + \cdots) + (a_0 + a_1 x + a_2 x^2 + \cdots) = 0$$

Collecting like powers of x shows that

$$(a_1 + a_0) + (2a_2 + a_1)x + (3a_3 + a_2)x^2 + \cdots = 0$$

To satisfy this equation, each coefficient must vanish. Thus

$$a_1 = -a_0$$
$$a_2 = -\tfrac{1}{2}a_1 = \tfrac{1}{2}a_0$$
$$a_3 = -\tfrac{1}{3}a_2 = -\frac{1}{(3)(2)}\, a_0$$

Substitution of these results into the original power series for y [Eq. (11.44)] gives

$$y = a_0\left(1 - x + \frac{1}{2!}x^2 - \frac{1}{3!}x^3 + \cdots\right) = a_0 e^{-x}$$

A major difficulty of the series method is the problem of determining beforehand whether or not a differential equation possesses a series solution. The following theorems have been developed for second-order differential equations.

Theorem I

If $f_1(x)$ and $f_2(x)$ can be expanded in a Taylor series about the point $x = x_0$, then each solution of the differential equation

$$y'' + f_1(x)y' + f_2(x)y = 0 \tag{11.45}$$

can be expressed as a Taylor series of the form

$$y = \sum_{k=0}^{\infty} a_k(x - x_0)^k = a_0 + a_1(x - x_0) + a_2(x - x_0)^2 + \cdots$$

For the usual case in which the series is taken about the origin ($x_0 = 0$), each solution has the form

$$y = \sum_{k=0}^{\infty} a_k x^k = a_0 + a_1 x + a_2 x^2 + \cdots \tag{11.46}$$

Theorem II

If $f_1(x)$ and $f_2(x)$ can be expanded in a Taylor series about the point $x = x_0$, then at least one solution of the differential equation

$$y'' + \frac{f_1(x)}{x - x_0}\, y' + \frac{f_2(x)}{(x - x_0)^2}\, y = 0 \tag{11.47}$$

possesses an expansion of the form

$$y = (x - x_0)^r \sum_{k=0}^{\infty} a_k (x - x_0)^k$$

where the exponent r is some constant.

When the expansion is about the point $x_0 = 0$, the solution has the form

$$y = x^r \sum_{k=0}^{\infty} a_k x^k = x^r (a_0 + a_1 x + a_2 x^2 + \cdots) \tag{11.48}$$

The general second-order differential equations given by Eqs. (11.45) and (11.47) have extensive applications to engineering situations.

To illustrate Theorem I, let it be desired to determine the solution of Legendre's differential equation. That is,

$$(1 - x^2)y'' - 2xy' + n(n + 1)y = 0 \tag{11.49}$$

Dividing through by $1 - x^2$ shows that this has the form of Eq. (11.45), in which $f_1(x) = -2x/(1 - x^2)$ and $f_2(x) = n(n + 1)/(1 - x^2)$.

Because of the singular points at $x = \pm 1$, the Taylor series expansion about $x = 0$ has a radius of convergence of 1. Thus, the solution converges for the interval $-1 < x < 1$. Using Eq. (11.46) to express each term in Legendre's equation as a series gives

$$n(n + 1)y = n(n + 1)(a_0 + a_1 x + a_2 x^2 + \cdots)$$
$$= n(n + 1) \sum_{k=0}^{\infty} a_k x^k$$
$$-2xy' = -2x(a_1 + 2a_2 x + 3a_3 x^2 + \cdots) = -2x \sum_{k=0}^{\infty} k a_k x^{k-1}$$
$$(1 - x^2)y'' = (1 - x^2)[2a_2 + (3)(2)a_3 x + (4)(3)a_4 x^2 + \cdots]$$
$$= (1 - x^2) \sum_{k=0}^{\infty} k(k - 1) a_k x^{k-2}$$

Adding each series and collecting coefficients of like powers of x gives

$$[n(n + 1)a_0 + 2a_2] + \{[n(n + 1) - 2]a_1 + (3)(2)a_3\}x$$
$$+ \{[n(n + 1) - 6]a_2 + (4)(3)a_4\}x^2 + \cdots = 0$$

Equating each coefficient to zero yields the following relationships:

$$a_0 = a_0$$

$$a_2 = -\frac{n(n + 1)}{2!} a_0$$

$$a_4 = -\frac{(n - 2)(n + 3)}{(4)(3)} a_2 = \frac{(n - 2)n(n + 1)(n + 3)}{4!} a_0$$

$$a_1 = a_1$$

$$a_3 = -\frac{(n - 1)(n + 2)}{3!} a_1$$

$$a_5 = -\frac{(n - 3)(n + 4)}{(5)(4)} a_3 = \frac{(n - 3)(n - 1)(n + 2)(n + 4)}{5!} a_1$$

Thus, the solution of Legendre's equation is

$$y = a_0 y_1(x) + a_1 y_2(x) \tag{11.50}$$

where

$$y_1(x) = 1 - \frac{n(n + 1)}{2!} x^2 + \frac{(n - 2)n(n + 1)(n + 3)}{4!} x^4 - \cdots$$

$$y_2(x) = x - \frac{(n - 1)(n + 2)}{3!} x^3$$
$$+ \frac{(n - 3)(n - 1)(n + 2)(n + 4)}{5!} x^5 - \cdots$$

For many engineering applications the parameter n is a positive integer. When n is an even integer, the series for $y_1(x)$ contains but the first $n/2 + 1$ terms. The remaining terms vanish because their coefficients are zero. For example, for $n = 2$ the series is $1 - 3x^2$.

For even values of n, the series $y_2(x)$ diverges at $x = \pm 1$. Thus, for the closed interval $-1 \le x \le 1$, Legendre's equation is satisfied by the first $(n/2) + 1$ terms of $y_1(x)$.

When n is an odd integer, the series for $y_2(x)$ contains only the first $(n + 1)/2$ terms. For odd values of n, the series $y_1(x)$ diverges at $x = \pm 1$. Thus, for $-1 \le x \le 1$, Legendre's equation is satisfied by the first $(n + 1)/2$ terms of $y_2(x)$.

The $y_1(x)$ polynomials that result for even values of n, and the $y_2(x)$ polynomials that result for odd values of n may be expressed in the same form. For even values of n, this form is obtained by multiplying $y_1(x)$ by

$$(-1)^{n/2} \frac{(1)(3)(5) \cdots (n - 1)}{(2)(4)(6) \cdots n} \qquad n \text{ even}$$

For odd values of n, this form is obtained by multiplying $y_2(x)$ by

$$(-1)^{(n-1)/2} \frac{(1)(3)(5) \cdots n}{(2)(4)(6) \cdots (n-1)} \qquad n \text{ odd}$$

The resulting polynomial obtained for each value of n is designated $P_n(x)$ and is called *Legendre's polynomial*. That is,

$$P_n(x) = \sum_{k=0}^{N} \frac{(-1)^k (2n - 2k)!}{2^n k! (n - k)! (n - 2k)!} x^{n-2k} \tag{11.51}$$

where $N = n/2$ when n is even and $N = (n - 1)/2$ when n is odd. Some specific values of $P_n(x)$ are

$$P_0(x) = 1 \qquad\qquad P_1(x) = x$$

$$P_2(x) = \frac{3x^2 - 1}{2} \qquad\qquad P_3(x) = \frac{5x^3 - 3x}{2} \tag{11.52}$$

$$P_4(x) = \frac{35x^4 - 30x^2 + 3}{8} \qquad P_5(x) = \frac{63x^5 - 70x^3 + 15x}{8}$$

It is to be noted that for each even value of n, then $P_n(x)$ is a constant times the first $(n/2) + 1$ terms of $y_1(x)$. Similarly, for each odd value of n, then $P_n(x)$ is a constant times the first $(n + 1)/2$ terms of $y_2(x)$. These constant multipliers are such that for $x = 1$, then $P_n(x) = 1$. Because $P_n(x)$ is merely a constant times the

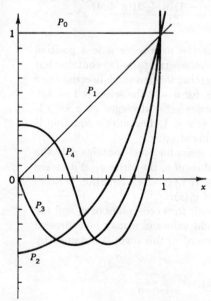

Fig. 11.6. Legendre polynomials.

corresponding $y_1(x)$ or $y_2(x)$ polynomial, then Legendre's equation is satisfied over the closed interval $-1 \leq x \leq 1$ by the Legendre polynomial appropriate to the particular value of n. In Fig. 11.6 is shown a plot of $P_n(x)$ over the interval $0 \leq x \leq 1$ for $n = 0, 1, 2, 3$, and 4. For even values, $P_n(x)$ is an even function. For odd values, $P_n(x)$ is an odd function.

To illustrate the application of Theorem II, consider Bessel's differential equation of order ν. That is,

$$x^2 y'' + xy' + (x^2 - \nu^2)y = 0$$

Dividing through by x^2 yields the same form as Eq. (11.47), in which $f_1(x) = 1$ and $f_2(x) = x^2 - \nu^2$. Because $f_1(x)$ and $f_2(x)$ have no singular points, the radius of convergence of each Taylor series is infinite. Thus, the series solution is valid for all values of x. Using Eq. (11.48) to express each term in the preceding equation as a series gives

$$(x^2 - \nu^2)y = (x^2 - \nu^2)(a_0 x^r + a_1 x^{r+1} + a_2 x^{r+2} + \cdots)$$
$$= (x^2 - \nu^2) \sum_{k=0}^{\infty} a_k x^{r+k}$$
$$xy' = x[ra_0 x^{r-1} + (r+1)a_1 x^r + (r+2)a_2 x^{r+1} + \cdots]$$
$$= x \sum_{k=0}^{\infty} (r+k)a_k x^{r+k-1}$$
$$x^2 y'' = x^2[r(r-1)a_0 x^{r-2} + (r+1)ra_1 x^{r-1}$$
$$+ (r+2)(r+1)a_2 x^r + \cdots]$$
$$= x^2 \sum_{k=0}^{\infty} (r+k)(r+k-1)a_k x^{r+k-2}$$

Adding these series and collecting coefficients of like powers of x gives

$$(r^2 - \nu^2)a_0 x^r + [(r+1)^2 - \nu^2]a_1 x^{r+1}$$
$$+ \{[(r+2)^2 - \nu^2]a_2 + a_0\}x^{r+2}$$
$$+ \cdots + \{[(r+k)^2 - \nu^2]a_k + a_{k-2}\}x^{r+k} + \cdots = 0$$

Because each of the preceding coefficients must be zero, then

$$(r^2 - \nu^2)a_0 = 0$$
$$[(r+1)^2 - \nu^2]a_1 = 0$$
$$[(r+2)^2 - \nu^2]a_2 + a_0 = 0 \qquad\qquad (11.53)$$
$$\cdots\cdots\cdots\cdots\cdots$$
$$[(r+k)^2 - \nu^2]a_k + a_{k-2} = 0$$

The coefficient a_0 of the first of the above relationships cannot be zero, for otherwise the series [Eq. (11.48)] would not begin with the term $a_0 x^r$ as assumed. Thus, it is necessary that

$$(r^2 - \nu^2) = (r - \nu)(r + \nu) = 0 \tag{11.54}$$

This equation, which is obtained from the coefficient of the lowest power of x, is called the *indicial equation*. The indicial equation yields two values of r. Substitution of each value of r into the coefficient relationships yields two separate sets of coefficients, which in turn result in two series which satisfy the differential equation. For $r = \nu$, the coefficient relationships given by Eq. (11.53) become

$$a_1 = 0$$

$$a_2 = \frac{-a_0}{(\nu + 2)^2 - \nu^2} = \frac{-a_0}{4(\nu + 1)} = \frac{-a_0}{2^2 1!(\nu + 1)}$$

$$a_3 = \frac{-a_1}{(\nu + 3)^2 - \nu^2} = 0$$

$$a_4 = \frac{-a_2}{(\nu + 4)^2 - \nu^2} = \frac{-a_2}{8(\nu + 2)} = \frac{a_0}{2^4 2!(\nu + 2)(\nu + 1)}$$

It is to be noted that all the odd coefficients vanish (that is, $a_1 = a_3 = a_5 = \cdots = 0$). By letting $a_0 = C_1/[2^\nu \Gamma(\nu + 1)]$, the preceding relationships become

$$a_2 = \frac{-C_1}{2^{\nu+2} 1!(\nu + 1)\Gamma(\nu + 1)} = \frac{-C_1}{2^{\nu+2} 1!\Gamma(\nu + 2)}$$

$$a_4 = \frac{C_1}{2^{\nu+4} 2!(\nu + 2)(\nu + 1)\Gamma(\nu + 1)} = \frac{C_1}{2^{\nu+4} 2!\Gamma(\nu + 3)}$$

and in general

$$a_{2k} = \frac{(-1)^k C_1}{2^{\nu+2k} k!\Gamma(\nu + k + 1)}$$

Substitution of these coefficients into Eq. (11.48) yields for the series solution corresponding to $r = \nu$

$$y_1(x) = C_1 \sum_{k=0}^{\infty} \frac{(-1)^k}{k!} \frac{1}{\Gamma(\nu + k + 1)} \left(\frac{x}{2}\right)^{\nu+2k} = C_1 J_\nu(x)$$

The second series which satisfies the differential equation is obtained from the second root of the indicial equation, $r = -\nu$. Thus, by letting $r = -\nu$ in the coefficient relationships, it is found that the second series is $y_2 = C_2 J_{-\nu}(x)$.

For a second-order differential equation, the indicial equation always yields two roots r_1 and r_2. Let r_1 denote the larger of the two roots and r_2 the smaller. A series solution y_1 is always obtained for the larger root r_1. However, the second series solution corresponding to r_2 is linearly independent of y_1 only if the difference $r_1 - r_2$ is not an integer, or if $r_1 \neq r_2$. For the case in which $r_1 - r_2$ is an integer, or $r_1 = r_2$, a different procedure must be employed to obtain the second independent solution. Note in the preceding example, if $\nu = n$ is an integer, then $r_1 - r_2 = n - (-n) = 2n$ is an integer. Hence, the second series $J_{-n}(x)$ is not independent of the first $J_n(x)$.

To obtain a second independent solution when $r_1 - r_2$ is an integer or when $r_1 = r_2$, it is necessary first to obtain the general form of the indicial equation. To do this, write $f_1(x)$ and $f_2(x)$ in their Taylor series forms.

$$f_1(x) = b_0 + b_1 x + b_2 x^2 + \cdots$$
$$f_2(x) = c_0 + c_1 x + c_2 x^2 + \cdots$$

Thus, the second-order differential equation becomes

$$x^2 y'' + x f_1(x) y' + f_2(x) y = x^2 y'' + x(b_0 + \cdots) y'$$
$$+ (c_0 + \cdots) y = 0 \quad (11.55)$$

Substitution of the series representations for y, y', and y'' into the preceding gives

$$x^2 [r(r-1) a_0 x^{r-2} + \cdots] + x(b_0 + \cdots)(r a_0 x^{r-1} + \cdots)$$
$$+ (c_0 + \cdots)(a_0 x^r + \cdots) = 0$$

The indicial equation which is the coefficient of the lowest-order x term (in this case the coefficient of x^r) is

$$[r(r-1) + b_0 r + c_0] a_0 = 0$$

Because a_0 cannot be zero, then

$$r^2 + (b_0 - 1) r + c_0 = 0$$

The roots of this general indicial equation are

$$r_1 = \frac{(1 - b_0) + \sqrt{(b_0 - 1)^2 - 4c_0}}{2}$$

$$r_2 = \frac{(1 - b_0) - \sqrt{(b_0 - 1)^2 - 4c_0}}{2} \quad (11.56)$$

For two equal roots, then $r = r_1 = r_2 = (1 - b_0)/2$. If the difference is equal to an integer m, then $r_1 - r_2 = \sqrt{(b_0 - 1)^2 - 4c_0} = m$. When the two solutions are linearly dependent, $y_2 = Cy_1$. The method of variation of parameters provides a general procedure for obtaining a second independent solution of the form

$$y_2 = u(x) y_1 \quad (11.57)$$

421

The first step in this method is to obtain successive derivatives of y_2. That is,

$$y_2' = u'y_1 + uy_1' \qquad y_2'' = u''y_1 + 2u'y_1' + uy_1''$$

Substitution of these values into Eq. (11.55) gives

$$x^2(u''y_1 + 2u'y_1' + uy_1'') + xf_1(x)(u'y_1 + uy_1') + f_2(x)uy_1 = 0$$

Because y_1 satisfies the original differential equation, then subtracting $u[x^2y_1'' + xf_1(x)y_1' + f_2(x)y_1]$ from the preceding result shows that

$$x^2y_1u'' + 2x^2y_1'u' + xf_1(x)y_1u' = 0$$

Dividing through by x^2y_1 and replacing $f_1(x)$ by its Taylor series gives

$$u'' + \left(2\frac{y_1'}{y_1} + \frac{b_0}{x} + b_1 + \cdots\right)u' = 0$$

From the series representation for y_1, it follows that y_1'/y_1 may be written in the form

$$\frac{y_1'}{y_1} = \frac{ra_0x^{r-1} + (r+1)a_1x^{r-2} + \cdots}{a_0x^r + a_1x^{r+1} + \cdots}$$
$$= \frac{x^{r-1}[ra_0 + (r+1)a_1x + \cdots]}{x^r(a_0 + a_1x + \cdots)} = \frac{1}{x}\left(r + \frac{a_1x}{a_0} + \cdots\right)$$

Substitution of this result into the above expression gives

$$u'' + \left(\frac{b_0 + 2r}{x} + b_1 + \frac{2a_1}{a_0} + \cdots\right)u' = 0$$

or

$$u'' + \left(\frac{b_0 + 2r}{x} + k_0 + k_1x + k_2x^2 + \cdots\right)u' = 0 \qquad (11.58)$$

where $k_0 = b_1 + 2a_1/a_0$. The r in Eq. (11.58) is r_1 corresponding to the series expansion for y_1. From the expressions for the roots of the indicial equation [Eq. (11.56)], it follows that $r_1 + r_2 = 1 - b_0$ and $r_1 - r_2 = m$, where $m = \sqrt{(b_0 - 1)^2 - 4c_0}$. Adding these relationships gives $b_0 + 2r_1 = m + 1$. Substitution of this last result into Eq. (11.58) gives

$$u'' + \left(\frac{m+1}{x} + k_0 + k_1x + \cdots\right)u' = 0$$

Integration shows that

$$\ln u' = -(m+1)\ln x - k_0x - k_1x^2/2 - \cdots$$

Thus,

$$u' = x^{-(m+1)}e^{-(k_0x+k_1x^2/2+\cdots)}$$

The preceding exponential may be expanded in a power series of the form

$$e^{-(k_0x+k_1x^2/2+\cdots)} = 1 + d_1x + d_2x^2 + \cdots$$

Thus u' becomes

$$u' = \frac{1}{x^{m+1}} + \frac{d_1}{x^m} + \cdots + \frac{d_m}{x} + d_{m+1} + \cdots$$

Integration gives

$$u = -\frac{1}{mx^m} - \cdots + d_m \ln x + d_{m+1}x + \cdots$$

Because $r_1 = r_2 + m$, the series solution y_1 may be written in the form

$$y_1 = x^{r_1}(a_0 + a_1x + \cdots) = x^{r_2+m}(a_0 + a_1x + \cdots)$$

The product uy_1 gives for the second independent solution

$$y_2 = d_my_1(x) \ln x + x^{r_2} \sum_{k=0}^{\infty} A_kx^k \tag{11.59}$$

The constant d_m and the A_k coefficients are evaluated by substitution of this general solution into the original differential equation.

When the roots are equal ($r_1 = r_2 = r$ and $m = 0$), then from the preceding analysis $u' = 1/x + \cdots$ and $u = \ln x + \cdots$. Thus the general form of the second independent solution is found to be

$$y_2(x) = y_1(x) \ln x + x^r \sum_{k=1}^{\infty} A_kx^k \tag{11.60}$$

To illustrate the application of this method, let it be desired to determine the second independent solution of Bessel's differential equation of order 0. For this case, the function $y_1(x)$ is known to be $J_0(x)$. Successive differentiation of Eq. (11.60), followed by the substitution of y_2, y_2', and y_2'' into Bessel's differential equation gives

$$\sum_{k=1}^{\infty} \frac{(-1)^kx^{2k-1}}{2^{2k-2}k!(k-1)!} + \sum_{k=1}^{\infty} k^2A_kx^{k-1} + \sum_{k=1}^{\infty} A_kx^{k+1} = 0$$

Expressing this result in terms of like powers of x gives

$$A_1 + (4A_2 - 1)x + (9A_3 + A_1)x^2 + (16A_4 + A_2 + \tfrac{1}{8})x^3$$
$$+ \cdots = 0$$

For the equality to hold, each coefficient must vanish. Thus, $A_1 = 0$, $A_3 = -A_1/9 = 0$, and similarly $A_5 = A_7 = \cdots = 0$. The values of the even coefficients are $A_2 = \frac{1}{4}$, $A_4 = -\frac{3}{128}$, and in general

$$A_{2k} = \frac{(-1)^{k-1}}{2^{2k}(k!)^2} h_k$$

where $h_0 = 0$, and for $k = 1, 2, 3, \ldots$, then

$$h_k = \frac{1}{1} + \frac{1}{2} + \frac{1}{3} + \cdots + \frac{1}{k}$$

Thus, the second independent solution is

$$y_2(x) = J_0(x) \ln x + \sum_{k=1}^{\infty} \frac{(-1)^{k-1} h_k}{(k!)^2} \left(\frac{x}{2}\right)^{2k} \tag{11.61}$$

By proceeding in a similar manner, a second independent solution $y_2(x)$ may be obtained for Bessel's differential equation of order n. This result is

$$y_2(x) = J_n(x) \ln x + \frac{1}{2} \sum_{k=0}^{\infty} \frac{(-1)^{k-1}(h_k + h_{k+n})}{k!(k+n)!} \left(\frac{x}{2}\right)^{2k+n}$$

$$- \frac{1}{2} \sum_{k=0}^{n-1} \frac{(n-k-1)!}{k!} \left(\frac{x}{2}\right)^{2k-n} \qquad n = 0, 1, 2, \ldots \tag{11.62}$$

For the case in which $n = 0$, the last summation vanishes.

It is desirable to have this second independent solution in a form which is applicable for integer as well as noninteger values of ν. Such a function is $Y_\nu(x)$ given by Eq. (11.34). That is,

$$Y_\nu(x) = \frac{1}{\sin \nu\pi} [J_\nu(x) \cos \nu\pi - J_{-\nu}(x)]$$

$$\lim_{\nu \to n} Y_\nu(x) = Y_n(x) \qquad n = 0, 1, 2, \ldots$$

The function $y_2(x)$ is related to $Y_n(x)$ by the linear combination $Y_n(x) = (2/\pi)[y_2(x) + (\gamma - \ln 2)J_n(x)]$. Thus,

$$Y_n(x) = \frac{2}{\pi} J_n(x) \left(\ln \frac{x}{2} + \gamma\right)$$

$$+ \frac{1}{\pi} \sum_{k=0}^{\infty} \frac{(-1)^{k-1}(h_k + h_{k+n})}{k!(k+n)!} \left(\frac{x}{2}\right)^{2k+n}$$

$$- \frac{1}{\pi} \sum_{k=0}^{n-1} \frac{(n-k-1)!}{k!} \left(\frac{x}{2}\right)^{2k-n} \qquad n = 0, 1, 2, \ldots \tag{11.63}$$

For $n = 0$, the last summation vanishes. The constant $\gamma = 0.5772$ is the Euler constant, which is defined by the limit

$$\gamma = \lim_{m \to \infty} 1 + \frac{1}{2} + \frac{1}{3} + \cdots + \frac{1}{m} - \ln m$$

For integer values, the function $Y_\nu(x)$ given by Eq. (11.34) becomes $Y_n(x)$ as given by Eq. (11.63). Because $Y_\nu(x)$ is independent of $J_\nu(x)$ for integer as well as noninteger values of ν, then

$$y(x) = C_1 J_\nu(x) + C_2 Y_\nu(x) \tag{11.64}$$

is a general solution of Bessel's equation for all values of ν.

11.5 CYLINDRICAL COORDINATES

In Fig. 11.7 is shown a point P in space. The relationships for transforming from rectangular to cylindrical coordinates are

$$r = \sqrt{x^2 + y^2}$$

$$\theta = \tan^{-1}\frac{y}{x}$$

The Laplacian operator is

$$\nabla^2 u = \frac{\partial^2 u}{\partial x^2} + \frac{\partial^2 u}{\partial y^2} + \frac{\partial^2 u}{\partial z^2} \tag{11.65}$$

This operator is transformed from rectangular to cylindrical coordinates as follows: First, apply the chain rule for partial differentiation to obtain $\partial u/\partial x$. That is,

$$\frac{\partial u}{\partial x} = \frac{\partial u}{\partial r}\frac{\partial r}{\partial x} + \frac{\partial u}{\partial \theta}\frac{\partial \theta}{\partial x} = u_r r_x + u_\theta \theta_x$$

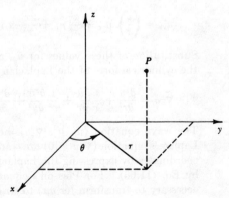

Fig. 11.7. Cylindrical coordinates.

where $u_r = \partial u/\partial r$, $r_x = \partial r/\partial x$, $u_\theta = \partial u/\partial \theta$, and $\theta_x = \partial\theta/\partial x$. Taking the partial derivative of the preceding expression with respect to x gives

$$\frac{\partial^2 u}{\partial x^2} = \frac{\partial}{\partial x}(u_r r_x + u_\theta \theta_x)$$
$$= (u_r)r_{xx} + (u_r)_x r_x + (u_\theta)\theta_{xx} + (u_\theta)_x \theta_x$$

The terms $(u_r)_x$ and $(u_\theta)_x$ may be further expanded,

$$(u_r)_x = u_{rr}r_x + u_{r\theta}\theta_x$$
$$(u_\theta)_x = u_{\theta r}r_x + u_{\theta\theta}\theta_x$$

Thus, the equation for u_{xx} becomes

$$u_{xx} = (r_x)^2 u_{rr} + (2r_x\theta_x)u_{r\theta} + (\theta_x)^2 u_{\theta\theta} + r_{xx}u_r + \theta_{xx}u_\theta$$

The terms r_x, r_{xx}, θ_x, and θ_{xx} are evaluated as follows:

$$r_x = \frac{1}{2}\frac{2x}{\sqrt{x^2+y^2}} = \frac{x}{r}$$
$$r_{xx} = \frac{r - xr_x}{r^2} = \frac{r - x(x/r)}{r^2} = \frac{y^2}{r^3}$$
$$\theta_x = \frac{1}{1+(y/x)^2}\left(\frac{1}{x}\frac{\partial y}{\partial x} - \frac{y}{x^2}\right) = -\frac{y}{r^2}$$
$$\theta_{xx} = -\frac{r^2\,\partial y/\partial x - 2ryr_x}{r^4} = \frac{2yr_x}{r^3} = \frac{2xy}{r^4}$$

Because x and y are independent variables, then $\partial y/\partial x = 0$. Substitution of the preceding results into the equation for u_{xx} gives

$$u_{xx} = \left(\frac{x}{r}\right)^2 u_{rr} + \frac{y^2}{r^3}u_r - \frac{2xy}{r^3}u_{r\theta} + \frac{2xy}{r^4}u_\theta + \left(\frac{-y}{r^2}\right)^2 u_{\theta\theta}$$

In a similar manner, $\partial^2 u/\partial y^2$ is found to be

$$u_{yy} = \left(\frac{y}{r}\right)^2 u_{rr} + \frac{x^2}{r^3}u_r + \frac{2xy}{r^3}u_{r\theta} - \frac{2xy}{r^4}u_\theta + \left(\frac{x}{r^2}\right)^2 u_{\theta\theta}$$

Substitution of these values for u_{xx} and u_{yy} into Eq. (11.65) yields the cylindrical form of the Laplacian ($\nabla^2 u$),

$$\nabla^2 u = \frac{\partial^2 u}{\partial r^2} + \frac{1}{r}\frac{\partial u}{\partial r} + \frac{1}{r^2}\frac{\partial^2 u}{\partial\theta^2} + \frac{\partial^2 u}{\partial z^2} \qquad (11.66)$$

The wave equation ($u_{tt} = c^2\nabla^2 u$), heat equation ($u_t = \alpha\nabla^2 u$), and Laplace's equation ($\nabla^2 u = 0$) are transformed directly to cylindrical coordinates by expressing the Laplacian operator in the form given by Eq. (11.66). For Poisson's equation [$\nabla^2 u = f(x,y,z)$], it is also necessary to transform $f(x,y,z)$ to $f(r,\theta,z)$. The role of Bessel functions in solving partial differential equations in cylindrical coordinates is illustrated in the following example.

Illustrative example In Fig. 11.8 is shown a cylinder whose outside radius is a. The top and bottom of this cylinder are insulated so that no heat flows in the z direction. The radial distance from the center of the cylinder to some internal point is designated r. Because of symmetry, the temperature $u(r,t)$ is a function of the radius r and time t. Determine the temperature $u(r,t)$ when the initial temperature distribution is $f(r)$ and the surface temperature is maintained at zero. That is,

$$u(r,0) = f(r)$$
$$u(a,t) = 0$$

Solution The partial differential equation is

$$\frac{\partial u}{\partial t} = \alpha \left(\frac{\partial^2 u}{\partial r^2} + \frac{1}{r} \frac{\partial u}{\partial r} \right) \tag{11.67}$$

To apply the method of separation of variables, it is assumed that the solution is the product of a function of the radius

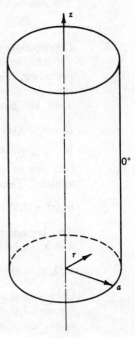

Fig. 11.8. Cylinder.

[that is, $R(r)$] and a function of time [that is, $T(t)$]. Thus

$$u(r,t) = R(r)T(t)$$

Differentiation shows that

$$\frac{\partial u}{\partial r} = R'T \qquad \frac{\partial^2 u}{\partial r^2} = R''T \qquad \frac{\partial u}{\partial t} = RT'$$

Substituting these results into Eq. (11.67) and dividing through by RT gives

$$\frac{T'}{T} = \frac{\alpha[R'' + (1/r)R']}{R} = k \tag{11.68}$$

Because T'/T is independent of R and similarly the next term is independent of T, each of these terms must equal a constant (k). Equation (11.68) may be written as two separate differential equations.

$$T' - kT = 0 \tag{11.69}$$
$$rR'' + R' - \frac{k}{\alpha}rR = 0 \tag{11.70}$$

The solution of Eq. (11.69) is

$$T = Ce^{kt}$$

It is necessary that k be negative (that is, $k < 0$), for otherwise the solution becomes infinite as time t becomes infinite. The preceding Bessel differential equation [Eq. (11.70)] has the same form as Eq. (11.36), in which ν is zero and $\lambda^2 = -k/\alpha$. Thus, the general solution of Eq. (11.70) is

$$R = C_1 J_0(\lambda r) + C_2 Y_0(\lambda r)$$

For $r = 0$, then $Y_0(\lambda r) = Y_0(0) = -\infty$. Because the temperature at $r = 0$ must remain finite, the constant C_2 must vanish. Thus, the product RT is

$$u(r,t) = RT = CC_1 J_0(\lambda r)e^{-\alpha\lambda^2 t}$$

The boundary condition $u(a,t) = 0$ is satisfied for values of $\lambda = \lambda_n$ such that

$$J_0(\lambda_n a) = 0 \qquad n = 1, 2, 3, \ldots \tag{11.71}$$

Only values of n for which λ_n is positive are used. This is because $J_0(\lambda r) = J_0(-\lambda r)$, and thus negative values yield the same result as positive values. The total response is the summation due to each value of λ_n; that is,

$$u(r,t) = \sum_{n=1}^{\infty} C_n J_0(\lambda_n r) e^{-\alpha \lambda_n^2 t} \qquad (11.72)$$

where $C_n = (CC_1)_n$.

The coefficient C_n is evaluated from the remaining boundary condition that $u(r,0) = f(r)$. Thus, letting $t = 0$ in the preceding equation gives

$$f(r) = \sum_{n=1}^{\infty} C_n J_0(\lambda_n r) \qquad (11.73)$$

This is a Fourier-Bessel series, which has properties similar to that of a Fourier series. The more general form for this series is

$$f(r) = \sum_{n=1}^{\infty} C_n J_\nu(\lambda_n r)$$

To evaluate the coefficient C_n, multiply each term by $r J_\nu(\lambda_m r)$, where m is a specific integer, and then integrate from 0 to a. When λ_n is such that $J_\nu(\lambda_n a) = 0$, then

$$\int_0^a r J_\nu(\lambda_m r) J_\nu(\lambda_n r) \, dr = \begin{cases} 0 & n \neq m \\ \dfrac{a^2}{2} [J_\nu'(\lambda_n a)]^2 & n = m \end{cases} \qquad (11.74)$$

Thus, the value of the coefficient C_n is

$$C_n = \frac{2}{a^2 [J_\nu'(\lambda_n a)]^2} \int_0^a r f(r) J_\nu(\lambda_n r) \, dr \qquad (11.75)$$

When $\nu = 0$, Eq. (11.26) shows that $J_0' = (J_{-1} - J_1)/2$. Because $J_{-1} = -J_1$, then $J_0' = -J_1$. For the particular case of Eq. (11.73), in which $\nu = 0$, the coefficient C_n is

$$C_n = \frac{2}{a^2 [J_1(\lambda_n a)]^2} \int_0^a r f(r) J_0(\lambda_n r) \, dr \qquad (11.76)$$

The solution is thus given by Eq. (11.72), in which the coefficient C_n is known from the preceding expression.

Transform solutions

Transformations are well suited to the solution of partial differential equations. In general, the transform of a partial differential equation is a simpler equation, which may be solved and then inverted to obtain the desired solution.

To illustrate this method, let it be desired to determine the temperature distribution $u(r,t)$ for a cylinder of radius a and infinite length. The boundary conditions are

$$u(r,0) = 0$$

$$u(a,t) = \begin{cases} 0 & t < 0 \\ U_0 & t > 0 \end{cases}$$

The first boundary condition shows that the initial temperature is zero throughout, and the second shows that the surface temperature is U_0 for $t > 0$. Taking the Laplace transform of Eq. (11.67) gives

$$sU(r,s) - u(r,0) = \alpha \left[\frac{\partial^2 U(r,s)}{\partial r^2} + \frac{1}{r} \frac{\partial U(r,s)}{\partial r} \right]$$

Inserting the initial condition and rearranging,

$$\frac{\partial^2 U(r,s)}{\partial r^2} + \frac{1}{r} \frac{\partial U(r,s)}{\partial r} - \frac{s}{\alpha} U(r,s) = 0 \tag{11.77}$$

This Bessel equation has the same form as Eq. (11.36), in which $\lambda^2 = -s/\alpha$ and $\nu = 0$. The corresponding solution is given by Eq. (11.38). Because the temperature is not infinite at $r = 0$, the term $Y_0(\lambda r)$, which is infinite at $r = 0$, is not applicable for this solution. Thus

$$U(r,s) = CJ_0(\lambda r) = CJ_0\left(jr \sqrt{\frac{s}{\alpha}} \right)$$

where $\lambda = j\sqrt{s/\alpha}$.

The constant C is evaluated from the boundary condition $u(a,t) = U_0$. The transform of this boundary condition is $U(a,s) = U_0/s$. For $r = a$, the preceding equation becomes

$$U(a,s) = CJ_0(\lambda a) = \frac{U_0}{s}$$

whence $C = U_0/[sJ_0(\lambda a)]$. Substitution of this value of C into the equation for $U(r,s)$ gives

$$U(r,s) = \frac{U_0 J_0(\lambda r)}{s J_0(\lambda a)} \tag{11.78}$$

The residue of $U(r,s)e^{st}$ due to the pole at $s = 0$ is

$$\lim_{s \to 0} \frac{U_0 J_0(jr\sqrt{s/\alpha})e^{st}}{J_0(ja\sqrt{s/\alpha})} = \frac{U_0 J_0(0)}{J_0(0)} = U_0$$

By designating λ_n as the value of λ such that $J_0(\lambda_n a) = 0$, then the corresponding value of s is $s_n = -\alpha\lambda_n{}^2$. The residue of $U(r,s)e^{st}$ due to the pole at $s = s_n$ is

$$\lim_{s \to s_n} \frac{U_0}{s} \frac{J_0(jr\sqrt{s/\alpha})e^{st}}{(d/ds)J_0(ja\sqrt{s/\alpha})} = \frac{U_0}{s_n} \frac{J_0(jr\sqrt{s_n/\alpha})e^{s_nt}}{ja/(2\sqrt{\alpha s_n})J_0'(ja\sqrt{s_n/\alpha})}$$

$$= -\frac{2U_0}{a\lambda_n} \frac{J_0(-\lambda_n r)}{J_0'(-\lambda_n a)} e^{-\alpha\lambda_n{}^2 t}$$

The inverse transform of Eq. (11.78) is the summation of all the residues. Because $J_0(-x) = J_0(x)$ and $J_0'(-x) = -J_1(-x) = J_1(x)$, then this result is

$$u(r,t) = U_0\left[1 - \frac{2}{a}\sum_n \frac{J_0(\lambda_n r)}{\lambda_n J_1(\lambda_n a)} e^{-\alpha\lambda_n{}^2 t}\right] \qquad (11.79)$$

This summation is over all the positive values of λ_n. Because $J_0(\lambda_n a) = J_0(-\lambda_{-n}a) = 0$, then $\lambda_n = -\lambda_{-n}$. Thus, positive or negative values of λ_n yield the same $s_n = -\alpha\lambda_n{}^2 = -\alpha(-\lambda_{-n})^2$. Therefore, it suffices to consider positive values of λ_n only.

As another example, let it be desired to determine the steady-state temperature distribution $u(r,z)$ in a cylinder of length L, as shown in Fig. 11.9. The temperature on the top and bottom is maintained at $0°$, and that around the circular sides is U_0.

$$u(r,0) = u(r,L) = 0$$
$$u(a,z) = U_0$$

When u is a function of r and z, the partial differential equation is

$$\frac{\partial^2 u(r,z)}{\partial r^2} + \frac{1}{r}\frac{\partial u(r,z)}{\partial r} + \frac{\partial^2 u(r,z)}{\partial z^2} = 0$$

Taking the sine transform with respect to the vertical coordinate z gives

$$\frac{\partial^2}{\partial r^2} U(r,n) + \frac{1}{r}\frac{\partial}{\partial r} U(r,n) - \left(\frac{\pi n}{L}\right)^2 U(r,n) = 0 \qquad (11.80)$$

This Bessel equation is the same as Eq. (11.36), in which $\lambda^2 = -(\pi n/L)^2$. Because the temperature at $r = 0$ must be finite, there can be no $Y_0(\lambda r)$ term in the solution. Thus

$$U(r,n) = CJ_0(\lambda r) = CJ_0\left(\frac{j\pi n r}{L}\right) = CI_0\left(\frac{\pi n r}{L}\right)$$

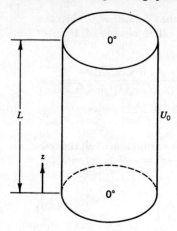

Fig. 11.9. Cylinder.

The sine transform of the boundary condition $u(a,z) = U_0$ is

$$U(a,n) = \frac{U_0 L}{\pi n} [1 - (-1)^n]$$

The constant C may now be evaluated. Thus, the transform $U(r,n)$ becomes

$$U(r,n) = \frac{U_0 L[1 - (-1)^n]}{\pi n I_0(\pi n a/L)} I_0(\pi n r/L)$$

Application of Eq. (8.3) to obtain the inverse gives

$$u(r,z) = \frac{2U_0}{\pi} \sum_{n=1}^{\infty} \frac{1 - (-1)^n}{n I_0(\pi n a/L)} I_0(\pi n r/L) \sin \frac{\pi n z}{L} \qquad (11.81)$$

When u is a function of r and θ, the partial differential equation is

$$r^2 \frac{\partial^2 u}{\partial r^2} + r \frac{\partial u}{\partial r} + \frac{\partial^2 u}{\partial \theta^2} = 0$$

Taking the sine transform with respect to θ reduces this partial differential equation to Euler's equation. The desired solution is obtained by solving this Euler equation and then inverting. Euler's equation is discussed in Sec. 11.7.

The preceding partial differential equations are functions of two variables. For equations which are functions of three variables, it is necessary to transform twice, and for four variables, it is necessary to transform three times, etc.

11.6 FINITE HANKEL TRANSFORMATION

This transformation is particularly well suited to the solution of partial differential equations in cylindrical coordinates.

The finite Hankel transform of a function $f(x)$ is defined by the integral equation

$$H_\nu[f(x)] = \int_0^a xf(x)J_\nu(sx)\, dx = F_\nu(s) \tag{11.82}$$

where $J_\nu(sx)$ is obtained by substituting sx for x in the Bessel function $J_\nu(x)$. After performing the integration in Eq. (11.82) and evaluating x at the upper and lower limits, the resulting function is some function of s [that is, $F_\nu(s)$]. This transformation is valid when $f(x)$ satisfies the Dirichlet conditions over the interval $0 \leq x \leq a$.

The corresponding finite Hankel transform with respect to r of the function $u(r,t)$ is

$$H_\nu[u(r,t)] = \int_0^a ru(r,t)J_\nu(sr)\, dr = F_\nu(s,t) \tag{11.83}$$

The parameter t is regarded as a constant in this transformation.

The finite Hankel transform is particularly well suited for transforming terms of the form

$$\frac{\partial^2 u(r,t)}{\partial r^2} + \frac{1}{r}\frac{\partial u(r,t)}{\partial r}$$

It may be shown (Ref. [6]) that

$$H_\nu\left(\frac{\partial^2 u}{\partial r^2} + \frac{1}{r}\frac{\partial u}{\partial r} - \frac{\nu^2}{r^2}\right)$$
$$= au'(a,t)J_\nu(sa) - sau(a,t)J_\nu'(sa) - s^2H_\nu[u(r,t)]$$

An important result follows for the case in which $\nu = 0$; that is,

$$H_0\left(\frac{\partial^2 u}{\partial r^2} + \frac{1}{r}\frac{\partial u}{\partial r}\right)$$
$$= au'(a,t)J_0(sa) - sau(a,t)J_0'(sa) - s^2H_0[u(r,t)] \tag{11.84}$$

The inversion formula for the finite Hankel transform depends on the value of s. Thus, it is necessary to distinguish two cases.

Case 1

For this case, $s = \lambda_n$ is such that

$$J_\nu(\lambda_n a) = 0 \tag{11.85}$$

When $s = \lambda_n$ is a root of this equation, the inversion formula is

$$u(r,t) = \frac{2}{a^2} \sum_n \frac{J_\nu(\lambda_n r)}{[J_\nu'(\lambda_n a)]^2} H_\nu[u(r,t)]$$

The summation is taken over all the positive roots λ_n of Eq. (11.85). For $\nu = 0$, then $J_0' = -J_1$. Thus the inversion formula becomes

$$u(r,t) = \frac{2}{a^2} \sum_n \frac{J_0(\lambda_n r)}{[J_1(\lambda_n a)]^2} H_0[u(r,t)] \tag{11.86}$$

Case 2

For this case, $s = \lambda_n$ is such that

$$J_\nu(\lambda_n a) + \frac{\lambda_n J_\nu'(\lambda_n a)}{c} = 0 \tag{11.87}$$

where c is a known constant. In the second illustrative example that follows, it is shown that for thermal problems, $c = h$ is the coefficient of surface heat transfer. From the theory of Bessel functions, the inversion formula for case 2 is found to be

$$u(r,t) = \frac{2}{a^2} \sum_n \frac{\lambda_n^2}{c^2 + (\lambda_n^2 - \nu^2/a^2)} \frac{J_\nu(\lambda_n r)}{[J_\nu(\lambda_n a)]^2} H_\nu[u(r,t)]$$

The summation is over all the positive roots λ_n of Eq. (11.87). For $\nu = 0$, the inversion formula becomes

$$u(r,t) = \frac{2}{a^2} \sum_n \frac{\lambda_n^2}{c^2 + \lambda_n^2} \frac{J_0(\lambda_n r)}{[J_0(\lambda_n a)]^2} H_0[u(r,t)] \tag{11.88}$$

When one of the given boundary conditions is $u(a,t)$, then case 1 ordinarily affords the most direct solution. The reason for this is that the term $au'(a,t)J_0(sa)$ in Eq. (11.84) vanishes because $J_0(sa)]_{s=\lambda_n} = 0$. When the boundary conditions are such as to give information about $u'(a,t)$, then case 2 usually provides the most direct solution.

In the limit as c becomes infinite, the last term in Eq. (11.87) vanishes. Thus, case 2 reduces to case 1 as c becomes infinite.

Illustrative example Use the Hankel transformation method to determine the temperature distribution in the cylinder of Fig. 11.8 when the boundary conditions are

$$u(r,0) = f(r)$$
$$u(a,t) = 0$$

This is the same problem as was solved by the method of separation of variables in the preceding section.

Solution   Multiplying each term in Eq. (11.67) by $rJ_0(\lambda_n r)$, where λ_n satisfies Eq. (11.85), and then integrating from 0 to a gives

$$\frac{\partial}{\partial t} H_0[u(r,t)] = \alpha H_0\left(\frac{\partial^2 u}{\partial r^2} + \frac{1}{r}\frac{\partial u}{\partial r}\right) \tag{11.89}$$

The right-hand side of this expression is evaluated by letting $s = \lambda_n$ in Eq. (11.84). Because $J_0(\lambda_n a) = u(a,t) = 0$, the first two terms on the right-hand side of Eq. (11.84) vanish. Thus

$$\left(\frac{\partial}{\partial t} + \alpha\lambda_n{}^2\right) H_0[u(r,t)] = 0$$

The solution of this first-order partial differential equation is

$$H_0[u(r,t)] = Ce^{-\alpha\lambda_n{}^2 t}$$

The constant C is evaluated from the boundary condition that $u(r,0) = f(r)$. Thus, evaluating the preceding equation at $t = 0$ gives

$$C = H_0[u(r,0)] = \int_0^a rf(r)J_0(\lambda_n r)\,dr$$

Substitution of this result for $H_0[u(r,t)]$ into the inversion formula [Eq. (11.86) yields the solution directly. That is,

$$u(r,t) = \frac{2}{a^2}\sum_n \frac{J_0(\lambda_n r)e^{-\alpha\lambda_n{}^2 t}}{[J_1(\lambda_n a)]^2} \int_0^a rf(r)J_0(\lambda_n r)\,dr$$

This verifies the solution obtained by the method of separation of variables [i.e., Eqs. (11.72) and (11.76)].

When the transfer of heat from the outer surface to the surrounding medium is primarily by convection, the temperature of the outer surface is different from that of the surrounding medium. The rate of heat flow from the surface to the surrounding medium for this case is

$$\frac{\partial u}{\partial r} = -hu$$

where h is the coefficient of surface heat transfer. As h becomes large, the temperature of the surface approaches that of the surrounding medium. For many problems h is sufficiently large so that the temperature of the outer surface may be regarded as being the same as that of the surrounding medium. When the effects of h must be considered, the technique is as described in the following.

Illustrative example Determine the solution of the preceding example when h is such that the temperature of the outer surface is significantly different from that of the surrounding medium. Thus, the boundary conditions are

$$\frac{\partial u(a,t)}{\partial r} = -hu(a,t)$$

$$u(r,0) = f(r)$$

Solution Transforming Eq. (11.67) yields the same result as that given by Eq. (11.89). The right-hand side of Eq. (11.89) is evaluated from Eq. (11.84) by using the information that $u'(a,t) = -hu(a,t)$ and that λ_n satisfies Eq. (11.87). Thus

$$\frac{\partial}{\partial t} H_0[u(r,t)] = -\alpha\lambda_n{}^2 H_0[u(r,t)]$$

This is the same first-order partial differential equation as was obtained in the preceding example and hence has the same solution.

$$H_0[u(r,t)] = Ce^{-\alpha\lambda_n{}^2 t}$$

Because the boundary condition $u(r,0) = f(r)$ is the same as in the preceding example, the value of C is also the same. Because λ_n satisfies Eq. (11.87), it is necessary to use the inversion formula given by Eq. (11.88). Thus, the solution is

$$u(r,t) = \frac{2}{a^2} \sum_n \frac{\lambda_n{}^2}{h^2 + \lambda_n{}^2} \frac{J_0(\lambda_n r)e^{-\alpha\lambda_n{}^2 t}}{[J_0(\lambda_n a)]^2} \int_0^a rf(r)J_0(\lambda_n r)\, dr$$

The preceding finite Hankel transforms apply to a function $u(r,t)$ for the interval $0 \le r \le a$. When the interval is $a \le r \le b$ (where a is the inside radius and b the outside radius), the finite Hankel transform is

$$H_\nu[u(r,t)] = \int_a^b ru(r,t)[J_\nu(sr)Y_\nu(sb) - J_\nu(sb)Y_\nu(sr)]\, dr$$

where J_ν is the Bessel function of the first kind and Y_ν is the Bessel function of the second kind.

It may be shown that

$$H_\nu\left(\frac{\partial^2 u}{\partial r^2} + \frac{1}{r}\frac{\partial u}{\partial r} - \frac{\nu^2}{r^2}\right) = \frac{J_\nu(\lambda_n b)}{J_\nu(\lambda_n a)}\, u(a,t) - u(b,t) - \lambda_n{}^2 H_\nu[u(r,t)]$$

For $\nu = 0$,

$$H_0\left(\frac{\partial^2 u}{\partial r^2} + \frac{1}{r}\frac{\partial u}{\partial r}\right) = \frac{J_0(\lambda_n b)}{J_0(\lambda_n a)}\, u(a,t) - u(b,t) - \lambda_n{}^2 H_0[u(r,t)]$$

(11.90)

When $s = \lambda_n$ is such that

$$J_\nu(\lambda_n a) Y_\nu(\lambda_n b) - J_\nu(\lambda_n b) Y_\nu(\lambda_n a) = 0 \tag{11.91}$$

then the inversion formula is

$$u(r,t) = \sum_n \frac{2\lambda_n{}^2 J_\nu{}^2(\lambda_n a)}{J_\nu{}^2(\lambda_n b) - J_\nu{}^2(\lambda_n a)}$$
$$[J_\nu(\lambda_n r) Y_\nu(\lambda_n b) - J_\nu(\lambda_n b) Y_\nu(\lambda_n r)] H_\nu[u(r,t)]$$

When $\nu = 0$, the inversion formula becomes

$$u(r,t) = \sum_n \frac{2\lambda_n J_0{}^2(\lambda_n a)}{J_0{}^2(\lambda_n b) - J_0{}^2(\lambda_n a)}$$
$$[J_0(\lambda_n r) Y_0(\lambda_n b) - J_0(\lambda_n b) Y_0(\lambda_n r)] H_0[u(r,t)] \quad (11.92)$$

Illustrative example   In Fig. 11.10 is shown a hollow cylinder of inside radius a and outside radius b.   Determine the tem-

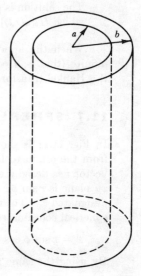

Fig. 11.10.   Hollow cylinder.

perature distribution $u(r,t)$ for the case in which the boundary conditions are

$$u(a,t) = u(b,t) = 0$$
$$u(r,0) = f(r)$$

Solution   Multiplying each term in Eq. (11.67) by

$$rJ_0(\lambda_n r) Y_0(\lambda_n b) - J_0(\lambda_n b) Y_0(\lambda_n r)$$

and integrating from a to b yields

$$\frac{\partial}{\partial t} H_0[u(r,t)] = \alpha H_0 \left(\frac{\partial^2 u}{\partial r^2} + \frac{1}{r} \frac{\partial u}{\partial r} \right)$$

The right-hand side of this expression is evaluated from Eq. (11.90). Because $u(a,t) = u(b,t) = 0$, then

$$\left(\frac{\partial}{\partial t} + \alpha \lambda_n{}^2 \right) H_0[u(r,t)] = 0$$

The solution of this first-order partial differential equation is

$$H_0[u(r,t)] = C e^{-\alpha \lambda_n{}^2 t}$$

The constant C is evaluated by letting $t = 0$ in the preceding expression and noting that $u(r,0) = f(r)$. Thus,

$$C = H_0[u(r,0)] = \int_a^b rf(r)[J_0(\lambda_n r) Y_0(\lambda_n b) - J_0(\lambda_n b) Y_0(\lambda_n r)] \, dr$$

The solution is given by the inversion formula [Eq. (11.92)], in which $H_0[u(r,t)]$ is known from the preceding expression.

When the upper limit of integration in Eq. (11.82) becomes infinite (that is, $a = \infty$), then the finite Hankel transform becomes the Hankel transform.

11.7 SPHERICAL COORDINATES

In Fig. 11.11 is shown a point P in space. The vector r extends from the origin to the point P. The angle from the z axis to the vector r is designated ϕ. The length of the projection of r on the xy plane is $r \sin \phi$. The angle from the x axis to this projection is designated θ. The rectangular coordinates (x,y,z) are related to the spherical coordinates (r,ϕ,θ) by the relationships

$$x = r \sin \phi \cos \theta \qquad y = r \sin \phi \sin \theta \qquad z = r \cos \phi$$

By using the same method described in Sec. 11.5 for transforming

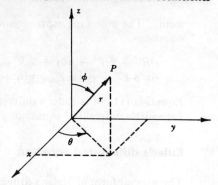

Fig. 11.11. Spherical coordinates.

the Laplacian $\nabla^2 u$ from rectangular to cylindrical coordinates, it is found that in spherical coordinates the Laplacian is

$$\nabla^2 u = \frac{\partial^2 u}{\partial r^2} + \frac{2}{r}\frac{\partial u}{\partial r} + \frac{1}{r^2}\frac{\partial^2 u}{\partial \phi^2} + \frac{\cos \phi}{r^2 \sin \phi}\frac{\partial u}{\partial \phi} + \frac{1}{r^2 \sin^2 \phi}\frac{\partial^2 u}{\partial \theta^2}$$

(11.93)

Often there is circular symmetry such that u is a function of r and ϕ only. For circular symmetry, the last term $(\partial^2 u/\partial \theta^2)$ vanishes. For this case, Laplace's equation is

$$\nabla^2 u = \frac{\partial^2 u}{\partial r^2} + \frac{2}{r}\frac{\partial u}{\partial r} + \frac{1}{r^2}\frac{\partial^2 u}{\partial \phi^2} + \frac{\cos \phi}{r^2 \sin \phi}\frac{\partial u}{\partial \phi} = 0 \qquad (11.94)$$

To solve this equation by the method of separation of variables, it is assumed that the solution is a product of a function of r [that is, $R(r)$] and a function of ϕ [that is, $F(\phi)$]. Thus,

$$u(r,\phi) = R(r)F(\phi) \qquad (11.95)$$

Differentiation shows that

$$\frac{\partial u}{\partial r} = R'F \qquad \frac{\partial^2 u}{\partial r^2} = R''F$$

$$\frac{\partial u}{\partial \phi} = RF' \qquad \frac{\partial^2 u}{\partial \phi^2} = RF''$$

Substituting these derivatives into Eq. (11.94) and rearranging gives

$$\frac{r^2 R'' + 2rR'}{R} = -\frac{F'' + (\cos \phi/\sin \phi)F'}{F} = n(n+1)$$

Because the first term is independent of ϕ and the middle term is independent of r, each is equal to a constant. This constant is written as $n(n+1)$ so that later results will appear in a convenient form. The preceding expression may be written as two differential equations:

$$r^2 R'' + 2rR' - n(n+1)R = 0 \tag{11.96}$$
$$\sin \phi \, F'' + \cos \phi \, F' + n(n+1) \sin \phi \, F = 0 \tag{11.97}$$

Equation (11.96) is Euler's differential equation, and Eq. (11.97) is Legendre's differential equation.

Euler's differential equation

The general form of Euler's differential equation is

$$a_n r^n \frac{d^n y}{dr^n} + a_{n-1} r^{n-1} \frac{d^{n-1} y}{dr^{n-1}} + \cdots + a_1 r \frac{dy}{dr} + a_0 y = 0$$

The solution of Euler's equation is simplified by transforming it to a linear differential equation with constant coefficients. This transformation is accomplished by making the change of variable

$$r = e^z \qquad \text{or} \qquad z = \ln r$$

Because $dz/dr = 1/r$, then

$$R' = \frac{dR}{dr} = \frac{dR}{dz}\frac{dz}{dr} = \frac{1}{r}\frac{dR}{dz}$$

$$R'' = \frac{d}{dr}\left(\frac{1}{r}\frac{dR}{dz}\right) = \frac{1}{r}\frac{d}{dr}\left(\frac{dR}{dz}\right) - \frac{1}{r^2}\frac{dR}{dz} = \frac{1}{r^2}\frac{d^2R}{dz^2} - \frac{1}{r^2}\frac{dR}{dz}$$

Substitution of these values into Eq. (11.96) gives

$$\frac{d^2R}{dz^2} + \frac{dR}{dz} - n(n+1)R = 0$$

By letting $D = d/dz$, then

$$[D^2 + D - n(n+1)]R = (D - n)[D + (n+1)]R = 0$$

Because the roots are n and $-(n+1)$, the solution is

$$R(z) = C_n e^{nz} + C_{n+1} e^{-(n+1)z}$$

Transforming $R(z)$ back to a function of r yields for the desired solution

$$R(r) = C_n r^n + \frac{C_{n+1}}{r^{n+1}} \tag{11.98}$$

Finite Mellin transformation

This transform is particularly well suited to transforming terms of the form $rf'(r), r^2f''(r), \ldots, r^nf^n(r)$. Thus, the finite Mellin transform lends itself very well to the solution of Euler's equation. This transform is defined by the equation

$$M[f(r)] = \int_0^a f(r)r^{s-1}\,dr = F(s)$$

The transform of $rf'(r)$ is obtained by integration by parts. That is,

$$M[rf'(r)] = \int_0^a f'(r)r^s\,dr = r^sf(r)\Big|_0^a - s\int_0^a f(r)r^{s-1}\,dr$$
$$= -sF(s) + a^sf(a)$$

Similarly, the transform of $r^2f''(r)$ is found to be

$$M[r^2f''(r)] = s(s+1)F(s) + [af'(a) - (s+1)f(a)]a^s$$

In solving Euler's equation, the function $(r/a)^n$ is frequently encountered. The finite Mellin transform of this function is

$$M\left[\left(\frac{r}{a}\right)^n\right] = \int_0^a \frac{r^{n+s-1}}{a^n}\,dr = \frac{a^s}{s+n}$$

To illustrate the application of the finite Mellin transform, let it be desired to determine the solution of Eq. (11.96). Transforming gives

$$[s^2 - s - n(n+1)]M[R(r)] = [(s-1)R(a) - aR'(a)]a^s$$
$$M[R(r)] = \frac{(n+1)R(a) + aR'(a)}{2n+1}\frac{a^s}{s+n}$$
$$+ \frac{nR(a) - aR'(a)}{2n+1}\frac{a^s}{s-(n+1)}$$

Inverting gives

$$R(r) = \frac{(n+1)R(a) + aR'(a)}{2n+1}\left(\frac{r}{a}\right)^n + \frac{nR(a) - aR'(a)}{2n+1}\left(\frac{r}{a}\right)^{-(n+1)}$$

This verifies the result of Eq. (11.98). It is to be noted that the finite Mellin transform method expresses the constants C_n and C_{n+1} in terms of the boundary conditions $R(a)$ and $R'(a)$.

The Mellin transform is the transform that results when the upper limit of integration in the equation for the finite Mellin transform becomes infinite (that is, $a = \infty$).

Legendre's differential equation

Legendre's equation [i.e., Eq. (11.97)] may be expressed in a more convenient algebraic form by letting $x = \cos \phi$. Because $dx/d\phi = -\sin \phi$, then

$$F' = \frac{dF}{d\phi} = \frac{dF}{dx}\frac{dx}{d\phi} = -\sin \phi \frac{dF}{dx}$$

$$F'' = \frac{d}{d\phi}\left(-\sin \phi \frac{dF}{dx}\right) = -\sin \phi \frac{d}{d\phi}\frac{dF}{dx} - \cos \phi \frac{dF}{dx}$$

$$= \sin^2 \phi \frac{d^2F}{dx^2} - \cos \phi \frac{dF}{dx}$$

Substituting these results into Eq. (11.97) and noting that $\sin^2 \phi = 1 - \cos^2 \phi = 1 - x^2$ gives

$$(1 - x^2)\frac{d^2F}{dx^2} - 2x\frac{dF}{dx} + n(n + 1)F = 0 \qquad (11.99)$$

This Legendre equation is satisfied by the Legendre polynomial $P_n(\dot{x})$ given by Eq. (11.51).

The trigonometric form of Legendre's equation [i.e., Eq. (11.97)] is satisfied by the trigonometric form of the Legendre polynomials. This form is obtained by replacing x by $\cos \phi$ in the Legendre polynomials [Eq. (11.52)]. That is,

$$P_0(\cos \phi) = 1$$
$$P_1(\cos \phi) = \cos \phi$$
$$P_2(\cos \phi) = \frac{3 \cos 2\phi + 1}{4}$$
$$P_3(\cos \phi) = \frac{5 \cos 3\phi + 3 \cos \phi}{8}$$
$$P_4(\cos \phi) = \frac{35 \cos 4\phi + 20 \cos 2\phi + 9}{64}$$
$$P_5(\cos \phi) = \frac{63 \cos 5\phi + 35 \cos 3\phi + 30 \cos \phi}{128}$$

An important property of the Legendre polynomials is that

$$\int_{-1}^{1} P_m(x)P_n(x)\,dx = \begin{cases} 0 & n \neq m \\ \dfrac{2}{2n + 1} & n = m \end{cases} \qquad (11.100)$$

where m is a specific integer.

The trigonometric form of the preceding equation is obtained by letting $x = \cos\phi$, $dx = -\sin\phi\,d\phi$ and noting that, when $x = -1$, then $\phi = \pi$ and, when $x = +1$, then $\phi = 0$. Thus

$$\int_0^\pi P_m(\cos\phi)P_n(\cos\phi)\cdot\sin\phi\,d\phi = \begin{cases} 0 & n \neq m \\ \dfrac{2}{2n+1} & n = m \end{cases}$$

$$(11.101)$$

The preceding concepts are now illustrated in the following example.

Illustrative example The temperature on the surface of a sphere of radius a is maintained at $f(\phi)$. Because of circular symmetry, the resulting temperature distribution $u(r,\phi)$ is a function of r and ϕ only (i.e., it is independent of θ). Use the method of separation of variables to determine the steady-state temperature distribution throughout this sphere.

Solution The product form of the solution of Laplace's equation [i.e., Eq. (11.94)] is $u(r,\phi) = R(r)F(\phi)$. The function $R(r)$ is given by Eq. (11.98), and similarly $F(\phi)$ has been found to be $P_n(\cos\phi)$. Thus, for a specific value of n the product $R(r)F(\phi)$ is

$$\left(C_n r^n + \frac{C_{n+1}}{r^{n+1}} \right) P_n(\cos\phi)$$

It is necessary that $C_{n+1} = 0$, for otherwise this solution becomes infinite at $r = 0$. Summing up all the terms which satisfy the differential equation gives, for the general solution,

$$u(r,\phi) = \sum_{n=0}^\infty C_n r^n P_n(\cos\phi) \qquad (11.102)$$

The constant C_n is determined from the boundary condition $u(a,\phi) = f(\phi)$ as follows: First, evaluate the preceding expression at $r = a$, then multiply each term by $\sin\phi\,P_m(\cos\phi)$, and integrate from 0 to π. Utilizing the result of Eq. (11.101), and then solving for C_n, gives

$$C_n = \frac{2n+1}{2a^n} \int_0^\pi f(\phi)\sin\phi\,P_n(\cos\phi)\,d\phi \qquad (11.103)$$

Thus, the solution is given by Eq. (11.102), in which the constant C_n is given by Eq. (11.103).

Transform solutions

To illustrate the application of transformation methods to partial differential equations in spherical coordinates, let it be desired to determine the temperature distribution $u(r,t)$ throughout a sphere when the boundary conditions are

$$u(r,0) = 0$$
$$u(a,t) = \begin{cases} 0 & t < 0 \\ U_0 & t > 0 \end{cases}$$

Thus, the initial temperature throughout the sphere is zero, and the surface temperature is U_0 for $t > 0$. The partial differential equation is

$$\frac{\partial u}{\partial t} = \alpha \left(\frac{\partial^2 u}{\partial r^2} + \frac{2}{r} \frac{\partial u}{\partial r} \right)$$

The Laplace transform is

$$sU(r,s) - u(r,0) = \alpha \left[\frac{\partial^2 U(r,s)}{\partial r^2} + \frac{2}{r} \frac{\partial U(r,s)}{\partial r} \right]$$

Inserting the initial condition and rearranging gives

$$\frac{\partial^2 U(r,s)}{\partial r^2} + \frac{2}{r} \frac{\partial U(r,s)}{\partial r} - \frac{s}{\alpha} U(r,s) = 0$$

This is transformed to Bessel's equation by making the change of variable $U = Y/\sqrt{r}$. Thus

$$r^2 \frac{\partial^2 Y(r,s)}{\partial r^2} + r \frac{\partial Y(r,s)}{\partial r} + \left(-\frac{s}{\alpha} r^2 - \frac{1}{4} \right) Y(r,s) = 0$$

This Bessel equation has the same form as Eq. (11.36), in which $\nu = \frac{1}{2}$ and $\lambda^2 = -s/\alpha$. The general form of the solution is

$$Y(r,s) = C_1 J_{\frac{1}{2}}(\lambda r) + C_2 J_{-\frac{1}{2}}(\lambda r)$$

Substitution of the trigonometric representations for $J_{\frac{1}{2}}$ and $J_{-\frac{1}{2}}$ given in Prob. 11.3 yields

$$Y(r,s) = \sqrt{\frac{2}{\pi \lambda r}} \left(C_1 \sin \lambda r + C_2 \cos \lambda r \right)$$

Converting back to $U(r,s)$ gives

$$U(r,s) = \sqrt{\frac{2}{\pi \lambda}} \left(\frac{C_1 \sin \lambda r}{r} + \frac{C_2 \cos \lambda r}{r} \right)$$

The constant C_2 must vanish, for otherwise, at $r = 0$, then $U(r,s)$ is infinite. Thus, $U(r,s)$ becomes

$$U(r,s) = C_1 \sqrt{\frac{2}{\pi\lambda}} \frac{\sin \lambda r}{r}$$

The transform of the boundary condition $u(a,t) = U_0$ is

$$U(a,s) = U_0/s$$

After evaluating C_1, the transform $U(r,s)$ becomes

$$U(r,s) = \frac{aU_0}{rs} \frac{\sin \lambda r}{\sin \lambda a} = \frac{aU_0}{rs} \frac{\sin (jr \sqrt{s/\alpha})}{\sin (ja \sqrt{s/\alpha})}$$

If λ_n is the value of λ such that $\sin \lambda a = 0$, then the corresponding value of s is $s_n = -\alpha\lambda_n^2$. The residues of $U(r,s)e^{st}$ due to the poles at $s = 0$ and at $s = s_n$ are

$$\frac{aU_0}{r} \frac{\sin (jr \sqrt{s/\alpha})e^{st}}{\sin (ja \sqrt{s/\alpha})}\bigg|_{s=0} = U_0$$

$$\frac{aU_0}{r} \frac{\sin (jr \sqrt{s/\alpha})e^{st}}{s(d/ds) \sin (ja \sqrt{s/\alpha})}\bigg|_{s=s_n} = \frac{2U_0}{r} \frac{\sin (jr \sqrt{s_n/\alpha})e^{s_n t}}{j \sqrt{s_n/\alpha} \cos (ja \sqrt{s_n/\alpha})}$$

$$= \frac{2U_0}{r} \frac{\sin \lambda_n r}{\lambda_n \cos \lambda_n a} e^{-\alpha\lambda_n^2 t}$$

Thus, the inverse is

$$u(r,t) = U_0 + \frac{2U_0}{r} \sum_n \frac{\sin \lambda_n r}{\lambda_n \cos \lambda_n a} e^{-\alpha\lambda_n^2 t}$$

Because $\sin \lambda_n a = \sin (-\lambda_{-n}a) = 0$, then $\lambda_n = -\lambda_{-n}$. Consequently, positive or negative values of λ_n yield the same

$$s_n = -\alpha\lambda_n^2 = -\alpha(-\lambda_{-n})^2$$

Thus, the preceding summation is taken over the positive values of λ_n only.

11.8 LEGENDRE TRANSFORMATION

This transformation is well suited to the solution of partial differential equations in spherical coordinates. The algebraic form of the Legendre transformation of a function $f(x)$ is obtained by multiplying the function by $P_n(x)$ and then integrating from -1 to $+1$.

That is,

$$L[f(x)] = \int_{-1}^{1} f(x)P_n(x)\, dx = F(n) \tag{11.104}$$

where $L[f(x)]$ indicates the Legendre transform of $f(x)$. It is to be noted that, after performing the preceding integration and evaluating x at the upper and lower limits, then the resulting expression is a function of n [that is, $F(n)$]. The inversion formula is obtained by first writing $f(x)$ in the form

$$f(x) = \sum_{n=0}^{\infty} C_n P_n(x) \tag{11.105}$$

To evaluate the coefficient C_n, multiply each term by $P_m(x)$, integrate from -1 to $+1$, and then utilize the result of Eq. (11.100). Thus

$$C_n = \frac{2n+1}{2} \int_{-1}^{1} f(x)P_n(x)\, dx = \frac{2n+1}{2} L[f(x)] \tag{11.106}$$

The desired inversion formula is now obtained by substituting this value of C_n into the above expression for $f(x)$; that is,

$$f(x) = \frac{1}{2} \sum_{n=0}^{\infty} (2n+1)P_n(x)L[f(x)] \tag{11.107}$$

A very important characteristic of this transformation will now be developed. Let $u = P_n(x)$, $v = (1 - x^2)\, \partial f/\partial x$, and then integrate by parts.

$$\int_{-1}^{1} \frac{\partial}{\partial x}\left[(1 - x^2)\frac{\partial f}{\partial x} \right] P_n(x)\, dx = \left[(1 - x^2)\frac{\partial f}{\partial x} P_n(x) \right]_{-1}^{1}$$
$$- \int_{-1}^{1} \frac{\partial f}{\partial x}(1 - x^2)P_n'(x)\, dx$$

The first term on the right-hand side vanishes at both the upper and lower limits. The last term is again integrated by parts, in which $u = (1 - x^2)P_n'(x)$ and $v = -f(x)$. This integration is simplified by noting that, because $P_n(x)$ satisfies Legendre's equation [i.e., Eq. (11.99)], then

$$\frac{du}{dx} = (1 - x^2)P_n''(x) - 2xP_n'(x) = -n(n+1)P_n(x)$$

This shows that $du = -n(n+1)P_n(x)\, dx$. Thus, integration by parts gives

$$\int_{-1}^{1} \frac{\partial}{\partial x}\left[(1-x^2)\frac{\partial f}{\partial x}\right]P_n(x)\,dx = -[f(x)(1-x^2)P_n'(x)]_{-1}^{1}$$
$$- n(n+1)\int_{-1}^{1} f(x)P_n(x)\,dx$$
$$= -n(n+1)L[f(x)] \quad (11.108)$$

This transform is thus seen to remove terms of the form

$$\frac{\partial}{\partial x}\left[(1-x^2)\frac{\partial f}{\partial x}\right] = (1-x^2)\frac{\partial^2 f}{\partial x^2} - 2x\frac{\partial f}{\partial x}$$

from partial differential equations. Such terms arise with oblate spheroidal coordinates. Of more immediate significance is the trigonometric form of this transformation, which is directly applicable for spherical coordinates. Thus, making the change of variable from x to $\cos\phi$ in Eq. (11.108) yields

$$\int_{0}^{\pi}\left[\frac{\partial^2 f(\phi)}{\partial \phi^2} + \frac{\cos\phi}{\sin\phi}\frac{\partial f(\phi)}{\partial\phi}\right]P_n(\cos\phi)\sin\phi\,d\phi$$
$$= -n(n+1)L[f(\phi)] \quad (11.109)$$

Similarly, the trigonometric form for Eq. (11.104) is found to be

$$L[f(\phi)] = \int_{0}^{\pi} f(\phi)P_n(\cos\phi)\sin\phi\,d\phi \quad (11.110)$$

The trigonometric form for the inversion formula [i.e., Eq. (11.107)] is

$$f(\phi) = \tfrac{1}{2}\sum_{n=0}^{\infty}(2n+1)L[f(\phi)]P_n(\cos\phi) \quad (11.111)$$

In applying the three preceding equations for the solution of partial differential equations such as Eq. (11.94), then $f(\phi)$ is replaced by $u(r,\phi)$.

Illustrative example Use the finite Legendre transform to verify the solution of the illustrative example of the preceding section.

Solution The finite Legendre transform of Eq. (11.94) is obtained by multiplying each term by $P_n(\cos\phi)\sin\phi$ and then integrating from 0 to π. Thus,

$$\frac{\partial^2}{\partial r^2}L[u(r,\phi)] + \frac{2}{r}\frac{\partial}{\partial r}L[u(r,\phi)] - \frac{n(n+1)L[u(r,\phi)]}{r^2} = 0$$

447

This transformation reduces the partial differential equation to Euler's equation. The solution of this Euler equation is

$$L[u(r,\phi)] = C_n r^n + \frac{C_{n+1}}{r^{n+1}} = C_n r^n$$

The constant C_{n+1} must vanish, for otherwise the transform "blows up" at $r = 0$. Using Eq. (11.111) to invert the preceding expression gives

$$u(r,\phi) = \tfrac{1}{2} \sum_{n=0}^{\infty} (2n + 1)L[u(r,\phi)]P_n(\cos \phi)$$

$$= \tfrac{1}{2} \sum_{n=0}^{\infty} (2n + 1)C_n r^n P_n(\cos \phi) \tag{11.112}$$

The coefficient C_n is evaluated from the boundary condition that $u(a,\phi) = f(\phi)$. That is, solving the $L[u(r,\phi)]$ expression for C_n and then letting $r = a$ gives

$$C_n = \frac{1}{a^n} L[f(\phi)] = \frac{1}{a^n} \int_0^\pi f(\phi)P_n(\cos \phi) \sin \phi \, d\phi$$

Thus, the solution is given by Eq. (11.112), in which the constant C_n is known from the preceding expression.

Even functions

The even-numbered Legendre polynomials (P_0, P_2, P_4, . . .) are even functions, and the odd-numbered Legendre polynomials (P_1, P_3, P_5, . . .) are odd functions over the interval $-1 \leq x \leq 1$. When $f(x)$ is an even function and n is odd, the product $f(x)P_n(x)$ is an odd function. Thus, Eq. (11.106) shows that the odd coefficients vanish when $f(x)$ is an even function. That is,

$$C_n = \frac{2n + 1}{2} \int_{-1}^{1} f(x)P_n(x) \, dx = 0 \qquad n \text{ odd}$$

The series given by Eq. (11.105) may now be written in the form

$$f(x) = \sum_{n=0}^{\infty} C_{2n}P_{2n}(x) \tag{11.113}$$

Replacing n by $2n$ and m by $2m$ in Eq. (11.100) gives

$$\int_{-1}^{1} P_{2m}(x)P_{2n}(x) \, dx = 2 \int_0^1 P_{2m}(x)P_{2n}(x) \, dx$$
$$= \begin{cases} 0 & n \neq m \\ \dfrac{2}{4n + 1} & n = m \end{cases}$$

Multiplying each term in Eq. (11.113) by P_{2m} and then integrating from 0 to 1 shows that

$$\int_0^1 f(x)P_{2n}(x)\, dx = \frac{C_{2n}}{4n + 1}$$

The Legendre transform for an even function is thus defined as follows:

$$L[f(x)] = \int_0^1 f(x)P_{2n}(x)\, dx = F(2n) \tag{11.114}$$

The preceding analysis shows that the corresponding inversion formula is

$$f(x) = \sum_{n=0}^{\infty} (4n + 1)L[f(x)]P_{2n}(x) \tag{11.115}$$

This transformation is applicable for any function $f(x)$ defined over the interval $0 \leq x \leq 1$ in that any such function may be regarded as an even function, as is illustrated in Fig. 8.1c. A most important property of this transformation is that

$$\int_0^1 \frac{\partial}{\partial x}\left[(1 - x^2)\,\frac{\partial f(x)}{\partial x} \right] P_{2n}(x)\, dx = -P_{2n}(0)\left[\frac{\partial f(x)}{\partial x} \right]_{x=0} - 2n(2n + 1)L[f(x)]$$

The trigonometric form of this result is obtained by letting $x = \cos \phi$; that is,

$$\int_0^{\pi/2} \left[\frac{\partial^2 f(\phi)}{\partial \phi^2} + \frac{\cos \phi}{\sin \phi}\,\frac{\partial f(\phi)}{\partial \phi} \right] P_{2n}(\cos \phi)\, \sin \phi \, d\phi$$
$$= P_{2n}(0)\left[\frac{1}{\sin \phi}\,\frac{\partial f(\phi)}{\partial \phi} \right]_{\phi=\pi/2} - 2n(2n + 1)L[f(\phi)]$$

Odd functions

By proceeding in the same manner as for even functions, the Legendre transform for an odd function is found to be

$$L[f(x)] = \int_0^1 f(x)P_{2n+1}(x)\, dx = F(2n + 1) \tag{11.116}$$

The corresponding inversion formula is

$$f(x) = \sum_{n=0}^{\infty} (4n + 3)L[f(x)]P_{2n+1}(x) \tag{11.117}$$

An important property of this transform is

$$\int_0^1 \frac{\partial}{\partial x}\left[(1 - x^2)\frac{\partial f(x)}{\partial x}\right]P_{2n+1}(x)\ dx = (2n + 1)P_{2n}(0)[f(x)]_{x=0}$$
$$- (2n + 1)(2n + 2)L[f(x)]$$

The trigonometric form of this property is

$$\int_0^{\pi/2}\left[\frac{\partial^2 f(\phi)}{\partial \phi^2} + \frac{\cos\ \phi}{\sin\ \phi}\frac{\partial f(\phi)}{\partial \phi}\right]P_{2n+1}(\cos\ \phi)\sin\ \phi\ d\phi$$
$$= (2n + 1)P_{2n}(0)[f(\phi)]_{\phi=\pi/2} - (2n + 1)(2n + 2)L[f(\phi)]$$

The choice as to whether to regard a function as an even or an odd function depends upon the given initial conditions. This is the same situation that exists with regard to the choice of the finite sine or cosine transformation, as is discussed in Chapter 8.

11.9   ORTHOGONAL FUNCTIONS

A set of functions

$$\theta_1(x),\ \theta_2(x),\ \ldots\ ,\ \theta_n(x),\ \ldots$$

is said to be orthogonal over the interval $a \le x \le b$ if

$$\int_a^b \theta_m(x)\theta_n(x)\ dx = \begin{cases} 0 & n \ne m \\ k_n & n = m \end{cases} \tag{11.118}$$

where m is a specific integer.   An example of a set of functions which is orthogonal over the interval $0 \le x \le L$ is

$$\theta_1 = \sin\ \omega_0 x \qquad \theta_2 = \sin\ 2\omega_0 x \qquad \theta_3 = \sin\ 3\omega_0 x$$

where $\omega_0 = \pi/L$.   For this set, it may be shown that

$$\int_0^L \theta_m(x)\theta_n(x)\ dx = \int_0^L \sin\ m\omega_0 x\ \sin\ n\omega_0 x\ dx$$
$$= \begin{cases} 0 & n \ne m \\ \dfrac{L}{2} & n = m \end{cases} \tag{11.119}$$

A set of functions is said to be *orthonormal* (i.e., the orthogonal set is normalized) if the value of the integral in Eq. (11.118) when $n = m$ is unity.   That is,

$$\int_a^b \phi_m(x)\phi_n(x)\ dx = \begin{cases} 0 & n \ne m \\ 1 & n = m \end{cases} \tag{11.120}$$

Any set of orthogonal functions [Eq. (11.118)] may be converted to an orthonormal set [Eq. (11.120)] by letting

$$\phi_1(x) = \frac{\theta_1(x)}{\sqrt{k_1}} \qquad \phi_2(x) = \frac{\theta_2(x)}{\sqrt{k_2}} \qquad \cdots$$

For the set of orthogonal functions given in Eq. (11.119), $k_1 = k_2 = \cdots = k_n = L/2$. Thus, this set is normalized by dividing each term by $\sqrt{L/2}$. That is,

$$\phi_1 = \frac{\sin \omega_0 x}{\sqrt{L/2}} \qquad \phi_2 = \frac{\sin 2\omega_0 x}{\sqrt{L/2}} \qquad \cdots$$

This set is recognized as the basic elements of the Fourier series expansion for a function $f(x)$ which is regarded as an odd function (see Fig. 8.1b) over the interval $0 \leq x \leq L$.

An example of another set which is orthonormal over the interval $0 \leq x \leq L$ [that is, which satisfies Eq. (11.120)] is

$$\phi_1 = \frac{1}{\sqrt{L}} \qquad \phi_2 = \frac{\cos \omega_0 x}{\sqrt{L/2}} \qquad \phi_3 = \frac{\cos 2\omega_0 x}{\sqrt{L/2}} \qquad \cdots$$

$$(11.121)$$

This set is recognized as the basic elements of the Fourier series for a function $f(x)$ which is regarded as an even function (see Fig. 8.1c) over the interval $0 \leq x \leq L$.

An example of a set which is orthonormal over the interval $a \leq x \leq b$ is

$$\phi_1 = \frac{1}{\sqrt{T}} \qquad \phi_2 = \frac{\sin \omega_0 x}{\sqrt{T/2}} \qquad \phi_3 = \frac{\cos \omega_0 x}{\sqrt{T/2}}$$

$$\phi_4 = \frac{\sin 2\omega_0 x}{\sqrt{T/2}} \qquad \phi_5 = \frac{\cos 2\omega_0 x}{\sqrt{T/2}} \qquad \cdots$$

where $\omega_0 = 2\pi/T$. This set is recognized as the basic elements of the Fourier series expansion for a function $f(x)$ of period T which extends from a to b (that is, $T = b - a$).

A function $f(x)$ may be expressed as a sum of orthonormal functions as follows:

$$f(x) = C_1\phi_1(x) + C_2\phi_2(x) + \cdots + C_n\phi_n(x) + \cdots$$

$$(11.122)$$

where C_1, C_2, $\ldots$ are constants. To evaluate the constants, multiply each term by $\phi_n(x)$, and then integrate over the interval from a to b.

$$\int_a^b f(x)\phi_n(x)\, dx = C_1 \int_a^b \phi_1(x)\phi_n(x)\, dx + C_2 \int_a^b \phi_2(x)\phi_n(x)\, dx$$
$$+ \cdots + C_n \int_a^b \phi_n{}^2(x)\, dx + \cdots$$

Thus, for an orthonormal set,

$$C_n = \int_a^b f(x)\phi_n(x)\, dx \tag{11.123}$$

Although the fact that the set is orthonormal is sufficient to evaluate the coefficients, another condition is needed to prove that the series $\sum\limits_{n=1}^{\infty} C_n\phi_n$ converges to $f(x)$. This condition is the property of *completeness*. For example, suppose that the $\phi_i(x)$ term had been omitted from the set. Thus, there would exist the function $\phi_i(x)$ such that

$$\int_a^b \phi_i(x)\phi_n(x)\, dx = 0$$

A set is thus said to be *complete* if there is no function $g(x)$ such that

$$\int_a^b g(x)\phi_n(x)\, dx = 0 \qquad n = 1, 2, 3, \ldots$$

To verify that a complete orthonormal set does converge to $f(x)$, let the function $g(x)$ represent the error over the interval $a \leq x \leq b$. That is,

$$g(x) = f(x) - \sum_{n=1}^{\infty} C_n\phi_n(x)$$

Multiplying each term by $\phi_n(x)$ and then integrating over the interval gives

$$\int_a^b g(x)\phi_n(x)\, dx = \int_a^b f(x)\phi_n(x)\, dx - C_n \int_a^b \phi_n{}^2(x)\, dx$$
$$= C_n - C_n = 0$$

Because the set is complete, the preceding equation can be satisfied only if the error $g(x)$ is indentically zero. Thus, for a complete orthonormal set $\Sigma C_n\phi_n$ converges to $f(x)$. If the first term were designated $C_0\phi_0(x)$, then the lower index would be $n = 0$.

Another interesting property of orthonormal sets is obtained by considering the integral of the square of the error E; that is,

$$E = \int_a^b [g(x)]^2 \, dx = \int_a^b [f(x) - (C_1\phi_1 + C_2\phi_2 + \cdots)]^2 \, dx$$

$$= \int_a^b \{[f(x)]^2 - 2f(x)(C_1\phi_1 + C_2\phi_2 + \cdots)$$

$$+ (C_1\phi_1 + C_2\phi_2 + \cdots)^2\} \, dx$$

For the middle terms, it is to be noted that $\int_a^b f(x)\phi_n(x) \, dx = C_n$. After squaring the last term, the integral of all the cross-product terms vanishes $\int_a^b \phi_m(x)\phi_n(x) \, dx = 0$. Thus, the preceding expression becomes

$$E = \int_a^b [f(x)]^2 \, dx - 2 \sum_{n=1}^{\infty} C_n{}^2 + \sum_{n=1}^{\infty} C_n{}^2$$

$$= \int_a^b [f(x)]^2 \, dx - \sum_{n=1}^{\infty} C_n{}^2 \qquad (11.124)$$

Because the integral of the mean squared error E must be greater than or equal to zero,

$$\sum_{n=1}^{\infty} C_n{}^2 \leq \int_a^b [f(x)]^2 \, dx$$

This relationship is called *Bessel's inequality*. The series converges to $f(x)$ when the integral of the mean squared error E is zero. Thus, letting E equal zero in Eq. (11.124) gives

$$\sum_{n=1}^{\infty} C_n{}^2 = \int_0^b [f(x)]^2 \, dx \qquad (11.125)$$

This result is called *Parseval's formula*. When the set of ortho-normal functions is such that Eq. (11.125) is satisfied, the set is said to be *closed*. It can be shown that every closed set is complete, and vice versa.

Sturm-Liouville equation

Orthogonal functions arise naturally in the solution of a large and important class of problems in advanced engineering mathematics. Consider the following differential equation, which is known as the

Sturm-Liouville equation,

$$\frac{d}{dx}\left[r(x)\frac{dy}{dx} \right] + [q(x) + \lambda p(x)]y = 0$$

or

$$(ry')' + (q + \lambda p)y = 0 \tag{11.126}$$

where the functions $r(x)$, $q(x)$, and $p(x)$ are continuous over the interval from a to b. The values of λ for which solutions exist are called *eigenvalues*. The eigenvalues are designated $\lambda_1, \lambda_2, \lambda_3, \ldots$

The Sturm-Liouville equation is a generalized formulation for many differential equations such as Bessel's equations, Legendre's equation, Laguerre's equation, Hermite's equation, and other differential equations commonly encountered in engineering practice. For the case in which $\lambda = \lambda_m$, the Sturm-Liouville equation is

$$(ry_m')' + (q + \lambda_m p)y_m = 0$$

Similarly, when $\lambda = \lambda_n$, then

$$(ry_n')' + (q + \lambda_n p)y_n = 0$$

Multiplying the first of the preceding equations by y_n and the second by y_m, and then subtracting the second from the first, gives

$$(\lambda_m - \lambda_n)py_my_n = y_m(ry_n')' - y_n(ry_m')'$$
$$= \frac{d}{dx}[y_m(ry_n') - y_n(ry_m')]$$

Integration over the interval from a to b shows that

$$(\lambda_m - \lambda_n)\int_a^b py_my_n\,dx = r(y_my_n' - y_ny_m')\Big|_a^b$$
$$= r(b)[y_m(b)y_n'(b) - y_n(b)y_m'(b)]$$
$$\quad - r(a)[y_m(a)y_n'(a) - y_n(a)y_m'(a)] \tag{11.127}$$

When the initial conditions are such that the right-hand side of the preceding equation is zero, then

$$\int_a^b py_my_n\,dx = 0 \tag{11.128}$$

To rule out the trivial solution $y_m = y_n = 0$, it is necessary to stipulate that

$$\int_a^b py_n{}^2\,dx \neq 0$$

By letting $\phi_m(x) = \sqrt{p}\,y_m(x)$ and $\phi_n(x) = \sqrt{p}\,y_n(x)$, then the solution forms a system of orthogonal functions. The functions $y_m(x)$

and $y_n(x)$ are said to be orthogonal with respect to the weight function $p(x)$.

Legendre's equation

This differential equation is

$$(1 - x^2) \frac{d^2 y}{dx^2} - 2x \frac{dy}{dx} + n(n + 1)y = 0$$

Legendre's equation may also be written in the form

$$[(1 - x^2)y']' + n(n + 1)y = 0 \tag{11.129}$$

Legendre's equation is a special case of the Sturm-Liouville equation, in which $r(x) = 1 - x^2$, $q(x) = 0$, $p(x) = 1$, and $\lambda = n(n + 1)$. Ordinarily, Legendre's equation is considered over the interval $-1 \leq x \leq 1$. For this interval $r(a) = 1 - (-1)^2 = 0$, and $r(b) = 1 - (1)^2 = 0$. Because $r(a) = r(b) = 0$, the right-hand side of Eq. (11.127) is zero regardless of the boundary conditions. For $p(x) = 1$, the solutions of Legendre's equation satisfy the orthogonality relationships.

$$\int_{-1}^{1} y_m(x)y_n(x)\, dx = \begin{cases} 0 & n \neq m \\ k_n & n = m \end{cases}$$

In the preceding section, it was shown that the Legendre polynomial $y_n(x) = P_n(x)$ satisfies Legendre's equation when $n = 0, 1, 2, \ldots$. The validity of the preceding orthogonality relationships is substantiated by Eq. (11.100).

The Legendre series for approximating a function $f(x)$ which is defined over the interval $-1 \leq x \leq 1$ is

$$f(x) = \sum_{n=0}^{\infty} C_n P_n(x)$$

The coefficient C_n is evaluated by multiplying each term in the preceding expression by $P_m(x)$ and then integrating from -1 to $+1$. Thus,

$$C_n = \frac{2n + 1}{2} \int_{-1}^{1} f(x)P_n(x)\, dx$$

Bessel's equation

This differential equation is

$$r^2 \frac{d^2 y(r)}{dr^2} + r \frac{dy(r)}{dr} + (r^2 - \nu^2)y(r) = 0$$

Replacing r by λx and noting that

$$\frac{dy}{dr} = \frac{dx}{dr}\frac{dy}{dx} = \frac{1}{\lambda}\frac{dy}{dx} \qquad \text{and} \qquad \frac{d^2y}{dr^2} = \frac{1}{\lambda^2}\frac{d^2y}{dx^2}$$

then Bessel's differential equation becomes

$$x^2 y''(\lambda x) + x y'(\lambda x) + (\lambda^2 x^2 - \nu^2)y(\lambda x) = 0$$

The preceding equation may be expressed in the form

$$[xy'(\lambda x)]' + \left(-\frac{\nu^2}{x} + \lambda^2 x\right) y(\lambda x) = 0 \tag{11.130}$$

This is recognized as a special case of the Sturm-Liouville equation, in which $r(x) = x$, $q(x) = -\nu^2/x$, and $p(x) = x$. The parameter λ^2 in the above equation corresponds to λ in the Sturm-Liouville equation.

For the interval $0 \leq x \leq a$, then $r(x) = 0$ at the lower limit. Thus, the last term on the right-hand side of Eq. (11.127) vanishes. The preceding Bessel differential equation is satisfied by the Bessel function $y(\lambda x) = J_\nu(\lambda x)$. For values of $\lambda = \lambda_n$ such that $y_n = J_\nu(\lambda_n a) = 0$, the first term on the right-hand side of Eq. (11.127) also vanishes. Thus, the orthogonality relationships are satisfied. That is,

$$\int_0^a p y_m y_n \, dx = \int_0^a x J_\nu(\lambda_m x) J_\nu(\lambda_n x) \, dx = \begin{cases} 0 & n \neq m \\ k_n & n = m \end{cases}$$

In particular, it may be shown that

$$\int_0^a x J_\nu(\lambda_m x) J_\nu(\lambda_n x) \, dx = \begin{cases} 0 & n \neq m \\ \dfrac{a^2}{2}\,[J_\nu'(\lambda_n a)]^2 & n = m \end{cases}$$

A function $f(x)$ which is defined over the interval $0 \leq x \leq a$ may be expressed in terms of the following Fourier-Bessel series:

$$f(x) = \sum_{n=0}^{\infty} C_n J_\nu(\lambda_n x)$$

Multiplying each term in the preceding series by $xJ_\nu(\lambda_n x)$, and then integrating from 0 to a, yields for the coefficient C_n

$$C_n = \frac{2}{a^2[J_\nu'(\lambda_n a)]^2} \int_0^a x f(x) J_\nu(\lambda_n a) \, dx$$

A similar Fourier-Bessel series may be developed for a function $f(x)$ which is defined over the interval $a \leq x \leq b$.

Laguerre's equation

Another differential equation which arises frequently in engineering practice is the Laguerre equation

$$xy'' + (1 - x)y' + ny = 0 \qquad n = 0, 1, 2, \ldots$$

or

$$(xe^{-x}y')' + ne^{-x}y = 0 \tag{11.131}$$

This differential equation is satisfied by the Laguerre polynomial

$$L_n(x) = e^x \frac{d^n}{dx^n} (x^n e^{-x})$$

The first few Laguerre polynomials are

$$L_0(x) = 1 \qquad L_2(x) = x^2 - 4x + 2$$
$$L_1(x) = -x + 1 \qquad L_3(x) = -x^3 + 9x^2 - 18x + 6$$

The Laguerre equation is a special case of the Sturm-Liouville equation, in which $r(x) = xe^{-x}$, $q(x) = 0$, $p(x) = e^{-x}$, and $\lambda = n$. For the interval $0 \leq x \leq \infty$, then $r(a) = r(0) = 0$, and $r(b) = r(\infty) = 0$. Because $r(a) = r(b) = 0$, the right-hand side of Eq. (11.127) is zero, regardless of the boundary conditions. Thus, the solutions of Laguerre's equation are orthogonal with respect to the weight function $p(x) = e^{-x}$. That is,

$$\int_0^\infty e^{-x}y_m(x)y_n(x)\, dx = \begin{cases} 0 & n \neq m \\ k_n & n = m \end{cases}$$

In particular, it may be shown that

$$\int_0^\infty e^{-x}L_m(x)L_n(x)\, dx = \begin{cases} 0 & n \neq m \\ (n!)^2 & n = m \end{cases}$$

A function $f(x)$ defined over the interval $0 \leq x \leq \infty$ may be expanded in terms of Laguerre polynomials as follows:

$$f(x) = \sum_{n=0}^\infty C_n L_n(x) \tag{11.132}$$

The coefficient C_n is obtained by multiplying each term by $e^{-x}L_m(x)$ and then integrating from 0 to ∞. Thus

$$C_n = \frac{1}{(n!)^2} \int_0^\infty e^{-x}f(x)L_n(x)\, dx \tag{11.133}$$

There are a great many sets of orthogonal functions. Such functions have numerous and varied applications to engineering problems. An obvious application is that each set may be regarded as forming the basis of a transformation. For example, Eq. (11.133) transforms a function $f(x)$ to a function of n [that is, $C_n = F(n)$]. The corresponding inversion formula for this Laguerre transformation is given by Eq. (11.132).

PROBLEMS

11.1   Determine the solution y for each of the following differential equations:

(a) $\dfrac{d^2y}{dt^2} + t\,\dfrac{dy}{dt} - 2y = 2$ $y(0) = y'(0) = 0$

(b) $t\,\dfrac{d^2y}{dt^2} + \dfrac{dy}{dt} - (t-3)y = e^{-t}$ $y(0) = 0,\ y'(0) = 1$

(c) $t\,\dfrac{d^2y}{dt^2} + 2\,\dfrac{dy}{dt} + ty = 0$ $y(0) = 1,\ y'(0) = 0$

11.2   The Laguerre differential equation is

$$xy'' + (1-x)y' + ny = 0$$

Show that

$$\frac{Y'(s)}{Y(s)} = \frac{n+1-s}{s(s-1)} = \frac{n}{s-1} - \frac{n-1}{s}$$

and thus

$$Y(s) = \frac{(s-1)^n}{s^{n+1}}$$

Finally, show that, for $n = 0, 1, 2, \ldots$, the solutions are the Laguerre polynomials $L_0(x) = 1$, $L_1(x) = 1 - x$, etc.

11.3   Write the series expansion for $J_{\frac{1}{2}}(x)$ and $J_{-\frac{1}{2}}(x)$, and then show that

$$J_{\frac{1}{2}}(x) = \sqrt{\frac{2}{\pi x}}\,\sin x$$

$$J_{-\frac{1}{2}}(x) = \sqrt{\frac{2}{\pi x}}\,\cos x$$

Note that $\Gamma(\frac{1}{2}) = \sqrt{\pi}$. Use these results and appropriate recurrence relationships to show that

$$J_{\frac{1}{2}}(x) = \sqrt{\frac{2}{\pi x}} \left(\frac{\sin x}{x} - \cos x \right)$$

$$J_{-\frac{1}{2}}(x) = -\sqrt{\frac{2}{\pi x}} \left(\frac{\cos x}{x} + \sin x \right)$$

11.4 (a) Show that $y = x^n J_n(x)$ satisfies the equation

$$xy'' + (1 - 2n)y' + xy = 0$$

(b) Show that $y = x^{-n} J_n(x)$ satisfies the equation

$$xy'' + (1 + 2n)y' + xy = 0$$

11.5 Make the indicated substitutions to show that each of the following differential equations reduces to Bessel's differential equation:

(a) $xy'' - y' + xy = 0$ $y = xu$
(b) $x^2 y'' - xy' + (x^2 - v^2 + 1)y = 0$ $y = xu$
(c) $x^2 y'' + (x^2 + \frac{1}{4})y = 0$ $y = \sqrt{x}\, u$
(d) $x^2 y'' + (1 + 2a)xy' + (x^2 - v^2 + a^2)y = 0$ $y = u/x^a$

11.6 Determine the general form of the solution for each of the differential equations given in Prob. 11.5.

11.7 Determine the resulting differential equation when Bessel's differential equation is transformed by each of the following transformations:

(a) $y = \lambda u$
(b) $y = x^a u$
(c) $y = e^x u$

11.8 (a) Use Eq. (11.11) to show that, for small values of x, then $J_0(x)$ may be approximated by $J_0(x) \approx (1 - x^2/4)$. Determine the error in using this approximation for $x = 0.1$, 0.5, and 1.0. The exact values are $J_0(0.1) = 0.9975$, $J_0(0.5) = 0.9385$, and $J_0(1.0) = 0.7652$.

(b) Use Eq. (11.63) to show that, for small values of x, then $Y_0(x) \approx (2/\pi)[\ln (x/2) + \gamma]$.

11.9 For large values of x it may be shown that

$$\lim_{x \to \infty} J_v(x) = \sqrt{\frac{2}{\pi x}} \cos \left(x - \frac{\pi}{4} - \frac{v\pi}{2} \right)$$

$$\lim_{x \to \infty} Y_v(x) = \sqrt{\frac{2}{\pi x}} \sin \left(x - \frac{\pi}{4} - \frac{v\pi}{2} \right)$$

Use these relationships to determine the first three values of x for which $J_0(x) = 0$, and similarly determine the first three values of x for which $Y_0(x) = 0$. Compare these approximate values with the exact values, which are 2.405, 5.520, and 8.654 for $J_0(x)$ and 0.89, 3.96, and 7.09 for $Y_0(x)$.

11.10 Make the change of variable $y = u/\sqrt{x}$ in Bessel's differential equation of order ν [that is, Eq. (11.29)] to show that

$$u'' + \left(1 - \frac{\nu^2 - \frac{1}{4}}{x^2}\right) u = 0$$

For large values of x the $(\nu^2 - \frac{1}{4})/x^2$ term may be neglected. Thus, show that for large values of x the solution is approximately

$$y = \frac{A \sin x + B \cos x}{\sqrt{x}}$$

Verify the preceding result by substituting the approximations for $J_\nu(x)$ and $Y_\nu(x)$ given in Prob. 11.9 for large values of x into the general solution [Eq. (11.64)].

11.11 A general transform relationship for Bessel functions is

$$\mathcal{L}[x^\nu J_\nu(\lambda x)] = \frac{\lambda^\nu \Gamma(2\nu + 1)}{2^\nu \Gamma(\nu + 1)(s^2 + \lambda^2)^{(2\nu+1)/2}} \qquad \nu \geq 0$$

Use this general result to determine the transform of each of the following functions:

(a) $J_0(x)$ (b) $xJ_1(x)$ (c) $J_1(x)$

11.12 Use the series method to determine the solution of each of the following differential equations:

(a) $y'' + y = 0$ (b) $(x - 2)y' - xy = 0$

11.13 Use the series method to determine the solution of the following differential equation:

$$y'' - y = 0$$

11.14 Transform each of the following from rectangular to cylindrical coordinates:

(a) The gradient, $\nabla \phi$
(b) The divergence, $\nabla \cdot \mathbf{F}$
(c) The curl, $\nabla \times \mathbf{F}$

11.15 A cylinder with insulated ends is initially at zero temperature throughout. Use the Hankel transform method to determine the temperature distribution $u(r,t)$ for the case in which the temperature on the surface becomes U_0 for $t > 0$.

11.16 Expand $(x^2 - 1)^n$ by the binomial expansion, and then differentiate n times to show that

$$\frac{d^n}{dx^n}(x^2 - 1)^n = \frac{d^n}{dx^n}\left[\sum_{k=0}^{n}(-1)^k \frac{n!}{k!(n-k)!}x^{2n-2k}\right]$$

$$= \sum_{k=0}^{N}(-1)^k \frac{n!}{k!(n-k)!}\frac{(2n-2k)!}{(n-2k)!}x^{n-2k}$$

where $N = n/2$ when n is even and $N = (n-1)/2$ when n is odd. Compare the preceding expression with Eq. (11.51) to show that

$$P_n(x) = \frac{1}{2^n n!}\frac{d^n}{dx^n}(x^2 - 1)^n$$

This result is known as *Rodrigues's formula*. Use this formula to verify the results obtained for $P_0(x)$, $P_1(x)$, $P_2(x)$, and $P_3(x)$.

11.17 In Fig. P11.17 is shown a flexible cable, or chain, which is hanging by its own weight. This is similar to the flexible string discussed in Chap. 7, except that the tension at a station x is equal to the weight wx rather than a constant T (w is the weight per unit length). Thus, from Eq. (7.13), it follows that for the case of a flexible cable

$$\frac{\partial^2 y}{\partial t^2} = \frac{1}{\rho}\frac{\partial[wx(\partial y/\partial x)]}{\partial x} = g\left[x\frac{\partial^2 y}{\partial x^2} + \frac{\partial y}{\partial x}\right]$$

where $\rho = w/g$ is the mass per unit length. Determine the motion of a flexible cable when the initial position $y(x,0) = f(x)$. (*Hint:* Make the change of variable $x = \frac{1}{4}gz^2$.)

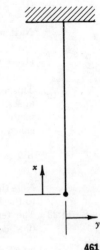

Fig. P11.17

11.18 The differential equation for the vibration of a circular membrane is

$$\frac{\partial^2 z}{\partial t^2} = c^2 \left(\frac{\partial^2 z}{\partial r^2} + \frac{1}{r} \frac{\partial z}{\partial r} \right)$$

where $c^2 = T/\rho$, in which T is the uniform tension (i.e., force per unit length) in the elastic membrane and ρ is the surface density (i.e., mass per unit area of the membrane). The radius of the membrane is a. The circumference is fixed so that $z(a,t) = 0$. The membrane is initially at rest so that $\partial z(r,0)/\partial t = 0$. The initial displacement is $z(r,0) = f(r)$. Use the Hankel transform method to determine the resulting motion $z(r,t)$.

11.19 Determine the steady-state temperature distribution in a semicircle for which the temperature on the circular portion is maintained at $0°$ and the temperature on the diameter is maintained at U_0. (*Hint:* Use the polar form of Laplace's equation.)

11.20 Determine the steady-state temperature distribution in a circular disk for which the temperature on the upper half of the circumference is maintained at $0°$ and the lower half of the circumference is maintained at U_0.

11.21 Transform each of the following from rectangular to spherical coordinates:

(a) The gradient, $\nabla \phi$
(b) The divergence, $\nabla \cdot \mathbf{F}$
(c) The curl, $\nabla \times \mathbf{F}$

11.22 Differentiate Legendre's equation m times with respect to x, and then let $w = d^m y/dx^m$ to show that

$$(1 - x)^2 \frac{d^2 w}{dx^2} - 2x(m + 1) \frac{dw}{dx} + (n - m)(n + m + 1)w = 0$$

Next make the change of variable $u = w(1 - x^2)^{m/2}$ to show that

$$(1 - x^2) \frac{d^2 u}{dx^2} - 2x \frac{du}{dx} + \left[n(n + 1) - \frac{m^2}{(1 - x^2)} \right] u = 0$$

This equation, which is called the *associated Legendre equation*, arises in solving Laplace's equation in spherical coordinates by the method of separation of variables. Show that the solution $P_n{}^m(x)$, called the *associated Legendre polynomial*, is

$$P_n{}^m(x) = (1 - x^2)^{m/2} \frac{d^m}{dx^m} P_n(x)$$

Note that for $m = 0$, then the associated Legendre equation reduces to the Legendre equation.

11.23 The temperature on the surface of the upper half of a sphere $(0 < \phi < \pi/2)$ is maintained at U_0, and that on the surface of the lower half $(\pi/2 < \phi < \pi)$ is maintained at $0°$. Determine the resulting steady-state temperature distribution throughout the sphere.

11.24   In Fig. P11.24 is shown a thin hemispherical shell of radius a.   The shell is sufficiently thin so that variations in the radial (r) direction may be neglected.   Initially the temperature of the shell is U_0 throughout.   At time $t = 0$, the temperature on the circular edge $(\phi = \pi/2)$ is changed to $0°$.   Thus, the boundary conditions are

$$u(\phi,0) = U_0$$

$$u\left(\frac{\pi}{2}, t\right) = \begin{cases} U_0 & t < 0 \\ 0 & t > 0 \end{cases}$$

Determine the equation for the resulting temperature distribution $u(\phi,t)$.

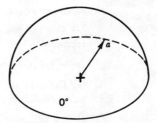

Fig. P11.24

11.25   Show that if the boundary conditions are

$$A_1 y(a) - A_2 y'(a) = 0$$
$$B_1 y(b) - B_2 y'(b) = 0$$

the right-hand side of Eq. (11.127) vanishes.   Thus, the solutions (y_m, y_n) of the Sturm-Liouville equation form orthogonal sets.

11.26   The generating function $g(x,t)$ for the Legendre polynomials, $P_n(x)$, is

$$g(x,t) = (1 - 2xt + t^2)^{-\frac{1}{2}} = \sum_{n=0}^{\infty} P_n(x)t^n$$

(*a*) Show that

$$P_n(x) = \frac{1}{n!}\left[\frac{\partial^n g(x,t)}{\partial t^n}\right]_{t=0} = \frac{1}{2^n n!}\frac{d^n(x^2 - 1)^n}{dx^n}$$

(*b*) Differentiate $g(x, t)$ with respect to t to show that

$$(1 - 2xt + t^2)\frac{\partial g(x,t)}{\partial t} = (x - t)g(x,t)$$

Express this result in the form

$$\sum_{n=1}^{\infty} nP_n(x)t^{n-1} - \sum_{n=0}^{\infty} (2n+1)xP_n(x)t^n + \sum_{n=0}^{\infty} (n+1)P_n(x)t^{n+1}$$
$$= 0$$

Change the index from $n = 0$ to $n = 1$ in the last two summations and note that $P_1(x) - xP_0(x) = 0$. Thus, verify the recurrence relationship

$$(n+1)P_{n+1}(x) - (2n+1)xP_n(x) + nP_{n-1}(x) = 0$$

(c) Differentiate $g(x,t)$ with respect to x to show that

$$t\frac{\partial g(x,t)}{\partial t} = (x - t)\frac{\partial g(x,t)}{\partial x}$$

Proceed as in part (b) to verify the recurrence relationship

$$nP_n(x) - xP'_n(x) + P'_{n-1}(x) = 0$$

(d) Use the preceding recurrence relationships to obtain the new relationship

$$(x^2 - 1)P'_n(x) + nP_{n-1}(x) - nxP_n(x) = 0$$

(e) Differentiate this last relationship and then proceed to obtain an expression in $P_n(x)$ and its derivatives only. That is,

$$(1 - x^2)P''_n(x) - 2xP'_n(x) + n(n+1)P_n(x) = 0$$

Hence, $P_n(x)$ satisfies Legendre's differential equation

$$(1 - x^2)y'' - 2xy' + n(n+1)y = 0$$

11.27 Use the recurrence relations developed in Prob. 11.26 to obtain the following additional recurrence relationships:

$$P'_{n+1}(x) - P'_{n-1}(x) - (2n+1)P_n(x) = 0$$
$$P'_{n+1}(x) - xP'_n(x) - (n+1)P_n(x) = 0$$

11.28 The generating function $g(x,t)$ for the Laguerre polynomials, $L_n(x)$, is

$$g(x,t) = \frac{e^{-xt/(1-t)}}{1 - t} = \sum_{n=0}^{\infty} \frac{L_n(x)}{n!} t^n$$

(a) Show that

$$L_n(x) = \left[\frac{\partial^n g(x,t)}{\partial t^n}\right]_{t=0} = e^x \frac{d^n}{dx^n}(x^n e^{-x})$$

(b) Differentiate $g(x,t)$ with respect to t to show that

$$(1 - t)^2 \frac{\partial g(x,t)}{\partial t} = (1 - t - x)g(x,t)$$

By successive differentiation of the preceding equation show that the nth partial derivative has the form

$$(1 - t^2) \frac{\partial^{n+1}g}{\partial t^{n+1}} - 2n(1 - t) \frac{\partial^n g}{\partial t^n} + \frac{2n(n - 1)}{2!} \frac{\partial^{n-1}g}{\partial t^{n-1}}$$
$$= (1 - t - x) \frac{\partial^n g}{\partial t} - n \frac{\partial^{n-1}g}{\partial t^{n-1}}$$

Evaluate this result at $t = 0$ to verify the recurrence relationship

$$L_{n+1}(x) - (2n + 1 - x)L_n(x) + n^2 L_{n-1}(x) = 0$$

(c) Differentiate $g(x,t)$ with respect to x to show that

$$(1 - t) \frac{\partial g(x,t)}{\partial x} = -tg$$

Proceed as in part (b) to verify the recurrence relationship

$$L_n'(x) - nL_{n-1}'(x) + nL_{n-1}(x) = 0$$

(d) Use the preceding recurrence relationships to obtain the new relationship

$$xL_n'(x) - nL_n(x) + n^2 L_{n-1}(x) = 0$$

(e) Use the preceding recurrence relationships to obtain an expression in $L_n(x)$ and its derivatives only. That is,

$$xL_n''(x) + (1 - x)L_n'(x) + nL_n(x) = 0$$

Hence, $L_n(x)$ satisfies the Laguerre differential equation

$$xy'' + (1 - x)y' + ny = 0$$

11.29 The generating function $g(x,t)$ for the Hermite polynomials $H_n(x)$ is

$$g(x,t) = e^{x^2}e^{-(t-x)^2} = \sum_{n=0}^{\infty} \frac{H_n(x)}{n!} t^n$$

(a) Show that

$$H_n(x) = \frac{\partial^n g(x,t)}{\partial t^n}\bigg|_{t=0} = (-1)^n e^{x^2} \frac{d^n}{dx^n}(e^{-x^2})$$

(b) Show that the first four Hermite polynomials are

$$H_0(x) = 1 \qquad H_2(x) = 4x^2 - 2$$
$$H_1(x) = 2x \qquad H_3(x) = 8x^3 - 12x$$

(c) Differentiate $g(x,t)$ with respect to x to show that

$$\frac{\partial g(x,t)}{\partial x} = 2tg(x,t)$$

and therefore

$$\sum_{n=0}^{\infty} \frac{H_n'(x)}{n!} t^n = 2t \sum_{n=0}^{\infty} \frac{H_n(x)}{n!} t^n$$

Thus, verify the recurrence relationship

$$H_n'(x) = 2nH_{n-1}(x)$$

(d) Differentiate $g(x,t)$ with respect to t to show that

$$\frac{\partial g(x,t)}{\partial t} = -2(t - x)g(x,t)$$

Proceed as in part (b) to verify the recurrence relationship

$$H_{n+1}(x) - 2xH_n(x) + 2nH_{n-1}(x) = 0$$

(e) Use the preceding recurrence relationships to show that

$$H_n''(x) - 2xH_n'(x) + 2nH_n(x) = 0$$

Hence, $H_n(x)$ satisfies the differential equation

$$y'' - 2xy' + 2ny = 0$$

(f) Show that the Hermite polynomials which satisfy the preceding differential equation form an orthogonal set with respect to the weight function e^{-x^2}.

11.30 A function $f(x)$ defined over the interval $-\infty < x < \infty$ may be represented by the series

$$f(x) = \sum_{n=0}^{\infty} C_n H_n(x)$$

Use the result that

$$\int_{-\infty}^{\infty} H_m(x) H_n(x) e^{-x^2}\, dx = \begin{cases} 0 & n \neq m \\ \sqrt{\pi}\, 2^n(n!) & n = m \end{cases}$$

to obtain a general equation for evaluating the coefficient, C_n.

REFERENCES

1 Watson, G. N.: "A Treatise on the Theory of Bessel Functions," 2d ed., The Macmillan Company, New York, 1944.
2 Whittaker, E. T., and G. N. Watson: "A Course of Modern Analysis," The Macmillan Company, New York, 1947.
3 Jahnke, E., F. Emde, and F. Lösch: "Tables of Higher Functions," 6th ed., McGraw-Hill Book Company, New York, 1960.

4 Erdêlyi, A., W. Magnus, F. Oberhettinger, and F. G. Tricomi: "Higher Transcendental Functions," McGraw-Hill Book Company, New York, 1953, 1955.

5 Erdêlyi, A., W. Magnus, F. Oberhettinger, and F. G. Tricomi: "Tables of Integral Transforms," McGraw-Hill Book Company, New York, 1954.

6 Sneddon, I. N.: "Fourier Transforms," McGraw-Hill Book Company, New York, 1951.

7 Tranter, C. J.: "Integral Transforms in Mathematical Physics," John Wiley & Sons, Inc., New York, 1951.

8 Kreysig, E.: "Advanced Engineering Mathematics," John Wiley & Sons, Inc., New York, 1962.

chapter 12

Matrices and numerical methods

Matrix calculus provides a powerful method for the analysis of linear equations, nonlinear equations, systems of differential equations, linear transformations, etc. A particularly valuable aspect of matrices is that they lend themselves to solution by modern computers. Matrix methods can often be combined with other engineering techniques to yield a better method of analysis. For example, new methods for investigating the behavior of complex control systems have resulted from a combination of the theory of matrices, differential equations, and dynamic programming.

12.1 MATRICES

Consider the two simultaneous equations

$$
\begin{aligned}
y_1 &= a_{11}x_1 + a_{12}x_2 \\
y_2 &= a_{21}x_1 + a_{22}x_2
\end{aligned}
\tag{12.1}
$$

where a_{11}, a_{12}, a_{21}, and a_{22} are coefficients. The independent variables are x_1 and x_2, and the dependent variables are y_1 and y_2. An example of a matrix is obtained by arranging the coefficients in the same manner in which they occur above. That is,

$$
\begin{bmatrix}
a_{11} & a_{12} \\
a_{21} & a_{22}
\end{bmatrix}
$$

A matrix is defined as a rectangular array of elements. Thus, an $m \times n$ (read "m by n") matrix is one that contains m rows and

n columns. That is,

$$
\begin{bmatrix}
a_{11} & a_{12} & \cdot \cdot \cdot & a_{1n} \\
a_{21} & a_{22} & \cdot \cdot \cdot & a_{2n} \\
\cdot & \cdot & \cdot \cdot \cdot \cdot \cdot \cdot & \cdot \\
a_{m1} & a_{m2} & \cdot \cdot \cdot & a_{mn}
\end{bmatrix} = [a_{ij}] = [a] = A \tag{12.2}
$$

The first subscript designates the row and the second subscript the column. Thus, a_{jk} is the element of the jth row and kth column. For convenience the entire matrix may be represented by $[a_{ij}]$ or $[a]$ or A.

Addition and subtraction

The sum $A + B$ is defined as the matrix that results by adding corresponding elements in A and B. Similarly, the difference A-B is defined as the matrix that results when corresponding elements in B are subtracted from A.

Example Let

$$
A = \begin{bmatrix} 2 & 5 \\ 7 & 6 \end{bmatrix} \quad \text{and} \quad B = \begin{bmatrix} 1 & 4 \\ 3 & 8 \end{bmatrix} \tag{12.3}
$$

Then

$$
A + B = \begin{bmatrix} 2+1 & 5+4 \\ 7+3 & 6+8 \end{bmatrix} = \begin{bmatrix} 3 & 9 \\ 10 & 14 \end{bmatrix}
$$

$$
A - B = \begin{bmatrix} 2-1 & 5-4 \\ 7-3 & 6-8 \end{bmatrix} = \begin{bmatrix} 1 & 1 \\ 4 & -2 \end{bmatrix}
$$

In general, it follows that

$$
\begin{aligned}
A + B &= [a_{ij} + b_{ij}] \\
A - B &= [a_{ij} - b_{ij}]
\end{aligned} \tag{12.4}
$$

It is to be noted that addition and subtraction are defined only for matrices which have the same number of rows m and the same number of columns n.

From the preceding definitions, it follows that

$$
\begin{aligned}
A + B &= B + A \\
c(A + B) &= cA + cB
\end{aligned} \tag{12.5}
$$

where c is a scalar quantity.

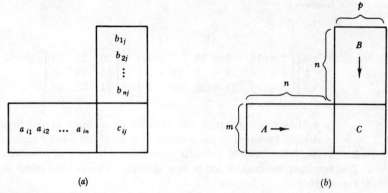

(a)   (b)

Fig. 12.1. Skeletal diagrams to illustrate matrix multiplication.

Multiplication

The product $C = AB$ is defined as the matrix whose element c_{ij} is

$$c_{ij} = a_{i1}b_{1j} + a_{i2}b_{2j} + \cdots + a_{in}b_{nj} \qquad (12.6)$$

As illustrated in Fig. 12.1a, the element c_{ij} is the sum of the individual products of corresponding terms in the ith row of A and the jth column of B. The matrix A is represented by the lower left rectangle and B by the upper right rectangle.

Example Determine the product AB for the two matrices A and B given by Eq. (12.3).

Solution

$$\begin{bmatrix} & 1 & & 4 \\ & 3 & & 8 \end{bmatrix}$$
$$\begin{bmatrix} 2 & 5 \\ 7 & 6 \end{bmatrix}\begin{bmatrix} 2+15 & 8+40 \\ 7+18 & 28+48 \end{bmatrix} = \begin{bmatrix} 17 & 48 \\ 25 & 76 \end{bmatrix}$$

Example Determine the product AB for the two matrices

$$A = \begin{bmatrix} 1 & 6 \\ 4 & 2 \\ 3 & 5 \end{bmatrix} \qquad B = \begin{bmatrix} 2 & 5 & 7 \\ 3 & 4 & 1 \end{bmatrix}$$

Solution

$$
\begin{bmatrix} 2 & 5 & 7 \\ 3 & 4 & 1 \end{bmatrix}
$$

$$
\begin{bmatrix} 1 & 6 \\ 4 & 2 \\ 3 & 5 \end{bmatrix}
\begin{bmatrix} 2+18 & 5+24 & 7+6 \\ 8+\ 6 & 20+\ 8 & 28+2 \\ 6+15 & 15+20 & 21+5 \end{bmatrix}
=
\begin{bmatrix} 20 & 29 & 13 \\ 14 & 28 & 30 \\ 21 & 35 & 26 \end{bmatrix}
$$

In matrix multiplication, the number of columns in the first matrix A must be equal to the number of rows of the second matrix B. Thus, if A is an $m \times n$ matrix, then B must be an $n \times p$ matrix. The resulting product is an $m \times p$ matrix. This is illustrated in Fig. 12.1b.

For the second of the foregoing examples, A is a 3×2 matrix, and B is a 2×3 matrix; hence the product is a 3×3 matrix.

When $m = p$, it is possible to obtain the product BA. For the matrices of the second example B is 2×3, and A is 3×2; hence the product BA is a 2×2 matrix. That is,

$$
\begin{bmatrix} 1 & 6 \\ 4 & 2 \\ 3 & 5 \end{bmatrix}
$$

$$
\begin{bmatrix} 2 & 5 & 7 \\ 3 & 4 & 1 \end{bmatrix}
\begin{bmatrix} 2+20+21 & 12+10+35 \\ 3+16+\ 3 & 18+\ 8+\ 5 \end{bmatrix}
=
\begin{bmatrix} 43 & 57 \\ 22 & 31 \end{bmatrix}
$$

The preceding skeletal forms are employed as an aid in visualizing the pattern of matrix multiplication. The usual form for expressing the preceding product is

$$
\begin{bmatrix} 2 & 5 & 7 \\ 3 & 4 & 1 \end{bmatrix}
\begin{bmatrix} 1 & 6 \\ 4 & 2 \\ 3 & 5 \end{bmatrix}
=
\begin{bmatrix} 43 & 57 \\ 22 & 31 \end{bmatrix}
$$

As illustrated in the preceding, matrix multiplication is not commutative. Thus,

$$
AB \neq BA \tag{12.7}
$$

It may be shown that matrix multiplication is associative and distributive. That is,

$$
\begin{aligned}
(AB)C &= A(BC) = ABC \\
A(B + C) &= AB + AC \\
(B + C)A &= BA + CA
\end{aligned} \tag{12.8}
$$

Transformation of coordinate systems

Suppose that the relationships for transforming from the x_1x_2 coordinate system to the y_1y_2 system are

$$y_1 = a_{11}x_1 + a_{12}x_2$$
$$y_2 = a_{21}x_1 + a_{22}x_2$$

Similarly, the transformation for going from this y_1y_2 coordinate system to a z_1z_2 system is

$$z_1 = b_{11}y_1 + b_{12}y_2$$
$$z_2 = b_{21}y_1 + b_{22}y_2$$

After much mathematical manipulation, it is found that the transformation for going directly from the x_1x_2 system to the z_1z_2 system is

$$z_1 = (b_{11}a_{11} + b_{12}a_{21})x_1 + (b_{11}a_{12} + b_{12}a_{22})x_2$$
$$z_2 = (b_{21}a_{11} + b_{22}a_{21})x_1 + (b_{21}a_{12} + b_{22}a_{22})x_2$$

(12.9)

With the aid of matrix notation, the preceding result may be obtained in a very direct manner. First of all, the transformation equations are written in the matrix form,

$$\begin{bmatrix} y_1 \\ y_2 \end{bmatrix} = \begin{bmatrix} a_{11} & a_{12} \\ a_{21} & a_{22} \end{bmatrix} \begin{bmatrix} x_1 \\ x_2 \end{bmatrix} = \begin{matrix} a_{11}x_1 + a_{12}x_2 \\ a_{21}x_1 + a_{22}x_2 \end{matrix}$$

$$\begin{bmatrix} z_1 \\ z_2 \end{bmatrix} = \begin{bmatrix} b_{11} & b_{12} \\ b_{21} & b_{22} \end{bmatrix} \begin{bmatrix} y_1 \\ y_2 \end{bmatrix} = \begin{matrix} b_{11}y_1 + b_{12}y_2 \\ b_{21}y_1 + b_{22}y_2 \end{matrix}$$

A matrix which has but one column is called a *column matrix*, or *column vector*. Similarly, a matrix which has but one row is called a *row matrix*, or *row vector*. The preceding matrix relationships may be written in the symbolic form

$$\mathbf{y} = A\mathbf{x} \tag{12.10}$$
$$\mathbf{z} = B\mathbf{y} \tag{12.11}$$

Column matrices, or vectors, are distinguished by boldface letters. Thus, **x**, **y**, and **z** designate the respective column matrices, and A and B designate the respective coefficient matrices. The substitution of **y** from Eq. (12.10) into Eq. (12.11) gives

$$\mathbf{z} = BA\mathbf{x}$$

The matrix product BA is

$$BA = \begin{bmatrix} b_{11} & b_{12} \\ b_{21} & b_{22} \end{bmatrix} \begin{bmatrix} a_{11} & a_{12} \\ a_{21} & a_{22} \end{bmatrix}$$

$$= \begin{bmatrix} b_{11}a_{11} + b_{12}a_{21} & b_{11}a_{12} + b_{12}a_{22} \\ b_{21}a_{11} + b_{22}a_{21} & b_{21}a_{12} + b_{22}a_{22} \end{bmatrix}$$

This then verifies the transformation given by Eq. (12.9).

12.2 SYSTEMS OF EQUATIONS

Consider the following set of n equations in n unknowns:

$$\begin{aligned} a_{11}x_1 + a_{12}x_2 + \cdots + a_{1n}x_n &= b_1 \\ a_{21}x_1 + a_{22}x_2 + \cdots + a_{2n}x_n &= b_2 \\ \cdots\cdots\cdots\cdots\cdots\cdots\cdots\cdots\cdots \\ a_{n1}x_1 + a_{n2}x_2 + \cdots + a_{nn}x_n &= b_n \end{aligned} \qquad (12.12)$$

Cramer's theorem states that, if the determinant D of a system of n linear equations in n unknowns is not zero, then

$$x_1 = \frac{D_1}{D} \qquad x_2 = \frac{D_2}{D} \qquad \cdots \qquad x_n = \frac{D_n}{D}$$

where D_k is the determinant obtained by replacing the kth column in D by the column of elements $b_1, b_2, \ldots, b_n$. To distinguish a determinant from a matrix, a determinant is enclosed by vertical lines, while a matrix is enclosed by brackets. The determinant D of the above system is

$$D = \begin{vmatrix} a_{11} & a_{12} & \cdots & a_{1n} \\ a_{21} & a_{22} & \cdots & a_{2n} \\ \cdots\cdots\cdots\cdots\cdots\cdots \\ a_{n1} & a_{n2} & \cdots & a_{nn} \end{vmatrix}$$

The determinant formed by crossing out the ith row and jth column is called the minor M_{ij}. The cofactor C_{ij} is

$$C_{ij} = (-1)^{i+j}M_{ij} \qquad (12.13)$$

The sign of the cofactor depends upon whether $i + j$ is odd or even. A determinant D may be expressed in terms of the cofactors of each element of the ith row as follows,

$$D = a_{i1}C_{i1} + a_{i2}C_{i2} + \cdots + a_{in}C_{in}$$

where $a_{i1}, a_{i2}, \ldots, a_{in}$ are the elements of the ith row. Similarly, a determinant D may be expressed in terms of the elements of the jth column as follows,

$$D = a_{1j}C_{1j} + a_{2j}C_{2j} + \cdots + a_{nj}C_{nj}$$

where $a_{1j}, a_{2j}, \ldots, a_{nj}$ are the elements of the jth column. With the preceding expansions, any $n \times n$ determinant may eventually be reduced to a sum of 3×3 determinants. Each 3×3 determinant may then be evaluated by the usual criss-cross pattern illustrated below:

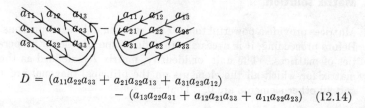

$$D = (a_{11}a_{22}a_{33} + a_{21}a_{32}a_{13} + a_{31}a_{23}a_{12})$$
$$- (a_{13}a_{22}a_{31} + a_{12}a_{21}a_{33} + a_{11}a_{32}a_{23}) \quad (12.14)$$

Example Let it be desired to determine the value of x_2 that satisfies the following system:

$$2x_1 + 5x_2 - x_3 = 4$$
$$x_1 + 3x_2 + 4x_3 = 6 \quad\quad (12.15)$$
$$5x_1 - x_2 + x_3 = 3$$

Solution The values of the determinants D_2 and D are

$$x_2 = \frac{D_2}{D} = \frac{\begin{vmatrix} 2 & 4 & -1 \\ 1 & 6 & 4 \\ 5 & 3 & 1 \end{vmatrix}}{\begin{vmatrix} 2 & 5 & -1 \\ 1 & 3 & 4 \\ 5 & -1 & 1 \end{vmatrix}} = \frac{91}{125}$$

Some operational properties of determinants are:

1. The value of a determinant is not altered if the rows and columns are interchanged. That is, the first row becomes the first column, the second row becomes the second column, etc. Because of this property, the following properties apply just as well to columns as to rows.
2. If each element of a row is multiplied by a constant k, the value of the determinant is multiplied by k (that is, kD).
3. Interchanging any two rows of a determinant changes the sign of the determinant.

4. Multiplying each element of a row by a constant k and then adding each resulting element to the corresponding element in another row does not affect the value of the determinant.

5. When all the elements of a row are zero, the value of the determinant is zero.

6. When each element of a row is equal to a constant k times each corresponding element in another row (i.e., the rows are proportional), the value of the determinant is zero.

Matrix solution

Matrices provide a powerful tool for the solution of linear equations. Before proceeding, it is necessary to define some additional properties of matrices. The unit, or identity, matrix I is defined as the matrix for which all the elements on the principal diagonal are 1 and all other elements are 0. That is,

$$I = \begin{bmatrix} 1 & & & 0 \\ & 1 & & \\ & & \cdot & \\ & & & \cdot \\ 0 & & & 1 \end{bmatrix} \tag{12.16}$$

Multiplying a matrix by the identity matrix does not change the value of the matrix. Thus,

$$I\mathbf{x} = \mathbf{x} \tag{12.17}$$

The inverse A^{-1} of a matrix A is the matrix such that

$$A^{-1}A = I \tag{12.18}$$

The matrix notation for Eq. (12.12) is

$$A\mathbf{x} = \mathbf{b}$$

Multiplying both sides by A^{-1} gives

$$A^{-1}A\mathbf{x} = A^{-1}\mathbf{b}$$

or

$$\mathbf{x} = A^{-1}\mathbf{b} \tag{12.19}$$

Thus, the solution is reduced to the problem of determining the inverse A^{-1}. The inverse of a square matrix exists only if the determinant of A is not zero (that is, $|A| \neq 0$). The inverse is given by the relationship

$$A^{-1} = \frac{\text{adj } A}{|A|} \tag{12.20}$$

The adjoint of a matrix A (adj A) is the matrix formed when each element is replaced by its cofactor and then the resulting matrix is transposed. A matrix is transposed by interchanging its rows and columns.

Example Determine the inverse for the coefficient matrix of Eq. (12.15). That is,

$$A = \begin{bmatrix} 2 & 5 & -1 \\ 1 & 3 & 4 \\ 5 & -1 & 1 \end{bmatrix} \tag{12.21}$$

Solution Replacing each element by its cofactor gives

$$\begin{bmatrix} 7 & 19 & -16 \\ -4 & 7 & 27 \\ 23 & -9 & 1 \end{bmatrix}$$

Because the value of the determinant is 125, then transposing and then dividing by $|A|$ gives

$$A^{-1} = \frac{1}{125} \begin{bmatrix} 7 & -4 & 23 \\ 19 & 7 & -9 \\ -16 & 27 & 1 \end{bmatrix} \tag{12.22}$$

In matrix form, the solution, $\mathbf{x} = A^{-1}\mathbf{b}$, is

$$\begin{bmatrix} x_1 \\ x_2 \\ x_3 \end{bmatrix} = \frac{1}{125} \begin{bmatrix} 7 & -4 & 23 \\ 19 & 7 & -9 \\ -16 & 27 & 1 \end{bmatrix} \begin{bmatrix} 4 \\ 6 \\ 3 \end{bmatrix} \tag{12.23}$$

The relationship for x_2 is

$$x_2 = \frac{(19)(4) + (7)(6) - (9)(3)}{125} = \frac{91}{125}$$

This verifies the result previously attained. In a similar manner, the values of x_1 and x_3 may also be computed.

The inverse of a matrix may be found by a method which involves elementary operations only. Such elementary operations include the interchanging of rows, multiplying a row by a constant, and adding one row to another. In this method, one begins by writing the unit matrix to the right of the given matrix. To obtain the inverse of Eq. (12.21), first write

$$
\begin{array}{ccc|ccc}
2 & 5 & -1 & 1 & 0 & 0 \\
1 & 3 & 4 & 0 & 1 & 0 \\
5 & -1 & 1 & 0 & 0 & 1
\end{array}
\tag{12.24}
$$

Elementary operations are now performed on the rows to convert the matrix on the left to the unit matrix. The resulting matrix on the right will be the desired inverse. Multiplying row 1 by $-\frac{1}{2}$ and adding to row 2, then multiplying row 1 by $-\frac{5}{2}$ and adding to row 3 gives

$$
\begin{array}{ccc|ccc}
2 & 5 & -1 & 1 & 0 & 0 \\
0 & \frac{1}{2} & \frac{9}{2} & -\frac{1}{2} & 1 & 0 \\
0 & -\frac{27}{2} & \frac{7}{2} & -\frac{5}{2} & 0 & 1
\end{array}
$$

Multiplying row 2 by -10 and adding to row 1, then multiplying row 2 by 27 and adding to row 3 gives

$$
\begin{array}{ccc|ccc}
2 & 0 & -46 & 6 & -10 & 0 \\
0 & \frac{1}{2} & \frac{9}{2} & -\frac{1}{2} & 1 & 0 \\
0 & 0 & 125 & -16 & 27 & 1
\end{array}
$$

Multiplying row 3 by $\frac{46}{125}$ and adding to row 1, then multiplying row 3 by $-\frac{9}{250}$ and adding to row 2 gives

$$
\begin{array}{ccc|ccc}
2 & 0 & 0 & \frac{14}{125} & -\frac{8}{125} & \frac{46}{125} \\
0 & \frac{1}{2} & 0 & \frac{19}{250} & \frac{7}{250} & -\frac{9}{250} \\
0 & 0 & 125 & -16 & 27 & 1
\end{array}
$$

Finally, multiply row 1 by $\frac{1}{2}$, row 2 by 2, and row 3 by $\frac{1}{125}$. Thus,

$$
\begin{array}{ccc|ccc}
1 & 0 & 0 & \frac{7}{125} & -\frac{4}{125} & \frac{23}{125} \\
0 & 1 & 0 & \frac{19}{125} & \frac{7}{125} & -\frac{9}{125} \\
0 & 0 & 1 & -\frac{16}{125} & \frac{27}{125} & \frac{1}{125}
\end{array}
$$

The matrix on the right is seen to be the desired inverse given by Eq. (12.22).

Rank

The rank of a matrix is a fundamental concept in the theory of matrices.

Consider the following 2×3 matrix:

$$
\begin{bmatrix}
a_{11} & a_{12} & a_{13} \\
a_{21} & a_{22} & a_{23}
\end{bmatrix}
$$

From any matrix, other submatrices may be formed by striking out rows or columns. The above matrix contains three 2 × 2 square matrices. That is,

$$\begin{bmatrix} a_{11} & a_{12} \\ a_{21} & a_{22} \end{bmatrix} \qquad \begin{bmatrix} a_{11} & a_{13} \\ a_{21} & a_{23} \end{bmatrix} \qquad \begin{bmatrix} a_{12} & a_{13} \\ a_{22} & a_{23} \end{bmatrix}$$

Because the original matrix has six elements, six 1 × 1 square matrices can be formed. It is possible to form the determinant for a square matrix.

A matrix is said to be of rank r if there exists at least one $r \times r$ square matrix whose determinant is not zero, whereas the determinants for all larger square matrices are zero.

Example Determine the rank of the following 3 × 4 matrix:

$$\begin{bmatrix} 3 & 5 & 0 & 1 \\ 1 & 0 & -4 & 3 \\ 2 & 5 & 4 & -2 \end{bmatrix} \qquad\qquad (12.25)$$

The determinant for each 3 × 3 matrix is

$$\begin{vmatrix} 3 & 5 & 0 \\ 1 & 0 & -4 \\ 2 & 5 & 4 \end{vmatrix} = 0 \qquad \begin{vmatrix} 3 & 5 & 1 \\ 1 & 0 & 3 \\ 2 & 5 & -2 \end{vmatrix} = 0$$

$$\begin{vmatrix} 3 & 0 & 1 \\ 1 & -4 & 3 \\ 2 & 4 & -2 \end{vmatrix} = 0 \qquad \begin{vmatrix} 5 & 0 & 1 \\ 0 & -4 & 3 \\ 5 & 4 & -2 \end{vmatrix} = 0$$

Because there is at least one 2 × 2 square matrix whose determinant is not 0, the rank of this matrix is 2.

The rank of a matrix may also be obtained by elementary operations. For example, in Eq. (12.25) multiplying column 1 by 4 and adding to column 3, then multiplying column 1 by −3 and adding to column 4 gives

$$\begin{bmatrix} 3 & 5 & 12 & -8 \\ 1 & 0 & 0 & 0 \\ 2 & 5 & 12 & -8 \end{bmatrix}$$

Multiplying row 2 by −3 and adding to row 1, then multiplying row 2 by −2 and adding to row 3 gives

$$\begin{bmatrix} 0 & 5 & 12 & -8 \\ 1 & 0 & 0 & 0 \\ 0 & 5 & 12 & -8 \end{bmatrix}$$

Multiply row 1 by -1 and add to row 3, then multiply column 2 by $-\frac{12}{5}$ and add to column 3, etc., and the matrix may be reduced to the following form, which is obviously of rank 2:

$$\begin{bmatrix} 0 & 1 & 0 & 0 \\ 1 & 0 & 0 & 0 \\ 0 & 0 & 0 & 0 \end{bmatrix}$$

A system of n equations with n unknowns has a solution if the determinant of the matrix coefficient is not zero. In other words, the rank of the coefficient matrix must be $r = n$.

12.3 HOMOGENEOUS EQUATIONS

A homogeneous equation is one for which the right-hand side is zero. For example, consider the following set of m homogeneous equations in n unknowns:

$$\begin{aligned} a_{11}x_1 + a_{12}x_2 + \cdots + a_{1n}x_n &= 0 \\ a_{21}x_1 + a_{22}x_2 + \cdots + a_{2n}x_n &= 0 \\ &\cdots \cdots \cdots \cdots \cdots \cdots \cdots \cdots \\ a_{m1}x_1 + a_{m2}x_2 + \cdots + a_{mn}x_n &= 0 \end{aligned} \tag{12.26}$$

If the coefficient matrix is of rank r, then r equations suffice to describe the system. Thus, there are m-r redundant equations. For r independent equations, there can be only r unknowns. Thus, n-r unknowns are arbitrary. By taking these n-r arbitrary unknowns to the right-hand side of the r equations, the resulting system has the form of Eq. (12.12). The solution is then ascertained by the application of Cramer's rule.

Example Determine the solution of the equations

$$\begin{aligned} 3x_1 + 5x_2 + 0x_3 + x_4 &= 0 \\ x_1 + 0x_2 - 4x_3 + 3x_4 &= 0 \\ 2x_1 + 5x_2 + 4x_3 - 2x_4 &= 0 \end{aligned}$$

Solution The coefficient matrix for this system is given by Eq. (12.25). Because this matrix is of rank 2, then two equations suffice to describe the system. That is, the solution for the first two equations automatically satisfies the third. The number of arbitrary unknowns is $n - r = 2$. By regarding x_3

and x_4 as the arbitrary unknowns, the first two equations become

$$3x_1 + 5x_2 = 0x_3 - x_4$$
$$x_1 + 0x_2 = 4x_3 - 3x_4$$

The system has now been reduced to two equations in the two unknowns x_1 and x_2. The corresponding solution is

$$x_1 = 4x_3 - 3x_4$$
$$x_2 = \frac{-12x_3 + 8x_4}{5}$$

Let us now consider a system of m nonhomogeneous equations in n unknowns. That is,

$$
\begin{aligned}
a_{11}x_1 + a_{12}x_2 + \cdots + a_{1n}x_n &= b_1 \\
a_{21}x_1 + a_{22}x_2 + \cdots + a_{2n}x_n &= b_2 \\
\cdots\cdots\cdots\cdots\cdots\cdots\cdots\cdots\cdots \\
a_{m1}x_1 + a_{m2}x_2 + \cdots + a_{mn}x_n &= b_m
\end{aligned}
\tag{12.27}
$$

This system is converted to a system of homogeneous equations by writing it in the form

$$
\begin{aligned}
a_{11}x_1 + a_{12}x_2 + \cdots + a_{1n}x_n + b_1x_{n+1} &= 0 \\
a_{21}x_1 + a_{22}x_2 + \cdots + a_{2n}x_n + b_2x_{n+1} &= 0 \\
\cdots\cdots\cdots\cdots\cdots\cdots\cdots\cdots\cdots\cdots \\
a_{m1}x_1 + a_{m2}x_2 + \cdots + a_{mn}x_n + b_mx_{n+1} &= 0
\end{aligned}
\tag{12.28}
$$

where $x_{n+1} = -1$. By setting $x_{n+1} = -1$, then the solution is obtained in the same manner as for a set of homogeneous equations. The matrix of the coefficients of Eq. (12.28), in which the last column is the b column, is called the *augmented matrix*.

Example Determine the solution of the following set of equations:

$$
\begin{aligned}
x_1 + 3x_2 - x_3 &= 4 \\
2x_1 - 4x_2 + x_3 &= 3 \\
5x_1 - 5x_2 + x_3 &= 10
\end{aligned}
$$

Solution Expressing this system in the form of Eq. (12.28) gives

$$
\begin{aligned}
x_1 + 3x_2 - x_3 + 4x_4 &= 0 \\
2x_1 - 4x_2 + x_3 + 3x_4 &= 0 \\
5x_1 - 5x_2 + x_3 + 10x_4 &= 0
\end{aligned}
\tag{12.29}
$$

The augmented matrix is

$$\begin{bmatrix} 1 & 3 & -1 & 4 \\ 2 & -4 & 1 & 3 \\ 5 & -5 & 1 & 10 \end{bmatrix}$$

The rank r of this matrix is 2. Thus, only the first two equations are needed. There are $n - r = 3 - 2 = 1$ arbitrary unknowns. Let x_3 be the arbitrary unknown. The term x_4 is not unknown, because it is necessary that $x_4 = -1$. To solve for the unknowns x_1 and x_2, the first two equations of Eq. (12.29) are written in the form

$$x_1 + 3x_2 = x_3 - 4x_4$$
$$2x_1 - 4x_2 = -x_3 - 3x_4$$

The coefficient matrix A is

$$A = \begin{bmatrix} 1 & 3 \\ 2 & -4 \end{bmatrix}$$

For a 2×2 matrix the cofactors are merely the opposite elements, with care taken to provide the correct sign. That is,

$$\begin{bmatrix} -4 & -2 \\ -3 & 1 \end{bmatrix}$$

Transposing this matrix and dividing by $|A| = -10$ gives

$$A^{-1} = -\frac{1}{10}\begin{bmatrix} -4 & -3 \\ -2 & 1 \end{bmatrix} = \begin{bmatrix} 0.4 & 0.3 \\ 0.2 & -0.1 \end{bmatrix}$$

Thus, in matrix form the solution $\mathbf{x} = A^{-1}\mathbf{b}$ is

$$\begin{bmatrix} x_1 \\ x_2 \end{bmatrix} = \begin{bmatrix} 0.4 & 0.3 \\ 0.2 & -0.1 \end{bmatrix}\begin{bmatrix} (x_3 - 4x_4) \\ (-x_3 - 3x_4) \end{bmatrix}$$

Performing the matrix multiplication, and then letting $x_4 = -1$, gives for the solution

$$x_1 = 0.4(x_3 - 4x_4) + 0.3(-x_3 - 3x_4) = 0.1x_3 + 2.5$$
$$x_2 = 0.2(x_3 - 4x_4) - 0.1(-x_3 - 3x_4) = 0.3x_3 + 0.5$$

12.4 EIGENVALUES

A linear system is one whose behavior is described by a differential equation or a set of simultaneous differential equations. Such systems may be represented by a set of n first-order differential equations of the form

$$\dot{y}_1 = a_{11}y_1 + a_{12}y_2 + \cdots + a_{1n}y_n + b_1f(t)$$
$$\dot{y}_2 = a_{21}y_1 + a_{22}y_2 + \cdots + a_{2n}y_n + b_2f(t)$$
$$\cdots\cdots\cdots\cdots\cdots\cdots\cdots\cdots\cdots\cdots\cdots\cdots$$
$$\dot{y}_n = a_{n1}y_1 + a_{n2}y_2 + \cdots + a_{nn}y_n + b_nf(t)$$
(12.30)

This is referred to as the *state-space representation* for the system. The corresponding matrix form is

$$\dot{\mathbf{y}}(t) = A\mathbf{y}(t) + \mathbf{b}f(t)$$
(12.31)

The vector $\mathbf{y}(t)$ is called the *state of the system*, and the components $(y_1, y_2, \ldots, y_n)$ are referred to as the *state variables*. The forcing function, or excitation, is $f(t)$. By letting $\mathbf{y}(t) = \mathbf{x}e^{\lambda t}$, where $\mathbf{x}$ is a constant vector and λ a scalar, then $\dot{\mathbf{y}}(t) = \lambda\mathbf{x}e^{\lambda t}$. Thus, Eq. (12.30) becomes

$$\lambda\mathbf{x} = A\mathbf{x} + \mathbf{b}f(t)e^{-\lambda t}$$
(12.32)

For the case in which there is no forcing function, the solution is called the *force-free solution*. In the following development, the method for obtaining the force-free solution is described first. It is then shown that the complete solution is obtained by an extension of the solution for the force-free case.

When there is no forcing function, Eqs. (12.31) and (12.32) become

$$\dot{\mathbf{y}}(t) = A\mathbf{y}(t)$$
(12.33)
$$\lambda\mathbf{x} = A\mathbf{x}$$
(12.34)

This last equation may be written in the form

$$(\lambda I - A)\mathbf{x} = 0$$

Expanding gives

$$\begin{bmatrix} (\lambda - a_{11}) & -a_{12} & \cdots & -a_{1n} \\ -a_{21} & (\lambda - a_{22}) & \cdots & -a_{2n} \\ \cdots & \cdots & \cdots & \cdots \\ \cdots & \cdots & \cdots & \cdots \\ \cdots & \cdots & \cdots & \cdots \\ -a_{n1} & -a_{n2} & \cdots & (\lambda - a_{nn}) \end{bmatrix} \begin{bmatrix} x_1 \\ x_2 \\ \cdot \\ \cdot \\ \cdot \\ x_n \end{bmatrix} = 0$$

In order that this system of n equations in n unknowns have a solution, it is necessary that the determinant of $\lambda I - A$ vanish. That is,

$$|\lambda I - A| = 0$$
(12.35)

This determinant is an nth-degree polynomial in λ. Equation (12.35) is referred to as the *characteristic equation*, or *characteristic polynomial*, for the system. The n values of λ which cause the determinant to vanish are the values of λ for which solutions of the system exist. The n values (that is, $\lambda_1, \lambda_2, \ldots, \lambda_n$) are called the *eigenvalues* of the system. Eigenvalues are also referred to as *characteristic values, characteristic roots, natural modes,* and *free frequencies*.

For $\lambda = \lambda_j$ ($j = 1, 2, \ldots, n$), then Eq. (12.34) becomes

$$\lambda_j \begin{bmatrix} x_{1j} \\ x_{2j} \\ \cdot \\ \cdot \\ \cdot \\ x_{nj} \end{bmatrix} = A \begin{bmatrix} x_{1j} \\ x_{2j} \\ \cdot \\ \cdot \\ \cdot \\ x_{nj} \end{bmatrix}$$

where $x_{1j}, x_{2j}, \ldots, x_{nj}$ are the components of the solution vector corresponding to λ_j. Each solution vector is a column matrix which is designated $\mathbf{x}_j$. That is,

$$\mathbf{x}_j = \begin{bmatrix} x_{1j} \\ x_{2j} \\ \cdot \\ \cdot \\ \cdot \\ x_{nj} \end{bmatrix} \tag{12.36}$$

A solution vector $\mathbf{x}_j$ is also called an *eigenvector*, or *modal column*. The entire set of solutions corresponding to the n eigenvalues is embodied in the following relationship

$$\begin{bmatrix} x_{11} & x_{12} & \cdots & x_{1n} \\ x_{21} & x_{22} & \cdots & x_{2n} \\ \cdot & \cdot & & \cdot \\ \cdot & \cdot & & \cdot \\ \cdot & \cdot & & \cdot \\ x_{n1} & x_{n2} & \cdots & x_{nn} \end{bmatrix} \begin{bmatrix} \lambda_1 & & & 0 \\ & \lambda_2 & & \\ & & \cdot & \\ & & & \cdot \\ 0 & & & \lambda_n \end{bmatrix}$$

$$= \begin{bmatrix} a_{11} & a_{12} & \cdots & a_{1n} \\ a_{21} & a_{22} & \cdots & a_{2n} \\ \cdot & \cdot & & \cdot \\ a_{n1} & a_{n2} & \cdots & a_{nn} \end{bmatrix} \begin{bmatrix} x_{11} & x_{12} & \cdots & x_{1n} \\ x_{21} & x_{22} & \cdots & x_{2n} \\ \cdot & \cdot & & \cdot \\ x_{n1} & x_{n2} & \cdots & x_{nn} \end{bmatrix}$$

or

$$X\Lambda = AX \tag{12.37}$$

where X is the matrix whose columns are the solution vectors $(x_1, x_2, \ldots, x_n)$ and Λ (lambda) is the diagonal matrix that displays the eigenvalues on its diagonal. The matrix X is called the *modal column matrix*. Multiplication of Eq. (12.37) by X^{-1} shows that

$$\Lambda = X^{-1}AX \qquad (12.38)$$

A transformation of the form $X^{-1}AX$ is called a *similarity transformation*.

Canonical form

The solution of Eq. (12.33) is most readily ascertained by transforming it to the canonical form. This is accomplished by letting $y = Xq$. Thus,

$$X\dot{q} = AXq$$

Multiplying by X^{-1} gives

$$X^{-1}X\dot{q} = X^{-1}AXq$$

Because $X^{-1}X = I$ and $X^{-1}AX = \Lambda$, then

$$\dot{q} = \Lambda q \qquad (12.39)$$

Expanding shows that then

$$
\begin{bmatrix} \dot{q}_1 \\ \dot{q}_2 \\ \cdot \\ \cdot \\ \cdot \\ \dot{q}_n \end{bmatrix} = \begin{bmatrix} \lambda_1 & & & 0 \\ & \lambda_2 & & \\ & & \cdot & \\ & & & \cdot \\ 0 & & & \lambda_n \end{bmatrix} \begin{bmatrix} q_1 \\ q_2 \\ \cdot \\ \cdot \\ q_n \end{bmatrix} \qquad (12.40)
$$

Thus, the original system has been transformed to a set of n differential equations of the form

$$\dot{q}_j = \lambda_j q_j \qquad j = 1, 2, \ldots, n \qquad (12.41)$$

The components $q_1, q_2, \ldots, q_n$ are called the *normal coordinates*, or *canonical variables*. Integrating Eq. (12.41) to obtain the solution gives

$$\ln q_j = \lambda_j t + \ln C$$

or

$$\frac{q_j}{C} = e^{\lambda_j t}$$

Evaluating the preceding expression at $t = 0$ shows that $C = q_j(0)$. Thus, the desired solution is

$$q_j = q_j(0)e^{\lambda_j t}$$

The matrix form of the solution is

$$\mathbf{q}(t) = \begin{bmatrix} e^{\lambda_1 t} & & & & 0 \\ & e^{\lambda_2 t} & & & \\ & & \cdot & & \\ & & & \cdot & \\ & & & & \cdot \\ 0 & & & & e^{\lambda_n t} \end{bmatrix} \mathbf{q}(0) = [e^{\lambda_i t}]\mathbf{q}^{(0)} \qquad (12.42)$$

This is the force-free solution expressed in terms of the canonical variables. This result is converted back to the original coordinates by application of the transform $\mathbf{y}(t) = X\mathbf{q}(t)$.

The complete solution is obtained by replacing $\mathbf{y}(t)$ by $X\mathbf{q}(t)$ in Eq. (12.31), and then multiplying through by X^{-1}. Thus,

$$\begin{aligned} \dot{\mathbf{q}}(t) &= X^{-1}AX\mathbf{q}(t) + X^{-1}\mathbf{b}f(t) \\ &= \Lambda\mathbf{q}(t) + X^{-1}\mathbf{b}f(t) \end{aligned} \qquad (12.43)$$

This is the canonical form of the complete solution. After solving this system of equations for the canonical variables $\mathbf{q}(t)$, then multiplication by X yields the desired solution: $\mathbf{y}(t) = X\mathbf{q}(t)$.

Illustrative example Let it be desired to determine the solution of the following differential equation:

$$(D + 1)(D + 2)(D + 5)c = (D^3 + 8D^2 + 17D + 10)c = f(t) \qquad (12.44)$$

The initial conditions are $c(0) = 1$ and $\dot{c}(0) = \ddot{c}(0) = 0$. The forcing function is

$$f(t) = \begin{cases} 10 & t < 0 \\ h + 10 & t > 0 \end{cases}$$

Solution By letting $c = y_1$, $\dot{c} = \dot{y}_1 = y_2$, and $\ddot{c} = \ddot{y}_1 = \dot{y}_2 = y_3$, the preceding differential equation may be written in the state-space form

$$\begin{aligned} \dot{y}_1 &= y_2 \\ \dot{y}_2 &= y_3 \\ \dot{y}_3 &= -10y_1 - 17y_2 - 8y_3 + f(t) \end{aligned}$$

The corresponding matrix representation is

$$\begin{bmatrix} \dot{y}_1 \\ \dot{y}_2 \\ \dot{y}_3 \end{bmatrix} = \begin{bmatrix} 0 & 1 & 0 \\ 0 & 0 & 1 \\ -10 & -17 & -8 \end{bmatrix} \begin{bmatrix} y_1 \\ y_2 \\ y_3 \end{bmatrix} + \begin{bmatrix} 0 \\ 0 \\ 1 \end{bmatrix} f(t)$$

This has the form $\dot{\mathbf{y}} = A\mathbf{y} + \mathbf{b}f(t)$. The eigenvalues are $\lambda_1 = -1$, $\lambda_2 = -2$, and $\lambda_3 = -5$. The modal column $\mathbf{x}_1$ corresponding to λ_1 is obtained by letting $\lambda = \lambda_1 = -1$ in the equation $(\lambda I - A)\mathbf{x} = 0$. Thus

$$\begin{bmatrix} -1 & -1 & 0 \\ 0 & -1 & -1 \\ 10 & 17 & 7 \end{bmatrix} \begin{bmatrix} x_{11} \\ x_{21} \\ x_{31} \end{bmatrix} = 0$$

This matrix yields the three algebraic equations

$$\begin{aligned} x_{11} + \quad x_{21} \qquad\qquad &= 0 \\ x_{21} + \quad x_{31} &= 0 \\ 10x_{11} + 17x_{21} + 7x_{31} &= 0 \end{aligned}$$

Because $c_j\mathbf{x}_j$ satisfies Eq. (12.34) as well as $\mathbf{x}_j$, then any modal column may be multiplied by an arbitrary constant c_j. Usually the constant c_j is taken such that the first component x_{1j} of each modal column is unity. Thus, taking $x_{11} = 1$, the first of the preceding equations shows that $x_{21} = -x_{11} = -1$. The second equation shows that $x_{31} = -x_{21} = 1$. The third relationship is automatically satisfied. By proceeding in this manner, the three modal columns are found to be

$$\mathbf{x}_1 = \begin{bmatrix} 1 \\ -1 \\ 1 \end{bmatrix} \qquad \mathbf{x}_2 = \begin{bmatrix} 1 \\ -2 \\ 4 \end{bmatrix} \qquad \mathbf{x}_3 = \begin{bmatrix} 1 \\ -5 \\ 25 \end{bmatrix}$$

Thus, the modal column matrix is

$$X = \begin{bmatrix} 1 & 1 & 1 \\ -1 & -2 & -5 \\ 1 & 4 & 25 \end{bmatrix} \tag{12.45}$$

The inverse X^{-1} is

$$X^{-1} = \frac{1}{12} \begin{bmatrix} 30 & 21 & 3 \\ -20 & -24 & -4 \\ 2 & 3 & 1 \end{bmatrix} \tag{12.46}$$

From the preceding results, the canonical form is found to be

$$
\begin{bmatrix} \dot{q}_1 \\ \dot{q}_2 \\ \dot{q}_3 \end{bmatrix} = \begin{bmatrix} -1 & 0 & 0 \\ 0 & -2 & 0 \\ 0 & 0 & -5 \end{bmatrix} \begin{bmatrix} q_1 \\ q_2 \\ q_3 \end{bmatrix} + \frac{1}{12} \begin{bmatrix} 3 \\ -4 \\ 1 \end{bmatrix} f(t)
$$

The corresponding differential equations are

$(D + 1)q_1(t) = \frac{3}{12}f(t)$
$(D + 2)q_2(t) = -\frac{4}{12}f(t)$
$(D + 5)q_3(t) = \frac{1}{12}f(t)$

The $\mathbf{q}(0)$ initial conditions are found by use of the relationship $\mathbf{q}(0) = X^{-1}\mathbf{y}(0)$. Thus, it is found that $q_1(0) = \frac{30}{12}$, $q_2(0) = -\frac{20}{12}$, and $q_3(0) = \frac{2}{12}$. With these initial conditions, the preceding set of differential equations may be solved for q_1, q_2, and q_3. Thus,

$q_1 = 3[(h + 10) - he^{-t}]/12$
$q_2 = -2[(h + 10) - he^{-2t}]/12$
$q_3 = 0.2[(h + 10) - he^{-5t}]/12$

Application of the transform $\mathbf{y}(t) = X\mathbf{q}(t)$ gives

$y_1 = q_1 + q_2 + q_3$
$y_2 = -q_1 - 2q_2 - 5q_3$
$y_3 = q_1 + 4q_2 + 25q_3$

In particular, the desired result $y_1 = c(t)$ is

$c(t) = (h + 10)/10 + h(-3e^{-t} + 2e^{-2t} - 0.2e^{-5t})/12.$

It is to be noted that with the use of matrices, the solution of this third-order differential equation has been reduced to the solution of three first-order differential equations.

The complete solution may be expressed in terms of the convolution integral. Thus, transforming Eq. (12.39) gives

$$sQ(s) - \mathbf{q}(0) = \Lambda Q(s)$$

or

$$Q(s) = [sI - \Lambda]^{-1}\mathbf{q}(0)$$

Inverting yields

$$q(t) = \mathcal{L}^{-1}\{[sI - \Lambda]^{-1}\}\mathbf{q}(0)$$

Comparison of this result with Eq. (12.42) shows that

$$\mathcal{L}^{-1}[sI - \Lambda] = [e^{\lambda_i t}] \tag{12.47}$$

Next, transforming Eq. (12.43) and solving for $Q(s)$ gives

$$Q(s) = [sI - \Lambda]^{-1}q(0) + [sI - \Lambda]^{-1}[X^{-1}\mathbf{b}F(s)]$$

Inverting yields the complete solution in terms of the canonical variables.

$$\mathbf{q}(t) = [e^{\lambda_i t}]\mathbf{q}(0) + [e^{\lambda_i t}] * [X^{-1}\mathbf{b}f(t)] \tag{12.48}$$

To obtain the complete solution in terms of the original variables, replace $\mathbf{q}(0)$ by $X^{-1}\mathbf{y}(0)$, and then multiply through by X. Thus,

$$\mathbf{y}(t) = X\mathbf{q}(t) = \Phi(t)\mathbf{y}(0) + \Phi(t) * \mathbf{b}f(t) \tag{12.49}$$

where

$$\Phi(t) = X[e^{\lambda_i t}]X^{-1}$$

The matrix $\Phi(t)$ is called the state transition matrix in that it operates on the initial state of the system $\mathbf{y}(0)$ and the forcing function $f(t)$ to yield the subsequent state of the system $\mathbf{y}(t)$. This convolution technique usually requires more computational effort than the method described in the preceding example in which the canonical form of the differential equations is solved directly.

12.5 COMPUTER SOLUTIONS

This technique yields not only the computer diagram but also an alternate method for obtaining the modal matrix X. Multiplying each column of X by an arbitrary constant has the effect of dividing each row of X^{-1} by the corresponding constant. Thus, if the first column of X in the preceding example [Eq. (12.45)] is multiplied by $\frac{3}{12}$, the second column by $-\frac{4}{12}$, and the third by $\frac{1}{12}$, then the corresponding X and X^{-1} matrices are

$$X = \frac{1}{12}\begin{bmatrix} 3 & -4 & 1 \\ -3 & 8 & -5 \\ 3 & -16 & 25 \end{bmatrix} \qquad X^{-1} = \begin{bmatrix} 10 & 7 & 1 \\ 5 & 6 & 1 \\ 2 & 3 & 1 \end{bmatrix} \tag{12.50}$$

Any corresponding set of X and X^{-1} matrices may be used for the solution of a problem. Thus, X and X^{-1} may be taken as the

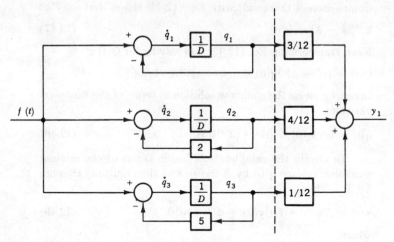

Fig. 12.2. Computer diagram for solution of Eq. (12.44).

matrices in which each row of X^{-1} is multiplied by a constant such that each coefficient of $X^{-1}\mathbf{b}$ is unity. For the preceding example, $\dot{\mathbf{q}} = \Lambda\mathbf{q} + X^{-1}\mathbf{b}f(t)$ may be written in the form

$$\dot{\mathbf{q}} = \begin{bmatrix} -1 & 0 & 0 \\ 0 & -2 & 0 \\ 0 & 0 & -5 \end{bmatrix} \mathbf{q} + \begin{bmatrix} 1 \\ 1 \\ 1 \end{bmatrix} f(t) \qquad (12.51)$$

The three corresponding equations are

$$\dot{q}_1 = -q_1 + f(t) \qquad \dot{q}_2 = -2q_2 + f(t) \qquad \dot{q}_3 = -5q_3 + f(t)$$

The portion of the computer diagram to the left of the dotted line in Fig. 12.2 may now be constructed. Replacing $\dot{q}_1$ by Dq_1 and solving for q_1, etc., gives

$$q_1 = \frac{f(t)}{D+1} \qquad q_2 = \frac{f(t)}{D+2} \qquad q_3 = \frac{f(t)}{D+5} \qquad (12.52)$$

The remaining portion of the computer diagram is now obtained by writing the partial fraction expansion for Eq. (12.44). That is,

$$y_1 = c(t) = \frac{3}{12}\frac{f(t)}{D+1} - \frac{4}{12}\frac{f(t)}{D+2} + \frac{1}{12}\frac{f(t)}{D+5}$$

$$= \frac{3}{12}q_1 - \frac{4}{12}q_2 + \frac{1}{12}q_3$$

Because $\mathbf{y} = X\mathbf{q}$, the coefficients ($\frac{3}{12}$, $-\frac{4}{12}$, and $\frac{1}{12}$) are the first row of the coefficients of the modal matrix X [Eq. (12.50)].

Multiplying Eq. (12.44) by D to obtain $y_2 = \dot{c}(t) = Dc(t)$, and then performing a partial fraction expansion, gives

$$y_2 = Dc(t) = -\frac{3}{12}\frac{f(t)}{D+1} + \frac{8}{12}\frac{f(t)}{D+2} - \frac{5}{12}\frac{f(t)}{D+5}$$

$$= -\frac{3}{12}q_1 + \frac{8}{12}q_2 - \frac{5}{12}q_3$$

These coefficients ($-\frac{3}{12}$, $\frac{8}{12}$, and $-\frac{5}{12}$) are the coefficients for the second row of X. The last row is similarly obtained from the partial fraction expansion for $y_3 = D^2c(t)$. Thus,

$$y_3 = D^2c(t) = \tfrac{3}{12}q_1 - \tfrac{16}{12}q_2 + \tfrac{25}{12}q_3$$

The only difference between the computer diagrams for y_1, y_2, and y_3 is the value of the coefficients for combining q_1, q_2, and q_3. These coefficients affect the portion of the computer diagram to the right of the dotted line of Fig. 12.2. The solution may be accomplished on either an analog or a digital computer.

Complex conjugate eigenvalues

When the eigenvalues are complex, then the matrix Λ and the modal matrix X also have complex elements. A major disadvantage in using complex elements is that it is not possible to construct a computer diagram for imaginary elements. Fortunately, there is an alternative matrix representation for complex conjugate eigenvalues that employs real elements only. This matrix form for a second order system is

$$\dot{\mathbf{q}} = \Lambda\mathbf{q} + X^{-1}\mathbf{b}f(t) = \begin{bmatrix} a & b \\ -b & a \end{bmatrix}\mathbf{q} + \begin{bmatrix} 0 \\ 1 \end{bmatrix}f(t) \qquad (12.53)$$

The corresponding differential equations are

$$\begin{aligned} \dot{q}_1 &= aq_1 + bq_2 \\ \dot{q}_2 &= -bq_1 + aq_2 + f(t) \end{aligned} \qquad (12.54)$$

The computer diagram for this set of equations is shown in Fig. 12.3. The quadratic form is revealed by replacing $\dot{q}_1$ by Dq_1 and $\dot{q}_2$ by Dq_2 in the preceding equations. This gives

$$(D^2 - 2aD + a^2 + b^2)q_1 = bf(t) \qquad (12.55)$$

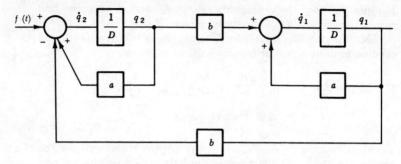

Fig. 12.3. Computer diagram for complex conjugate eigenvalues.

To illustrate this method, let it be desired to determine the computer diagram for solving the following differential equation:

$$(D + 1)(D^2 + 6D + 25)c(t) = 20(D + 3)f(t) \qquad (12.56)$$

The roots of the characteristic equation are -1 and $-3 \pm j4$. Thus, the canonical form $\dot{\mathbf{q}} = \Lambda\mathbf{q} + X^{-1}\mathbf{b}f(t)$ is

$$\dot{\mathbf{q}} = \begin{bmatrix} -1 & 0 & 0 \\ \hdashline 0 & -3 & 4 \\ 0 & -4 & -3 \end{bmatrix} \mathbf{q} + \begin{bmatrix} 1 \\ 0 \\ 1 \end{bmatrix} f(t) \qquad (12.57)$$

The dotted partitioning in Λ is used to distinguish the real eigenvalue from the complex conjugate pair. The three corresponding equations are

$$\dot{q}_1 = -q_1 + f(t)$$
$$\dot{q}_2 = -3q_2 + 4q_3$$
$$\dot{q}_3 = -4q_2 - 3q_3 + f(t)$$

The portion of the computer diagram to the left of the dotted line of Fig. 12.4 may now be constructed. By replacing $\dot{q}_1$ by Dq_1, the first equation may be written in the form

$$q_1 = \frac{f(t)}{D + 1}$$

Similarly, upon replacing $\dot{q}_2$ by Dq_2 and $\dot{q}_3$ by Dq_3, then q_3 may be eliminated from the last two equations. Thus

$$q_2 = \frac{4f(t)}{D^2 + 6D + 25}$$

The remaining portion of the computer diagram may now be ascertained by writing the partial fraction expansion for Eq. (12.56). That is,

$$y_1 = c(t) = \left(\frac{2}{D+1} - \frac{2D-10}{D^2 + 6D + 25} \right) f(t)$$

$$= 2q_1 - \frac{2Dq_2}{4} + \frac{10q_2}{4} = 2q_1 - 0.5\dot{q}_2 + 2.5q_2 \quad (12.58)$$

The right side of this equation yields directly the portion of the computer diagram to the right of the dotted line in Fig. 12.4.

Because $\dot{q}_2 = -3q_2 + 4q_3$, then Eq. (12.58) may be written in the form

$$y_1 = c(t) = 2q_1 + 4q_2 - 2q_3 \quad (12.59)$$

Equation (12.59) yields directly the first row of the modal column matrix. That is,

$$\begin{bmatrix} y_1 \\ y_2 \\ y_3 \end{bmatrix} = \begin{bmatrix} 2 & 4 & -2 \\ x_{21} & x_{22} & x_{23} \\ x_{31} & x_{32} & x_{33} \end{bmatrix} \begin{bmatrix} q_1 \\ q_2 \\ q_3 \end{bmatrix}$$

By similarly writing y_2 and y_3 as partial fractions, then the remaining rows of the modal matrix can be determined.

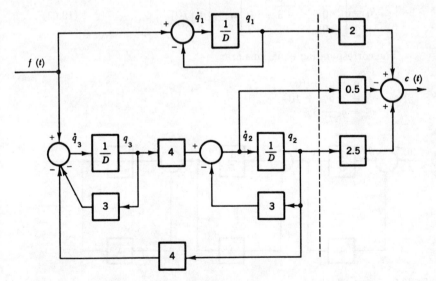

Fig. 12.4. Computer diagram for Eq. (12.56).

Repeated eigenvalues

For repeated eigenvalues the canonical form is

$$\dot{\mathbf{q}} = \begin{bmatrix} \lambda & 0 & 0 \\ 1 & \lambda & 0 \\ 0 & 1 & \lambda \end{bmatrix} \mathbf{q} + \begin{bmatrix} 1 \\ 0 \\ 0 \end{bmatrix} f(t) \tag{12.60}$$

The corresponding differential equations are

$$\begin{aligned}
\dot{q}_1 &= \lambda q_1 + f(t) \\
\dot{q}_2 &= q_1 + \lambda q_2 \\
\dot{q}_3 &= q_2 + \lambda q_3
\end{aligned} \tag{12.61}$$

The computer diagram for this set of equations is shown in Fig. 12.5. Let us now consider the differential equation

$$(D + 1)(D + 3)^2 c(t) = 8(D + 2) f(t) \tag{12.62}$$

The eigenvalues are -1, -3, and -3. Thus, the canonical form $\dot{\mathbf{q}} = \Lambda \mathbf{q} + X^{-1} \mathbf{b} f(t)$ is

$$\dot{\mathbf{q}} = \begin{bmatrix} -1 & 0 & 0 \\ \hline 0 & -3 & 0 \\ 0 & 1 & -3 \end{bmatrix} \mathbf{q} + \begin{bmatrix} 1 \\ 1 \\ 0 \end{bmatrix} f(t) \tag{12.63}$$

The corresponding equations are

$$\begin{aligned}
\dot{q}_1 &= -q_1 + f(t) \\
\dot{q}_2 &= -3q_2 + f(t) \\
\dot{q}_3 &= q_2 - 3q_3
\end{aligned}$$

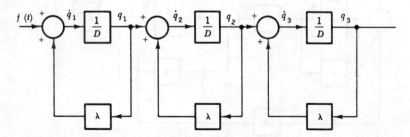

Fig. 12.5. Computer diagram for repeated eigenvalues.

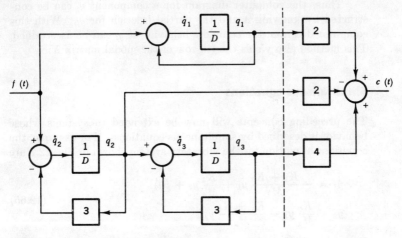

Fig. 12.6. Computer diagram for Eq. (12.62).

The portion of the computer diagram to the left of the dotted line in Fig. 12.6 may now be constructed. Replacing $\dot{q}$ by Dq in each of these equations and eliminating q_2 from the last equation shows that

$$q_1 = \frac{f(t)}{D + 1}$$

$$q_2 = \frac{f(t)}{D + 3}$$

$$q_3 = \frac{f(t)}{(D + 3)^2}$$

The remaining portion of the computer diagram is obtained from the partial fraction expansion. That is,

$$y_1 = c(t) = \left[\frac{2}{D + 1} - \frac{2}{D + 3} + \frac{4}{(D + 3)^2} \right] f(t)$$

$$= 2q_1 - 2q_2 + 4q_3 \qquad (12.64)$$

The preceding coefficients are the first row of the modal matrix X. That is,

$$\begin{bmatrix} y_1 \\ y_2 \\ y_3 \end{bmatrix} = \begin{bmatrix} 2 & -2 & 4 \\ x_{21} & x_{22} & x_{23} \\ x_{31} & x_{32} & x_{33} \end{bmatrix} \begin{bmatrix} q_1 \\ q_2 \\ q_3 \end{bmatrix}$$

495

Thus, the computer diagram for a component y_j can be constructed by knowing Λ and the partial fraction for y_j. With this computer diagram, the complete behavior of y_j can be ascertained. This method also yields the jth row of the modal matrix X.

Simultaneous equations

The preceding concepts will now be extended to systems whose behavior is described by simultaneous equations. For example, the equations of motion for the vibratory system shown in Fig. 12.7 are

$$\ddot{y}_1 = -\frac{K_1 + K_2}{M_1} y_1 + \frac{K_2}{M_1} y_2 + f(t)$$
$$\ddot{y}_2 = \frac{K_2}{M_2} y_1 - \frac{K_1 + K_2}{M_2} y_2 \tag{12.65}$$

The general form of these equations is

$$\ddot{\mathbf{y}} = A\mathbf{y} + \mathbf{b}f(t) \tag{12.66}$$

By letting $\mathbf{y} = \mathbf{x}e^{\lambda t}$, the general form becomes

$$\lambda^2 \mathbf{x} = A\mathbf{x} + \mathbf{b}f(t)e^{-\lambda t}$$

Thus, the characteristic equation is

$$\lambda^2 I - A = 0 \tag{12.67}$$

Illustrative example For the system of Fig. 12.7, determine the computer diagrams for y_1 and y_2. The system parameters are $K_1 = K_2 = K_3 = M_1 = M_2 = 1$.

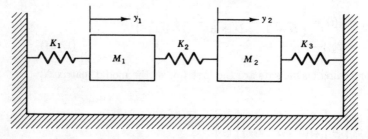

Fig. 12.7. Vibratory system described by simultaneous equations.

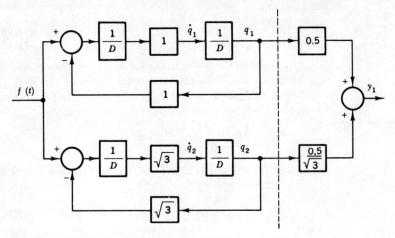

Fig. 12.8. Computer diagram for system of Fig. 12.7.

Solution For these values, the equations of motion are

$$\ddot{y}_1 = -2y_1 + y_2 + f(t)$$
$$\ddot{y}_2 = y_1 - 2y_2$$

(12.68)

By letting $\dot{y}_1 = y_a$ and $\dot{y}_2 = y_b$, then the state-space representation for this system is

$$\dot{y}_1 = 0 + y_a + 0 + 0$$
$$\dot{y}_a = -2y_1 + 0 + y_2 + 0 + f(t)$$
$$\dot{y}_2 = 0 + 0 + 0 + y_b$$
$$\dot{y}_b = y_1 + 0 - 2y_2 + 0$$

The determinant of the characteristic equation is

$$\begin{vmatrix} \lambda^2 + 2 & 1 \\ 1 & \lambda^2 + 2 \end{vmatrix} = \lambda^4 + 4\lambda^2 + 3 = (\lambda^2 + 1)(\lambda^2 + 3)$$

Thus, the four eigenvalues are

$$\lambda = \pm j \quad \text{and} \quad \lambda = \pm j\sqrt{3}$$

The corresponding canonical form $\dot{\mathbf{q}} = \Lambda\mathbf{q} + X^{-1}\mathbf{b}f(t)$ is

$$\begin{bmatrix} \dot{q}_1 \\ \dot{q}_a \\ \dot{q}_2 \\ \dot{q}_b \end{bmatrix} = \begin{bmatrix} 0 & 1 & 0 & 0 \\ -1 & 0 & 0 & 0 \\ 0 & 0 & 0 & \sqrt{3} \\ 0 & 0 & -\sqrt{3} & 0 \end{bmatrix} \begin{bmatrix} q_1 \\ q_a \\ q_2 \\ q_b \end{bmatrix} + \begin{bmatrix} 0 \\ 1 \\ 0 \\ 1 \end{bmatrix} f(t)$$

The portion of the computer diagram to the left of the dotted line in Fig. 12.8 may now be constructed. The equations for

497

q_1 and q_2 are

$$(D^2 + 1)q_1 = f(t) \tag{12.69}$$
$$(D^2 + 3)q_2 = \sqrt{3}\,f(t) \tag{12.70}$$

After solving these equations for q_1 and q_2, then q_a and q_b may be determined from the relationships $q_a = \dot{q}_1$ and $q_b = \dot{q}_2$.

Replacing $\ddot{y}_1$ by $D^2 y_1$ and $\ddot{y}_2$ by $D^2 y_2$ in Eq. (12.68), solving for y_1, and then performing a partial fraction expansion gives

$$y_1 = \frac{(D^2 + 2)f(t)}{(D^2 + 1)(D^2 + 3)} = \frac{0.5f(t)}{D^2 + 1} + \frac{0.5f(t)}{D^2 + 3}$$

Using Eqs. (12.69) and (12.70) to express this result in terms of q_1 and q_2 gives

$$y_1 = 0.5q_1 + \frac{0.5}{\sqrt{3}}q_2 \tag{12.71}$$

The computer diagram may now be completed.

By similarly solving Eq. (12.68) for y_2 and expanding, then

$$y_2 = \frac{f(t)}{(D^2 + 1)(D^2 + 3)} = \frac{0.5f(t)}{D^2 + 1} - \frac{0.5f(t)}{D^2 + 3} = 0.5q_1 - \frac{0.5}{\sqrt{3}}q_2 \tag{12.72}$$

The computer diagram for y_2 is the same as that for y_1 except that the $(0.5/\sqrt{3})q_2$ term should be subtracted at the final summing point in Fig. 12.8.

Differentiating y_1 and y_2 shows that

$$\dot{y}_1 = 0.5\dot{q}_1 + (0.5/\sqrt{3})\dot{q}_2 \quad \text{and} \quad \dot{y}_2 = 0.5\dot{q}_1 - (0.5/\sqrt{3})\dot{q}_2$$

Because $\dot{y}_1 = y_a$, $\dot{y}_2 = y_b$, $\dot{q}_1 = q_a$, and $\dot{q}_2 = \sqrt{3}\,q_b$, then

$$y_a = 0.5q_a + 0.5q_b$$
$$y_b = 0.5q_a - 0.5q_b$$

Thus, the $\mathbf{y} = X\mathbf{q}$ transformation is

$$\begin{bmatrix} y_1 \\ y_a \\ y_2 \\ y_b \end{bmatrix} = \begin{bmatrix} 0.5 & 0 & \dfrac{0.5}{\sqrt{3}} & 0 \\ 0 & 0.5 & 0 & 0.5 \\ 0.5 & 0 & -\dfrac{0.5}{\sqrt{3}} & 0 \\ 0 & 0.5 & 0 & -0.5 \end{bmatrix} \begin{bmatrix} q_1 \\ q_a \\ q_2 \\ q_b \end{bmatrix}$$

Because y_1 and y_2 are independent of q_a and q_b, the simplified matrix representation for y_1 and y_2 is

$$\begin{bmatrix} y_1 \\ y_2 \end{bmatrix} = \begin{bmatrix} 0.5 & 0.5/\sqrt{3} \\ 0.5 & -0.5/\sqrt{3} \end{bmatrix} \begin{bmatrix} q_1 \\ q_2 \end{bmatrix}$$

For a given $f(t)$, then q_1 and q_2 may be obtained from Eqs. (12.69) and (12.70), respectively. For the force-free case, these equations reduce to

$$(D^2 + 1)q_1 = 0$$
$$(D^2 + 3)q_2 = 0$$

The corresponding solutions are

$$
\begin{aligned}
q_1 &= A_1 \sin (\omega_1 t + \psi_1) & \omega_1 &= \sqrt{1} \\
q_2 &= A_2 \sin (\omega_2 t + \psi_2) & \omega_2 &= \sqrt{3}
\end{aligned}
\qquad (12.73)
$$

The solutions q_1 and q_2 for the force-free case are referred to as the *principal modes*. It is to be noted that the frequency of oscillation for each mode is determined by the eigenvalues for the system. The remaining constants are determined from the initial conditions for the particular problem.

The general force-free solutions y_1 and y_2 are linear combinations of the principal modes q_1 and q_2. That is,

$$\begin{bmatrix} y_1 \\ y_2 \end{bmatrix} = \begin{bmatrix} x_{11} & x_{12} \\ x_{21} & x_{22} \end{bmatrix} \begin{bmatrix} q_1 \\ q_2 \end{bmatrix}$$

Thus, y_1 and y_2 may be expressed in the general form

$$
\begin{aligned}
y_1 &= A_{11} \sin (\omega_1 t + \psi_1) + A_{12} \sin (\omega_2 t + \psi_2) \\
y_2 &= A_{21} \sin (\omega_1 t + \psi_1) + A_{22} \sin (\omega_2 t + \psi_2)
\end{aligned}
$$

where $A_{11} = x_{11}A_1$, $A_{12} = x_{12}A_2$, $A_{21} = x_{21}A_1$, and $A_{22} = x_{22}A_2$.

Consider now the following set of simultaneous differential equations:

$$
\begin{aligned}
\ddot{y}_1 &= -32y_1 - 8\dot{y}_1 + 39y_2 + f(t) \\
\ddot{y}_2 &= 5y_1 - 10y_2
\end{aligned}
\qquad (12.74)
$$

By letting $y = xe^{\lambda t}$, then

$$
\begin{aligned}
\lambda^2 x_1 &= -(8\lambda + 32)x_1 + 39x_2 + f(t)e^{-\lambda t} \\
\lambda^2 x_2 &= -5x_1 - 10x_2
\end{aligned}
$$

This has the form $\lambda^2 \mathbf{x} = A\mathbf{x} + \mathbf{b}f(t)e^{-\lambda t}$. The determinant of the characteristic equation is

$$\begin{vmatrix} (\lambda^2 + 8\lambda + 32) & 39 \\ 5 & (\lambda^2 + 10) \end{vmatrix}$$

The roots of the characteristic equation are $\lambda = -1 \pm j2$ and $\lambda = -3 \pm j4$. By letting $y_a = \dot{y}_1$ and $y_b = \dot{y}_2$, then the state-space representation is

$$\dot{y}_1 = 0 + y_a + 0 + 0$$
$$\dot{y}_a = -32y_1 - 8y_a - 39y_2 + 0 + f(t)$$
$$\dot{y}_2 = 0 + 0 + 0 + y_b$$
$$\dot{y}_b = 5y_1 + 0 + 0 - 10y_2 + 0$$

The corresponding canonical form is

$$\begin{bmatrix} \dot{q}_1 \\ \dot{q}_a \\ \dot{q}_2 \\ \dot{q}_b \end{bmatrix} = \begin{bmatrix} -1 & 2 & 0 & 0 \\ -2 & -1 & 0 & 0 \\ 0 & 0 & -3 & 4 \\ 0 & 0 & -4 & -3 \end{bmatrix} \begin{bmatrix} q_1 \\ q_a \\ q_2 \\ q_b \end{bmatrix} + \begin{bmatrix} 0 \\ 1 \\ 0 \\ 1 \end{bmatrix} f(t)$$

The differential equations for q_1 and q_2 are

$$(D^2 + 2d + 5)q_1 = 2f(t)$$
$$(D^2 + 6D + 25)q_2 = 4f(t)$$

After solving these differential equations for q_1 and q_2, then q_a and q_b may be determined from the relationships

$$q_a = (\dot{q}_1 + q_1)/2 \qquad \text{and} \qquad q_b = (\dot{q}_2 + 3q_2)/4$$

The first row of the modal matrix X is obtained by solving the original set of simultaneous equations for y_1 and then performing a partial fraction expansion. Thus,

$$y_1 = \frac{(D^2 + 10)f(t)}{(D^2 + 2D + 5)(D^2 + 6D + 25)}$$
$$= \left(\frac{aD + b}{D^2 + 2D + 5} + \frac{cD + d}{D^2 + 6D + 25} \right) f(t)$$

Multiplying through by the quadratic $(D^2 + 2D + 5)$, and then taking the limit as D approaches the root $-1 + j2$, gives

$$(b - a) + j2a = \frac{D^2 + 10}{D^2 + 2D + 25} \Big|_{D=-1+j2} = \frac{10 - j15}{40}$$

Equating real and imaginary parts shows that $a = -\frac{3}{16}$ and $b = \frac{1}{16}$. Similarly, it is found that $c = \frac{3}{16}$ and $d = \frac{27}{16}$. Thus,

$$y_1 = \frac{1}{16} \left(\frac{-3D + 1}{D^2 + 2D + 5} + \frac{7D + 27}{D^2 + 6D + 25} \right) f(t)$$
$$= \frac{(-3D + 1)}{16} \frac{q_1}{2} + \frac{(7D + 27)}{16} \frac{q_2}{4}$$

Because $Dq_1 = -q_1 + 2q_a$ and $Dq_2 = -3q_2 + 4q_b$, then the preceding becomes

$$y_1 = \tfrac{1}{32}(4q_1 - 6q_a + 9q_2 + 6q_b)$$

Thus, the coefficients of the first row of the modal matrix X are $\tfrac{4}{32}$, $-\tfrac{6}{32}$, $\tfrac{9}{32}$, and $\tfrac{6}{32}$. Differentiating the preceding equation and noting that $\dot{y}_1 = y_a$, $\dot{q}_1 = -q_1 + 2q_a$, $\dot{q}_a = -2q_1 - q_a + f(t)$, $\dot{q}_2 = -3q_2 + 4q_b$, and $\dot{q}_b = -4q_2 - 3q_b + f(t)$ gives

$$y_a = \tfrac{1}{32}(8q_1 + 14q_a - 51q_2 + 18q_b)$$

The remaining rows of X are determined in a similar manner. Thus, it is found that

$$X = \frac{1}{32} \begin{bmatrix} 4 & -6 & 9 & 6 \\ 8 & 14 & -51 & 18 \\ 4 & -2 & -1 & 2 \\ 0 & 10 & -5 & -10 \end{bmatrix}$$

With knowledge of X and the canonical form, the computer diagram may be constructed for any of the variables, or the solution may be completed by matrix methods.

12.6 NUMERICAL METHODS

When a problem cannot be solved by the ordinary analytical methods of mathematics, the engineer turns to numerical methods. With numerical methods, data are fed into a digital computer in terms of numbers. Likewise, the solution which is obtained from the computer is in the form of numbers. For example, if the answer for a particular problem were $y = e^{-t}$, the computer would yield corresponding numerical values of t and y (not the equation $y = e^{-t}$).

When possible, an analytical solution is preferred so that the engineer can see in equation form the role of the various system parameters in the solution. With numerical methods, the value of each parameter must be specified. Thus, to investigate the effect of changing a parameter, it is necessary to obtain the solution for each value of the parameter to be considered. Because of the complexity of many engineering situations, analytical solutions are not always feasible. In such cases, numerical analysis provides a means for obtaining concrete results.

Many specialized techniques and methods of numerical analysis have been developed. The method to be used depends upon the particular problem, the information sought, the required accuracy, etc. Some of the basic concepts and fundamental ideas employed in numerical analysis are now described.

Euler's method

To illustrate this method, let it be desired to determine by numerical methods the solution of the following differential equation over the interval $1.0 \leq x \leq 1.5$:

$$\frac{dy}{dx} = 2x \qquad (12.75)$$

The initial conditions are $x_0 = y_0 = 1.0$. By taking equal increments for Δx, then successive values of x are

$$x_1 = x_0 + \Delta x, \, x_2 = x_1 + \Delta x, \, \ldots , \, x_{n+1} = x_n + \Delta x$$

An equation for y_{n+1} may be obtained from the Taylor series expansion about y_n; that is,

$$y_{n+1} = y_n + \Delta x \frac{dy_n}{dx} + \frac{\Delta x^2}{2!} \frac{d^2 y_n}{dx^2} + \cdots \qquad (12.76)$$

In the Euler method, y_{n+1} is approximated by the first two terms of this series. Thus,

$$y_{n+1} = y_n + y_n' \, \Delta x \qquad (12.77)$$

The value of the derivative term in the preceding equation is obtained from the original differential equation. That is,

$$y_n' = \frac{dy_n}{dx} = 2x_n \qquad (12.78)$$

A chain of equations has now been established for computing corresponding values of x_n and y_n. The results for increments of $\Delta x = 0.1$ are given in Table 12.1. This particular differential equation may be integrated directly to obtain the exact solution $y = x^2$. Thus, the solution y_n in Table 12.1 obtained by numerical methods may be compared with the exact solution x^2. It is to be noted that

Table 12.1

x_n	y'_n	y_n	$y = x^2$
$x_0 = 1.0$	$y'_0 = 2.0$	$y_0 = 1.00$	1.00
$x_1 = 1.1$	$y'_1 = 2.2$	$y_1 = 1.20$	1.21
$x_2 = 1.2$	$y'_2 = 2.4$	$y_2 = 1.42$	1.44
$x_3 = 1.3$	$y'_3 = 2.6$	$y_3 = 1.66$	1.69
$x_4 = 1.4$	$y'_4 = 2.8$	$y_4 = 1.92$	1.96
$x_5 = 1.5$	$y'_5 = 3.0$	$y_5 = 2.20$	2.25

the error becomes progressively larger with each successive point. Accuracy can be improved by using smaller increments for Δx.

The preceding concepts may be extended to higher-order differential equations. For example, consider the equation

$$\frac{d^3y}{dx^3} + \frac{d^2y}{dx^2} + \frac{dy}{dx} = 2x \tag{12.79}$$

The initial conditions are y_0, y'_0, y''_0, and x_0. Application of Eq. (12.77) to obtain y_1 gives

$$y_1 = y_0 + y'_0 \Delta x \tag{12.80}$$

Successive differentiation yields equations for evaluating y'_1 and y''_1. Thus

$$y'_1 = y'_0 + y''_0 \Delta x \tag{12.81}$$
$$y''_1 = y''_0 + y'''_0 \Delta x \tag{12.82}$$

The only unknown in these equations is y'''_0 which may be ascertained from Eq. (12.79). That is,

$$y'''_0 = 2x_0 - y'_0 - y''_0 \tag{12.83}$$

Replacing the subscript 1 by n and the subscript 0 by $n - 1$ in Eqs. (12.80) to (12.83) yields a chain process for numerically determining the solution of the differential equation.

Equation (12.77) shows that, in the Euler method, the solution points y_n and y_{n+1} are connected by a straight line (i.e., the tangent y'_n at point y_n). A closer approximation is afforded by the Adams method, which connects these points by a parabola.

Adams' method

The equation for a parabola through the point y_n is

$$y = y_n + a(x - x_n) + b(x - x_n)^2 \tag{12.84}$$

To evaluate the constants a and b, first differentiate

$$\frac{dy}{dx} = a + 2b(x - x_n)$$

When $x = x_n$, then $dy/dx = y_n'$ is the slope at x_n. Similarly, when $x = x_{n-1}$, then $dy/dx = y_{n-1}'$ is the slope at x_{n-1}. Substitution of these boundary conditions into the preceding equation gives $a = y_n'$ and $b = (y_{n-1}' - y_n')/2(x_{n-1} - x_n)$. Because $x_{n-1} - x_n = -\Delta x$, then $b = (y_n' - y_{n-1}')/2\,\Delta x$. Substitution of these values of a and b into Eq. (12.84) gives

$$y = y_n + y_n'(x - x_n) + (y_n' - y_{n-1}') \frac{(x - x_n)^2}{2\,\Delta x}$$

For $y = y_{n+1}$, then $x = x_{n+1}$ and $x_{n+1} - x_n = \Delta x$; thus

$$y_{n+1} = y_n + \left(y_n' + \frac{y_n' - y_{n-1}'}{2} \right) \Delta x \tag{12.85}$$

The technique for solving differential equations by the Adams method is the same as that for the Euler method except that in the Adams method Eq. (12.85) is used in place of Eq. (12.77).

12.7 FINITE DIFFERENCES

An important concept in numerical analysis is that of finite differences. To illustrate this, first consider the equation for a straight line that passes through two points (x_0, y_0) and (x_1, y_1) as shown in Fig. 12.9a. The general equation for a straight line is

$$y = f(x) = a + bx$$

By letting $x = x_0 + k\,\Delta x$, then, when $k = 1$, $x = x_0 + \Delta x = 1.0$. Similarly, when $k = 0.6$, then $x = x_0 + 0.6\,\Delta x$ is six-tenths of the

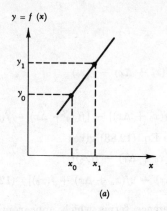

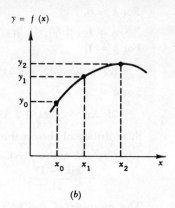

(a)   (b)

Fig. 12.9. Finite differences. (a) Straight line that passes through two points; (b) parabola that passes through three points.

distance from x_0 to x_1. Thus, the general equation for the straight line is

$$y = f(x_0 + k\,\Delta x) = a + b(x_0 + k\,\Delta x)$$
$$= (a + bx_0) + kb\,\Delta x \qquad (12.86)$$

For $k = 0$, the preceding expression shows that $f(x_0) = a + bx_0$.

Subtracting the value of Eq. (12.86) for $k = 0$ from that for $k = 1$ shows that

$$\begin{aligned} f(x_0 + \Delta x) &= (a + bx_0) + b\,\Delta x \\ -f(x_0) \quad &= -(a + bx_0) \\ \hline f(x_0 + \Delta x) - f(x_0) &= b\,\Delta x \end{aligned}$$

Substitution of these results for $b\,\Delta x$ and $a_0 + bx_0$ into Eq. (12.86) gives

$$f(x_0 + k\,\Delta x) = f(x_0) + k[f(x_0 + \Delta x) - f(x_0)] \qquad (12.87)$$

Because $f(x_0 + \Delta x) - f(x_0) = y_1 - y_0$, this is recognized as the equation for linear or straight-line interpolation between two points (i.e., the method of proportional parts).

Improved accuracy can be obtained by putting a parabola through three successive points, as shown in Fig. 12.9b. The general equation for a parabola is $y = f(x) = a + bx + cx^2$. Replacing x by $x_0 + k\,\Delta x$ gives

$$\begin{aligned} f(x_0 + k\,\Delta x) &= a + b(x_0 + k\,\Delta x) + c(x_0 + k\,\Delta x)^2 \\ &= (a + bx_0 + cx_0^2) + k[(b + 2cx_0)\,\Delta x + c\,\Delta x^2] \\ &\qquad + k(k - 1)c\,\Delta x^2 \quad (12.88) \end{aligned}$$

For $k = 0$,

$$a + bx_0 + cx_0{}^2 = f(x_0)$$

For $k = 1$,

$$[(b + 2cx_0) \Delta x + c \Delta x^2] = f(x_0 + \Delta x) - f(x_0)$$

For $k = 2$,

$$2!c \Delta x^2 = [f(x_0 + 2 \Delta x) - f(x_0 + \Delta x)] - [f(x_0 + \Delta x) - f(x_0)]$$

Substitution of these values into Eq. (12.88) gives

$$f(x_0 + k \Delta x) = f(x_0) + k[f(x_0 + \Delta x) - f(x_0)]$$
$$+ \frac{k(k - 1)}{2!} [f(x_0 + 2 \Delta x) - 2f(x_0 + \Delta x) + f(x_0)] \quad (12.89)$$

The meaning of the finite-difference terms which appear in the brackets of the preceding equation is best understood by referring to a difference table. Such a table for the function $f(x) = x^3$ is shown in Table 12.2. The term $\Delta f(x)$ midway to the right of two $f(x)$ terms is the difference between these two terms. Similarly, the difference $\Delta^2 f(x)$ midway to the right of two $\Delta f(x)$ terms is the difference between these terms, etc. A general difference table is illustrated in Table 12.3. It is to be noted that

$$\Delta f_0 = f_1 - f_0$$
$$\Delta^2 f_0 = \Delta f_1 - \Delta f_0 = (f_2 - f_1) - (f_1 - f_0) = f_2 - 2f_1 + f_0$$

Thus, Eqs. (12.87) and (12.89) may be written in the form

$$f(x + k \Delta x) = f_0 + k \Delta f_0$$
$$f(x + k \Delta x) = f_0 + k\Delta f_0 + \frac{k(k - 1)}{2!} \Delta^2 f_0$$

Table 12.2 Difference table for $f(x) = x^3$

x	$f(x)$	$\Delta f(x)$	$\Delta^2 f(x)$	$\Delta^3 f(x)$	$\Delta^4 f(x)$
0	0				
		1			
1	1		6		
		7		6	
2	8		12		0
		19		6	
3	27		18		
		37			
4	64				

Table 12.3   General difference table

x	f	Δf	$\Delta^2 f$	$\Delta^3 f$	$\Delta^4 f$
x_0	f_0				
		Δf_0			
x_1	f_1		$\Delta^2 f_0$		
		Δf_1		$\Delta^3 f_0$	
x_2	f_2		$\Delta^2 f_1$		$\Delta^4 f_0$
		Δf_2		$\Delta^3 f_1$	
x_3	f_3		$\Delta^2 f_2$		
		Δf_3			
x_4	f_4				

By extending this method to an nth-order polynomial which passes through n points, the general form of the interpolation formula becomes

$$f(x + k\,\Delta x) = f_0 + k\,\Delta f_0 + \frac{k(k-1)}{2!}\,\Delta^2 f_0$$

$$+ \frac{k(k-1)(k-2)}{3!}\,\Delta^3 f_0 + \cdots \quad (12.90)$$

This is referred to as the *Gregory-Newton interpolation formula.*

Milne's method

To develop this method, first differentiate Eq. (12.90) with respect to k,

$$\Delta x f'(x_0 + k\,\Delta x) = \Delta f_0 + \frac{2k-1}{2}\,\Delta^2 f_0$$

$$+ \frac{3k^2 - 6k + 2}{6}\,\Delta^3 f_0 + \cdots$$

For $k = 1$, 2, and 3, this expression becomes

$$f'(x + \Delta x) = \frac{1}{\Delta x}\left(\Delta f_0 + \tfrac{1}{2}\Delta^2 f_0 - \tfrac{1}{6}\Delta^3 f_0 + \tfrac{1}{12}\Delta^4 f_0 + \cdots\right)$$

$$f'(x + 2\,\Delta x) = \frac{1}{\Delta x}\left(\Delta f_0 + \tfrac{3}{2}\,\Delta^2 f_0 + \tfrac{1}{3}\,\Delta^3 f_0 - \tfrac{1}{12}\,\Delta^4 f_0 + \cdots\right)$$

$$f'(x + 3\,\Delta x) = \frac{1}{\Delta x}\left(\Delta f_0 + \tfrac{5}{2}\,\Delta^2 f_0 + \tfrac{11}{6}\,\Delta^3 f_0 + \tfrac{1}{4}\,\Delta^4 f_0 + \cdots\right)$$

Neglecting differences beyond the fourth, and replacing the remaining differences by their equivalent functional values (that is,

$\Delta f_0 = f_1 - f_0 = y_1 - y_0, \Delta^2 f_0 = f_2 - 2f_1 + f_0 = y_2 - 2y_1 + y_0$, etc.), gives

$$y_1' = \frac{1}{12\,\Delta x}\,(-3y_0 - 10y_1 + 18y_2 - 6y_3 + y_4)$$

$$y_2' = \frac{1}{12\,\Delta x}\,(y_0 - 8y_1 + 8y_3 - y_4)$$

$$y_3' = \frac{1}{12\,\Delta x}\,(-y_0 + 6y_1 - 18y_2 + 10y_3 + 3y_4)$$

where $y_1' = f'(x + \Delta x)$, $y_2' = f'(x + 2\,\Delta x)$, and $y_3' = f'(x + 3\,\Delta x)$. Multiplication of the sum of the first and last equations by 2, and then subtracting the middle equation, yields the basic equation for the Milne method. That is,

$$y_4 = y_0 + \frac{4\,\Delta x}{3}\,(2y_1' - y_2' + 2y_3')$$

The general form of this expression is

$$y_{n+1} = y_{n-3} + \frac{4\,\Delta x}{3}\,(2y_{n-2}' - y_{n-1}' + 2y_n') \tag{12.91}$$

The Milne method is the same as Euler's or Adams' except that the preceding expression is employed rather than Eq. (12.77) or (12.85). In the Milne equation, it is necessary to know values at four preceding stations. Thus, to get this process under way, the first few stations (y_0, y_1, y_2, and y_3) must be obtained by a method other than Milne's (e.g., Euler's method or the Runge-Kutta method).

Runge-Kutta method

This is one of the most accurate methods of numerical analysis. The basic Runge-Kutta relationship for solving a first-order differential equation of the form $y' = f(x,y)$ is

$$y_{n+1} = y_n + \tfrac{1}{6}(A_n + 2B_n + 2C_n + D_n) \tag{12.92}$$

The constants A_n, B_n, C_n, and D_n are determined as follows:

$$A_n = f(x_n,y_n)\,\Delta x$$

$$B_n = f\!\left(x_n + \frac{\Delta x}{2}, y_n + \frac{A_n}{2}\right)\Delta x$$

$$C_n = f\!\left(x_n + \frac{\Delta x}{2}, y_n + \frac{B_n}{2}\right)\Delta x$$

$$D_n = f(x_{n+1}, y_n + C_n)\,\Delta x$$

Inspection of the preceding relationships shows that the Runge-Kutta method is a self-starting method. That is, the value y_1 can be determined from y_0, and thus y_2 from y_1, etc. In the past, a disadvantage of the Runge-Kutta method has been the laborious computational effort required. However, the advent of the high-speed digital computer has now made the Runge-Kutta method very popular.

Partial differential equations

The partial derivatives $\partial u/\partial x$ and $\partial^2 u/\partial x^2$ are defined by the relationships

$$\frac{\partial u}{\partial x} = \lim_{\Delta x \to 0} \frac{u(x + \Delta x, y) - u(x,y)}{\Delta x}$$

$$\frac{\partial^2 u}{\partial x^2} = \lim_{\Delta x \to 0} \frac{1}{\Delta x}\left[\frac{u(x + \Delta x, y) - u(x,y)}{\Delta x} - \frac{u(x,y) - u(x - \Delta x, y)}{\Delta x} \right]$$

$$= \lim_{\Delta x \to 0} \frac{u(x + \Delta x, y) - 2u(x,y) + u(x - \Delta x, y)}{\Delta x^2}$$

A basic approach in the numerical solution of partial differential equations is to construct a grid of points which covers the region of interest. Such a grid is illustrated in Fig. 12.10. The closer the grid spacing, the more accurate the approximation of the preceding partial derivatives. In accordance with the numbering of Fig. 12.10, let $u(x,y) = u_0$, $u(x + \Delta x, y) = u_1$, and $u(x - \Delta x, y) = u_3$.

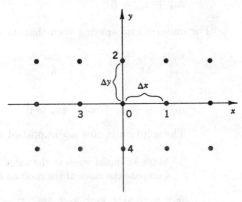

Fig. 12.10. Rectangular grid.

Thus, the difference equations are

$$\frac{\partial u}{\partial x} \approx \Delta_x = \frac{1}{\Delta x}(u_1 - u_0) \tag{12.93}$$

$$\frac{\partial^2 u}{\partial x^2} \approx \Delta_{xx} = \frac{1}{\Delta x^2}(u_1 - 2u_0 + u_3) \tag{12.94}$$

At any point 0, the values of Δ_x and Δ_{xx} are determined by adjacent horizontal values of u.

Partial derivatives with respect to y are defined by the relationships

$$\frac{\partial u}{\partial y} = \lim_{\Delta y \to 0} \frac{u(x, y + \Delta y) - u(x,y)}{\Delta y}$$

$$\frac{\partial^2 u}{\partial y^2} = \lim_{\Delta y \to 0} \frac{u(x, y + \Delta y) - 2u(x,y) + u(x, y - \Delta y)}{\Delta y^2}$$

The corresponding difference equations are

$$\frac{\partial u}{\partial y} \approx \Delta y = \frac{1}{\Delta y}(u_2 - u_0) \tag{12.95}$$

$$\frac{\partial^2 u}{\partial y^2} \approx \Delta_{yy} = \frac{1}{\Delta y^2}(u_2 - 2u_0 + u_4) \tag{12.96}$$

At any point 0, the values of Δ_y and Δ_{yy} are determined by adjacent vertical values of u.

To illustrate the preceding concepts, let it be desired to solve Laplace's equation for two dimensions over some region when the values on the boundary are known. The difference form of Laplace's equation is

$$\Delta_{xx} + \Delta_{yy} = 0$$

For uniform grid spacing such that $\Delta x = \Delta y = h$, then

$$\frac{u_1 - 2u_0 + u_3}{h^2} + \frac{u_2 - 2u_0 + u_4}{h^2} = 0$$

or

$$u_1 + u_2 + u_3 + u_4 - 4u_0 = 0 \tag{12.97}$$

The solution is now accomplished as follows:

1. Make an initial guess at the value of $u(x,y)$ at each grid point.
2. Compute the value of the residual R_0 at each grid point. That is,

$$R_0 = u_1 + u_2 + u_3 + u_4 - 4u_0 \tag{12.98}$$

Note that Eq. (12.97) is not satisfied, for the initial value of u at each point is merely a guess.

3. Readjust the value of u at each point in accordance with the relationship

$$\text{New } u_0 = \text{old } u_0 + \frac{R_0}{4}$$

In systematically proceeding from one grid point to another, new values of u are progressively obtained. Thus, the new value of u for some of the adjacent grid points will be known. Use these latest values, whenever available, in calculating R_0 in accordance with Eq. (12.98).

4. Repeat the preceding step until no residual is found which is greater than some predetermined value. Thus, the solution is within some predetermined degree of accuracy.

This method may be extended for other coordinate systems. For example, for cylindrical coordinates the grid is as shown in Fig. 12.11. The technique described in this section for partial differential equations is referred to as the *relaxation method*. There are numerous methods for solving various types of differential equations. As a detailed discussion of these methods is beyond the scope of this chapter, the reader is referred to the extensive literature on numerical analysis.

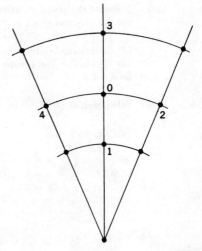

Fig. 12.11. Cylindrical grid.

PROBLEMS

12.1 For the matrices,

$$A = \begin{bmatrix} 3 & 2 & 4 \\ -1 & 5 & 7 \\ 2 & -3 & 6 \end{bmatrix} \quad \text{and} \quad B = \begin{bmatrix} 2 & 4 & -3 \\ 5 & 1 & 2 \\ 4 & 6 & 5 \end{bmatrix}$$

Determine (a) $A + B$; (b) $A - B$; (c) $B - A$; (d) AB; (e) BA.

12.2 The transformation which translates the axes so that the point (h,k) of the old (x_1,x_2) coordinate system becomes the origin of the new (y_1,y_2) coordinate system is

$$y_1 = x_1 - h$$
$$y_2 = x_2 - k$$

The transformation which rotates the axes about a fixed origin so that the new (z_1,z_2) coordinate system is rotated an angle θ with respect to the old (y_1,y_2) coordinate system is

$$z_1 = y_1 \cos \theta + y_2 \sin \theta$$
$$z_2 = -y_1 \sin \theta + y_2 \cos \theta$$

Write each of these transformations in matrix form, and then use matrix methods to determine the transformation for going directly from the (x_1,x_2) system to the (z_1,z_2) system.

12.3 Show that $(AB)^T = B^T A$, where the capital T designates the transpose.

12.4 For each of the matrices A and B given in Prob. 12.1, determine A^{-1} and B^{-1}.

12.5 Consider the transformations $y = Ax$ and $z = By$, where A and B are the matrices given in Prob. 12.1.

(a) Determine the transformation which expresses z in terms of x.
(b) Determine the inverse transformation which expresses x in terms of z.

12.6 Solve each of the following systems of linear equations:

(a) $\quad x - y = 2$
$\quad 2x - 3y = 1$
(b) $\quad x + y + 2z = 1$
$\quad x - y - z = 0$
$\quad 4x + y + 3z = 4$
(c) $\quad x + 4y + 2z + w = 2$
$\quad 3x - y \quad\quad - 4w = 5$
$\quad 2x \quad\quad - 3z + w = 10$
$\quad\quad\quad y + 5z - w = 0$

12.7 Solve each of the following systems of homogeneous linear equations:

(a) $\quad x + 3y + z = 0$
$\quad\quad 2x - y - 5z = 0$
(b) $\quad x + 2y - w = 0$
$\quad\quad x - y + w = 0$
$\quad\quad 2x + 4y - 2w = 0$

12.8 Solve each of the following systems of nonhomogeneous linear equations:

(a) $\quad x + y - 3z + w = 2$
$\quad\quad x - y + z + 3w = 4$
(b) $\quad x + y + z - 2w = 2$
$\quad\quad 2x - y + z + w = 3$
$\quad\quad x - y - z = 0$
$\quad\quad 4x - y + z - w = 5$

12.9 Consider the differential equation

$$(D + 1)(D + 2)c(t) = f(t)$$

The initial conditions are $c(0) = \dot{c}(0) = 0$, and the forcing function is

$$f(t) = \begin{cases} 0 & t < 0 \\ h & t > 0 \end{cases}$$

Let $y_1(t) = c(t)$, and then determine;

(a) The state-space representation for this differential equation
(b) The modal column matrix X
(c) The solution $c(t)$ by employing the canonical form
(d) The state-transition matrix $\Phi(t)$
(e) The solution $c(t)$ by using the state-transition matrix

12.10 For the differential equation of Prob. 12.9, determine the computer diagram for obtaining the component:

(a) $y_1(t) = c(t)$
(b) $y_2(t)$

12.11 In Fig. P12.11 is shown a block diagram for a feedback control system. Determine:

(a) The state-space and the canonical representations
(b) The computer diagram for $c(t)$ and the modal matrix

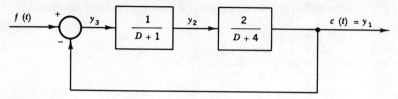

Fig. P12.11

12.12 In Fig. P12.12 is shown a block diagram for a feedback control system. Determine:

(a) The state-space and the canonical representations
(b) The computer diagram for $c(t)$ and the modal matrix

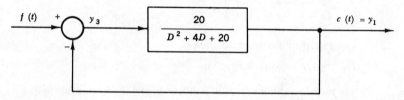

Fig. P12.12

12.13 For the differential equation

$$(D + 1)(D + 2)^2(D^2 + 6D + 25)c(t) = (139D + 159)f(t)$$

(a) Show that the partial fraction expansion is

$$c(t) = \left(\frac{1}{D + 1} + \frac{7}{(D + 2)^2} - \frac{2}{D + 2} + \frac{D - 4}{D^2 + 6D + 25} \right) f(t)$$

(b) Determine the computer diagram for obtaining $c(t)$.

12.14 For the system of Fig. 12.7, let the initial conditions be $y_1(0) = 1$, $y_2(0) = 2$, and $y_1'(0) = y_2'(0) = 0$. Determine the force-free solutions for the motions y_1 and y_2 when $K_1 = K_2 = K_3 = M_1 = M_2 = 1$.

12.15 Suppose that the force acts on the mass M_2 rather than M_1 of Fig. 12.7. Determine the corresponding computer diagrams for y_1 and y_2.

12.16 In Fig. P12.16 is shown a two-degree-of-freedom torsional system. For $J_1 = 1$, $J_2 = 5$, $K_1 = 20$, and $K_2 = 25$, determine:

(a) The equations of motion
(b) The computer diagrams for θ_1 and θ_2
(c) The general form of the principal modes of vibration

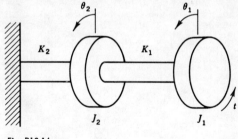

Fig. P12.16

12.17 Determine the response y_1 of Eqs. (12.74) when all the initial conditions are zero and $f(t) = h$ for $t > 0$.

12.18 For the differential equation

$$y'' + 3y' + 2y = t$$

determine the solution over the interval $0 \le t \le 1$ in increments of $\Delta t = 0.2$ for the case in which all the initial conditions are zero, by using the:

(a) Euler method
(b) Adams method
(c) Milne method
(d) Runge-Kutta method

For methods which are not self-starting, use the results of Euler's method for the starting values.

REFERENCES

1 Frazer, R. A., W. J. Duncan, and A. R. Collar: "Elementary Matrices," Cambridge University Press, New York, 1938.
2 Perlis, S.: "Theory of Matrices," Addison-Wesley Publishing Company, Inc., Reading, Mass., 1952.
3 Bellman, R.: "Introduction to Matrix Analysis," McGraw-Hill Book Company, New York, 1960.
4 Peschon, J.: "Disciplines and Techniques of Systems Control," Blaisdell Publishing Co., New York, 1965.
5 Scarborough, J. B.: "Numerical Mathematical Analysis," 4th ed., The John Hopkins Press, Baltimore, 1958.
6 Hildebrand, F. B.: "Introduction to Numerical Analysis," McGraw-Hill Book Company, New York, 1956.

Appendix A

LAPLACE TRANSFORMS

$$\mathcal{L}[f(t)] = \int_0^\infty f(t)e^{-st}\,dt = F(s)$$

$F(s)$	$f(t)$
1. $\dfrac{1}{(s-a)^2 + b^2}$	$\dfrac{e^{at}}{b}\sin bt$
2. $\dfrac{s-a}{(s-a)^2 + b^2}$	$e^{at}\cos bt$
3. $\dfrac{1}{[(s-a)^2 + b^2]^2}$	$\dfrac{e^{at}}{2b^3}(\sin bt - bt\cos bt)$
4. $\dfrac{s-a}{[(s-a)^2 + b^2]^2}$	$\dfrac{te^{at}}{2b}\sin bt$
5. $e^{-k\sqrt{s}}\ (k>0)$	$\dfrac{k}{2\sqrt{\pi t^3}}e^{-(k^2/4t)}$
6. $\dfrac{1}{\sqrt{s}}e^{-k\sqrt{s}}\ (k \geq 0)$	$\dfrac{1}{\sqrt{\pi t}}e^{-(k^2/4t)}$
7. $\dfrac{1}{s}e^{-k\sqrt{s}}\ (k \geq 0)$	$\operatorname{erfc}\left(\dfrac{k}{2\sqrt{t}}\right)$
8. $\dfrac{1}{s^{\frac{3}{2}}}e^{-k\sqrt{s}}\ (k \geq 0)$	$2\sqrt{\dfrac{t}{\pi}}e^{-(k^2/4t)}$
9. $\dfrac{1}{s^{\frac{1}{2}}}$	$\dfrac{1}{\sqrt{\pi t}}$
10. $\dfrac{1}{s^{(n+\frac{1}{2})}}\ (n = 1, 2, \ldots)$	$\dfrac{2^n t^{n-\frac{1}{2}}}{(1)(3)(5)\cdots(2n-1)\sqrt{\pi}}$
11. $\dfrac{\Gamma(k)}{s^k}\ (k>0)$	t^{k-1}
12. $\dfrac{\Gamma(k)}{(s-a)^k}\ (k>0)$	$t^{k-1}e^{at}$
13. $\dfrac{1}{\sqrt{s}+a}$	$\dfrac{1}{\sqrt{\pi t}} - ae^{a^2 t}\operatorname{erfc}(a\sqrt{t})$
14. $\dfrac{\sqrt{s}}{s-a^2}$	$\dfrac{1}{\sqrt{\pi t}} + ae^{a^2 t}\operatorname{erf}(a\sqrt{t})$
15. $\dfrac{1}{\sqrt{s}(s-a^2)}$	$\dfrac{e^{a^2 t}}{a}\operatorname{erf}(a\sqrt{t})$
16. $\dfrac{1}{s\sqrt{s+a^2}}$	$\dfrac{1}{a}\operatorname{erf}(a\sqrt{t})$

LAPLACE TRANSFORMS (*Continued*)

$$\mathcal{L}[f(t)] = \int_0^\infty f(t)e^{-st}\,dt = F(s)$$

$F(s)$	$f(t)$
17. $\dfrac{1}{\sqrt{s^2 + a^2}}$	$J_0(at)$
18. $\dfrac{1}{\sqrt{s^2 - a^2}}$	$I_0(at) = J_0(jat)$
19. $\dfrac{(\sqrt{s^2 + a^2} - s)^\nu}{\sqrt{s^2 + a^2}}$ $(\nu > -1)$	$a^\nu J_\nu(at)$
20. $\dfrac{(s - \sqrt{s^2 - a^2})^\nu}{\sqrt{s^2 - a^2}}$ $(\nu > -1)$	$a^\nu I_\nu(at)$
21. $\dfrac{1}{s}\ln s$	$\gamma - \ln t(\gamma = 0.5772)$
22. $\ln\dfrac{s - a}{s - b}$	$\dfrac{1}{t}(e^{bt} - e^{at})$
23. $\tan^{-1}\dfrac{k}{s}$	$\dfrac{1}{t}\sin kt$
24.† $\dfrac{1}{s}\left(\dfrac{s - 1}{s}\right)^n$	$L_n(t) = \dfrac{e^t}{n!}\dfrac{d^n}{dt^n}(t^n e^{-t})$
25.‡ $\dfrac{(1 - s)^n}{s^{n+\frac{1}{2}}}$	$\dfrac{n!}{(2n)!\,\sqrt{\pi t}}H_{2n}(\sqrt{t})$

† $L_n(t)$ is the Laguerre polynomial $L_n(t) = e^t(d^n/dt^n)(t^n e^{-t})$.
‡ $H_n(t)$ is the Hermite polynomial $H_n(t) = e^{t^2}(d^n/dt^n)(e^{-t^2})$.

Appendix B

FINITE SINE TRANSFORMS

$$S[f(x)] = \int_0^L f(x) \sin \frac{\pi n x}{L}\, dx = F(n)$$

	$S[f(x)]$	$f(x)$
1.	$\dfrac{L}{\pi n}[1 - (-1)^n]$	1
2.	$\dfrac{L}{\pi n}(-1)^{n+1}$	$\dfrac{x}{L}$
3.	$\dfrac{L}{\pi n}\left\{(-1)^{n-1} - \dfrac{2[1-(-1)^n]}{\pi^2 n^2}\right\}$	$\left(\dfrac{x}{L}\right)^2$
4.	$\dfrac{L}{\pi n}$	$\dfrac{L-x}{L}$
5.	$2\left(\dfrac{L}{\pi n}\right)^2 \sin \dfrac{\pi n}{2}$	$\begin{cases} x & \left(0 \le x \le \dfrac{L}{2}\right) \\ L-x & \left(\dfrac{L}{2} \le x \le L\right) \end{cases}$
6.	$2\left(\dfrac{L}{\pi n}\right)^3 [1 - (-1)^n]$	$x(L-x)$
7.	$6\left(\dfrac{L}{\pi n}\right)^3 (-1)^{n+1}$	$\dfrac{x(L^2 - x^2)}{L}$
8.	$\dfrac{\pi n L}{\pi^2 n^2 + k^2 L^2}[1 - (-1)^n e^{kL}]$	e^{kx}
9.	$\begin{cases} 0 & (n \ne m) \\ \dfrac{L}{2} & (n = m) \end{cases}$	$\sin \dfrac{\pi m x}{L}$ $(m = 1, 2, \ldots)$
10.	$\begin{cases} \dfrac{nL}{\pi(n^2 - m^2)}[1 - (-1)^{n+m}] & (n \ne m) \\ 0 & (n = m) \end{cases}$	$\cos \dfrac{\pi m x}{L}$ $(m = 1, 2, \ldots)$

Appendix C

FINITE COSINE TRANSFORMS

$$C[f(x)] = \int_0^L f(x) \cos \frac{\pi n x}{L}\, dx = F(n)$$

	$C[f(x)]$	$f(x)$
1.	$\begin{cases} L\ (n = 0) \\ 0\ (n = 1, 2, \ldots) \end{cases}$	1
2.	$\begin{cases} \frac{1}{2}L^2\ (n = 0) \\ \left(\dfrac{L}{\pi n}\right)^2 [(-1)^n - 1]\ (n = 1, 2, \ldots) \end{cases}$	x
3.	$\begin{cases} \frac{1}{3}L^3\ (n = 0) \\ \dfrac{2L^3}{\pi^2 n^2}\ (n = 1, 2, \ldots) \end{cases}$	x^2
4.	$\begin{cases} 0\ (n = 0) \\ \dfrac{2L}{\pi n} \sin \dfrac{\pi n}{2}\ (n = 1, 2, \ldots) \end{cases}$	$1 \left(0 < x < \dfrac{L}{2}\right)$ $-1 \left(\dfrac{L}{2} < x < L\right)$
5.	$\dfrac{kL^2}{\pi^2 n^2 + k^2 L^2} [(-1)^n e^{kL} - 1]$	e^{kx}
6.	$\begin{cases} 0\ (n \neq m) \\ \dfrac{L}{2}\ (n = m) \end{cases}$	$\cos \dfrac{\pi m x}{L}$
7.	$\begin{cases} \dfrac{mL}{\pi(n^2 - m^2)} [(-1)^{n+m} - 1]\ (n \neq m) \\ 0\ (n = m) \end{cases}$	$\sin \dfrac{\pi m x}{L}$

Index